Impacts of Climate Change on Agriculture and Aquaculture

Ahmed Karmaoui
Southern Center for Culture and Sciences, Morocco

Kirby Barrick
University of Florida, USA

Michael Reed
University of Kentucky, USA

Mirza Barjees Baig
King Saud University, Saudi Arabia

A volume in the Advances in Environmental Engineering and Green Technologies (AEEGT) Book Series

IGI Global
PUBLISHER of TIMELY KNOWLEDGE

Published in the United States of America by
IGI Global
Engineering Science Reference (an imprint of IGI Global)
701 E. Chocolate Avenue
Hershey PA, USA 17033
Tel: 717-533-8845
Fax: 717-533-8661
E-mail: cust@igi-global.com
Web site: http://www.igi-global.com

Library of Congress Cataloging-in-Publication Data

Names: Karmaoui, Ahmed, 1982- editor. | Barrick, R. Kirby, editor. | Reed, Michael R., editor. | Baig, Mirza Barjees, 1960- editor.
Title: Impacts of climate change on agriculture and aquaculture / Ahmed Karmaoui, Kirby Barrick, Michael Reed, and Mirza Barjees Baig, editors.
Description: Hershey PA : Engineering Science Reference, [2020] | Includes bibliographical references and index. | Summary: "This book provides an integrated assessment of climate change impacts on agriculture, aquaculture, and fisheries and explores a set of strategies to secure sustainable food security"-- Provided by publisher.
Identifiers: LCCN 2019051958 (print) | LCCN 2019051959 (ebook) | ISBN 9781799833437 (hardcover) | ISBN 9781799833444 (paperback) | ISBN 9781799833451 (ebook)
Subjects: LCSH: Crops and climate. | Aquaculture--Climatic factors. | Fisheries--Climatic factors. | Climatic changes. | Food security--Climatic factors. | Sustainable agriculture. | Sustainable aquaculture. | Sustainable fisheries.
Classification: LCC S600.5 .I466 2020 (print) | LCC S600.5 (ebook) | DDC 338.1/4--dc23
LC record available at https://lccn.loc.gov/2019051958
LC ebook record available at https://lccn.loc.gov/2019051959

This book is published in the IGI Global book series Advances in Environmental Engineering and Green Technologies (AEEGT) (ISSN: 2326-9162; eISSN: 2326-9170)

British Cataloguing in Publication Data
A Cataloguing in Publication record for this book is available from the British Library.

All work contributed to this book is new, previously-unpublished material.
The views expressed in this book are those of the authors, but not necessarily of the publisher.

For electronic access to this publication, please contact: eresources@igi-global.com.

Advances in Environmental Engineering and Green Technologies (AEEGT) Book Series

ISSN:2326-9162
EISSN:2326-9170

Editor-in-Chief: Sang-Bing Tsai, Zhongshan Institute, University of Electronic Science and Technology of China, China & Wuyi University, China & Ming-Lang Tseng, Lunghwa University of Science and Technology, Taiwan & Yuchi Wang, University of Electronic Science and Technology of China Zhongshan Institute, China

MISSION

Growing awareness and an increased focus on environmental issues such as climate change, energy use, and loss of non-renewable resources have brought about a greater need for research that provides potential solutions to these problems. Research in environmental science and engineering continues to play a vital role in uncovering new opportunities for a "green" future.

The **Advances in Environmental Engineering and Green Technologies (AEEGT)** book series is a mouthpiece for research in all aspects of environmental science, earth science, and green initiatives. This series supports the ongoing research in this field through publishing books that discuss topics within environmental engineering or that deal with the interdisciplinary field of green technologies.

COVERAGE

- Sustainable Communities
- Renewable Energy
- Water Supply and Treatment
- Contaminated Site Remediation
- Biofilters and Biofiltration
- Industrial Waste Management and Minimization
- Electric Vehicles
- Green Technology
- Air Quality
- Policies Involving Green Technologies and Environmental Engineering

IGI Global is currently accepting manuscripts for publication within this series. To submit a proposal for a volume in this series, please contact our Acquisition Editors at Acquisitions@igi-global.com or visit: http://www.igi-global.com/publish/.

The Advances in Environmental Engineering and Green Technologies (AEEGT) Book Series (ISSN 2326-9162) is published by IGI Global, 701 E. Chocolate Avenue, Hershey, PA 17033-1240, USA, www.igi-global.com. This series is composed of titles available for purchase individually; each title is edited to be contextually exclusive from any other title within the series. For pricing and ordering information please visit http://www.igi-global.com/book-series/advances-environmental-engineering-green-technologies/73679. Postmaster: Send all address changes to above address.

Titles in this Series

For a list of additional titles in this series, please visit:
http://www.igi-global.com/book-series/advances-environmental-engineering-green-technologies/73679

Solar Concentrating Modules With Louvered Heliostats Emerging Research and Opportunities
Dmitry Strebkov (Federal State Budget Scientific Institution, Russia & Federal Scientific Agroengineering Center VIM, Russia) Natalya Filippchenkova (Federal State Budget Scientific Institution, Russia & Federal Scientific Agroengineering Center VIM, Russia) and Anatoly Irodionov (Federal State Budget Scientific Institution, Russia & Federal Scientific Agroengineering Center VIM, Russia)
Engineering Science Reference • © 2021 • 200pp • H/C (ISBN: 9781799842767) • US $190.00

Recent Advancements in Bioremediation of Metal Contaminants
Satarupa Dey (Shyampur Siddheswari Mahavidyalaya, India) and Biswaranjan Acharya (School of Computer Engineering, KIIT University (Deemed), India)
Engineering Science Reference • © 2021 • 300pp • H/C (ISBN: 9781799848882) • US $195.00

Global Environmental Impacts on Food Security, Nutrition, and Human Health
Hicham Chatoui (Private University of Marrakech, (UPM), Morocco & Cady Ayyad University, Morocco) Mohamed Merzouki (Faculty of Science and Technology, Sultan Moulay Slimane University, Beni Mellal, Morocco) Hanane Moummou (Private University of Marrakech (UPM), Marrakech, Morroco) Mounir Tilaoui (Private University of Marrakech (UPM), Sultan Moulay Slimane University, Beni Mellal, Morocco) and Nabila Saadaoui (Private University of Marrakech (UPM), Marrakech, Morroco)
Engineering Science Reference • © 2020 • 300pp • H/C (ISBN: 9781799837480) • US $195.00

Spatial Information Science for Natural Resource Management
Suraj Kumar Singh (Suresh Gyan Vihar University, Jaipur, India) Shruti Kanga (Suresh Gyan Vihar University, Jaipur, India) and Varun Narayan Mishra (Suresh Gyan Vihar University, Jaipur, India)
Engineering Science Reference • © 2020 • 355pp • H/C (ISBN: 9781799850274) • US $195.00

For an entire list of titles in this series, please visit:
http://www.igi-global.com/book-series/advances-environmental-engineering-green-technologies/73679

IGI Global
PUBLISHER of TIMELY KNOWLEDGE

701 East Chocolate Avenue, Hershey, PA 17033, USA
Tel: 717-533-8845 x100 • Fax: 717-533-8661
E-Mail: cust@igi-global.com • www.igi-global.com

Table of Contents

Detailed Table of Contents

In this chapter, the authors propose a novel statistical model with a residual correction of downscaling coarse precipitation TRMM 3B43 product. The presented study was carried out over Morocco, and the objective is to improve statistical downscaling for TRMM 3B43 products using a machine learning algorithm. Indeed, the statistical model is based on the Transformed Soil Adjusted Vegetation Index (TSAVI), elevation, and distance from the sea. TSAVI was retrieved using the quantile regression method. Stepwise regression was implemented with the minimization of the Akaike information criterion and Mallows' Cp indicator. The model validation is performed using ten in-situ measurements from rain gauge stations (the most available data). The result shows that the model presents the best fit of the TRMM 3B43 product and good accuracy on estimating precipitation at 1km according to

Chapter 8

WFE nexus is an important aspect in building sustainable economies. Water is used in food production while water supply and food processing require energy. Understanding the interrelationships of the nexus components is a growing interest for researchers and policymakers towards sustainable development. This chapter analyses the in-depth meaning of the WFE nexus, its importance, and its involved processes. The chapter also evaluates the effects of climate change on the nexus using case examples in South Africa. It also proposes a road map to facilitate better management of the nexus by recommending useful action plans. These action plans prioritize on baseline data collection, optimization of WFE nexus processes and cooperative management of resources, and climate change adaptation.

Chapter 9

Climatic changes can cause severe food and water shortages, and desert nations such as Namibia can be challenged more than other countries for obvious reasons. Dependency on imports for food and electricity in Namibia is continuous in recent times. However, Industry 4.0-based large-scale symbiotic systems can potentially help in achieving a sustainable food security regime, as they operate under controlled conditions. Namibia is blessed with abundant sunshine and land availability, and hence, ample opportunities do exist for producing solar energy, which is used to meet the energy requirements of symbiotic systems. This chapter examines typical local operating conditions and then makes a strong case for fully automated symbiotic systems that use low-cost desalination and renewable energies.

Chapter 10

This chapter examines climate change and variability emergency disaster risks on agricultural food security of the local communities in Africa with a focus on gender equality lens in Uganda. Ugandan women contribute up to 75% of domestic food production and yet they are often overburdened with reproduction, household management, gender-specific discrimination, and adverse climate change effects like

Preface

Climate change is expected to influence several productive sectors, but especially agriculture. This important sector contributes to a large part of the global economy and includes crops, livestock, and seafood. Agriculture and aquaculture are closely linked to the climate. Changes in climatic conditions can affect animal and plant productivity, which in turn influence human well-being. The book provides an integrated assessment of climate change impacts on agriculture and aquaculture, and explores a set of strategies to secure sustainable food security. Information is synthesized from the scientific literature on the theory, models, and decision support tools. The main objectives of this edited book are:

- To explore the associations among climate change, agriculture, aquaculture, food security, and socio-economic development.
- To highlight good agricultural practices for the adaptation to climate change.
- To make a useful reference document for the development of recommendations for sustainable management of soil, water, and food.

Several researchers advance that the accumulation of greenhouse gases was considered the main cause of climate change. This later threatens the production systems (aquatic and terrestrial ecosystems), which raises disruptions of food availability. Consequently, dangerous levels of food security can be recorded in vulnerable societies, mainly for children and women in Africa and Asia. Reduced water quality and quantity, decreased agricultural production, raised sea-levels, and acidified oceans are the most serious examples of climate change impacts.

The current edited book deals with two productive systems that are linked to climate change and anthropogenic impacts: agriculture and aquaculture with a focus on agricultural activity. Agriculture is a system of food production by humans using and controlling the physical and biological components of ecosystems. Increased temperatures, more variable precipitation, heightened levels of pests, and higher concentrations of carbon dioxide and methane in the atmosphere may greatly influence agricultural production and food quality as well as deteriorate soil and water quality.

These changes will have major implications for the stability of poor communities and the biological diversity of these areas. For the most part, agriculture is one of the factors that exacerbate climate change through human activity, especially turning the forest into agricultural land. Poor land management and overuse can accelerate the desertification process, which is among the most devastating threats to food security.

The last special report of the Intergovernmental Panel on Climate Change (IPCC) (2019: Desertification. In: Climate Change and Land: an IPCC special report on climate change, desertification, land degradation, sustainable land management, food security, and greenhouse gas fluxes in terrestrial ecosystems) advanced the important action of the desertification and land degradation in the arid and semi-arid regions on agriculture and, consequently, poverty and food security. The report showed that desertification has decreased agricultural productivity and caused the loss of biodiversity in some arid zones and its risk will increase due to climate change.

Several solutions and sustainable technologies were developed and adopted in many countries for climate-friendly farming. Symbiotic systems such as aquaponic, aeroponic, and hydroponic systems are some interesting examples of these smart technologies. Some of these systems, mainly aquaponic systems, are used both for plants and fish production and result in much less water use. The orientation toward smart agriculture and the use of renewable energies to support the associated activities can be used as strategies to cope with climate change and extreme events impacts.

Aquaculture activity refers to production activities of organisms (aquatic plants, fish, crustaceans, etc.) in an aquatic environment. The scientific community considers aquaculture as an important solution to food insecurity. This farming activity is subject to several climate impacts such as temperature, salinity, and sea level rise, which subsequently influence food security for a large part of the world population. These changes affect the ecosystems, influencing their productivity. Several hundreds of people throughout the world depend on aquaculture, among the most important sources of protein in the whole world, making it an essential economic resource, mainly in poor societies.

ORGANIZATION OF THE BOOK

The eleven chapters of the collective book are organized into three sections: Section 1 deals with climate change and agriculture; Section 2 highlights the productive ecosystems, challenges and solutions; and Section 3 investigates Climate change and aquaculture. Within these three sections, Chapter 1 explores the downscaling of open coarse precipitation data using a machine learning algorithm. In this chapter, the authors propose a novel statistical model with a residual correction of downscaling coarse precipitation TRMM 3B43 product. The presented study was carried out over

Morocco, and the objective is to improve statistical downscaling for TRMM 3B43 products using a machine learning algorithm. Indeed, the statistical model is based on the Transformed Soil Adjusted Vegetation Index (TSAVI), elevation, and distance from the sea. TSAVI was retrieved using the quantile regression method. Stepwise regression was implemented with the minimization of the Akaike information criterion and Mallows' Cp indicator. The model validation is performed using ten in-situ measurements from rain gauge stations (the most available data). The result shows that the model presents the best fit of the TRMM 3B43 product and good accuracy on estimating precipitation at 1km according to *R*2, RMSE, bias, and MAE. Besides, TSAVI improved the model accuracy in the humid bioclimatic stage, and in the Saharan region to some extent due to its capacity to reduce soil brightness.

Chapter 2 highlights the designing a Framework for Agrisector Considering Disaster and Climatic Change. Change of technology, utilization of genetically modified crops, and organic farming can be taken care to improve the fertility of land and to increase production in farming, but the effect of climate change is a big problem for the Indian farmers. Farmers have to face extreme weather conditions, the heavy workload during the fieldwork like weeding, harvesting, etc. The conventional method of farming, lack of advanced technology makes farming too difficult. Due to climate change high heat, heavy rain, frost not only productivity decreases lands became barren, but farmer also suffers from a mental and physical disorder. Hence an effort is taken to design a framework for the agriculture sector keeping climate change factors in view. To improve the agricultural system in Indian Agrisector some implications must be done to avoid wastage of rainwater. By Creating an efficient drainage system in the crop field the extra water can be reused. To avoid loss due to heavy rain, Greenhouse or poly house can act as an effective way during natural calamities.

Chapter 3 investigates the mitigation of climate change through carbon sequestration. Climate change is considered one of the burning issues and threats to the globe. One of the most important drivers of climate change is the increasing concentration of atmospheric greenhouse gases that triggers the altered climate pattern. Therefore, researchers are now very much concerned to tackle the adverse impact of climate change as well in minimizing the concentration and emissions of atmospheric greenhouse gases. i.e. carbon sequestration seems one of the most attractive choices for the researchers that could be useful for mitigating climate change conditions. The process deals with the capturing of atmospheric carbon dioxide (CO2) in various segments of natural ecosystems like soil, biological systems, oceans and thereby offsetting the adverse impact of climate change. In this review, the author examines the current scenario of carbon sequestration in the biosphere and how it can be effectively implied over a specific ecosystem.

Chapter 4 reviews the vulnerability of oasis agriculture to climate change in Morocco that is considered as one of the most threatened countries. Over the last century, oases ecosystems in this country showed a high vulnerability to climate variation which has led to water scarcity, an increase in land salinity, and therefore, a decrease in agricultural production. Conscious of these issues, several solutions are initiated by the government to cope with climate change adverse effects. Many programs of rehabilitation were launched and advanced researches are in progress in order to use some biofertilizers to improve tolerance of oasis crops to drought and salinity toward sustainable agriculture. The aim of this chapter is to give an overview of the impacts of climate change on oasis agriculture in Morocco and to provide potentially effective strategies to promote oasis agriculture under climate change. As a conclusion, the authors found that the use of different biofertilizers could be a potential strategy to mitigate climate change effects on oasis agriculture in Morocco.

Chapter 5 explores the Impact of climate change through the thermal stress on the Moroccan Argan Agroecosystem. The argan tree is an exclusively endemic in the drylands of Southwest Morocco, an agroecosystem of great ecological, cultural, and economic importance. The argan agroecosystem is already damaged. It is particularly vulnerable to climate change as well as the harsh natural conditions aggravated by the current population growth and the exploitation in excess of the production capacities. Unfortunately, during the twentieth century, its area has been reduced by half. Current projections indicate an increase in temperature under climate scenarios. Anticipated climate change could accelerate this trend resulting in the argan tree degradation. To assess the climate change impact, we used the SDSM model at the argan agroecosystem scale and the thermal stress mode in order to assess the vulnerability of the argan agroecosystem and estimate its tolerance response in relation to temperature stress for a projected climate in the near term (2010-2025 years). In this chapter, the authors explored the impact of climate change on the argan tree regeneration.

Chapter 6 focuses on sustainable ecosystem management, challenges, and solutions. Ecosystems are part of human wellbeing and their sustainable management is essential for the survival of the human race and biodiversity. This chapter explores the concept of sustainable ecosystem management (SEM), its principles, elements, faces, and implementation. SEM is defined as environmentally sensitive, ecosystem-based, and eco-regional based. Its successful implementation is therefore complex due to the different priorities of stakeholders, the scope of ecosystems, some of which are transboundary, and the ever-changing nature of these areas amidst environmental uncertainties. These aspects are vulnerable to political changes and reconciling them is difficult. This chapter proposes a 5-step implementation plan on SEM that is pegged on adaptive management and holistic consideration of ecological resources.

Using documented case studies, SEM is a proposed solution to ecosystem challenges of modern-day amidst hindrances of rising resource demand, population increase, and climate variability.

Chapter 7 explores the climate change and its impact on terrestrial ecosystem. It provides a general overview of the effects of climate change on the terrestrial ecosystem and is meant to set the stage for the specific papers. The discussion in this chapter focuses basically on the effects of climatic disturbances on terrestrial flora and fauna, including increasing global temperature and changing climatic patterns of terrestrial areas of the globe. Basically climate disturbances derived increasing temperature and greenhouse gases have the ability to induce this phenomenon. Greenhouse gases emitted by a number of sources in the atmosphere such as urbanization, industrialization, transportation, and population growth, so these contributing factors and its effects on climatic events like temperature rise, change precipitation pattern, extreme weather events, survival and shifting of biodiversity, seasonal disturbances and effects on glaciers are relatively described in this chapter.

Chapter 8 investigates the water-food-energy nexus in the climate change era and highlights the roadmap to implementation in South Africa. The WFE nexus is an important aspect in building sustainable economies. Water is used in food production while water supply and food processing require energy. Understanding the interrelationships of the nexus components is a growing interest for researchers and policymakers towards sustainable development. This chapter analyses the in-depth meaning of the WFE nexus, its importance, and its involved processes. The chapter also evaluates the effects of climate change on the nexus using case examples in South Africa. It also proposes a road map to facilitate better management of the nexus by recommending useful action plans. These action plans prioritize on baseline data collection, optimization of WFE nexus processes and cooperative management of resources, and climate change adaptation.

Chapter 9 analyses the industry 4.0 based large scale symbiotic systems for sustainable food security in Namibia. Climatic changes can cause severe food and water shortage and desert nations such as Namibia can be challenged more than other countries for obvious reasons. Dependency on imports for food and electricity in Namibia is continuously in recent times. However, industry 4.0 based large scale symbiotic systems can potentially help in achieving a sustainable food security regime, as they operate under controlled conditions. Namibia is blessed with abundant sunshine and land availability and hence ample opportunities do exist for producing solar energy, which is used to meet the energy requirements of symbiotic systems. This chapter examines typical local operating conditions and then makes a strong case for fully automated symbiotic systems that use low-cost desalination and renewable energies.

Chapter 10 assesses gender equality and climate change advocacy for sustainable community agricultural food security in South Western Uganda. This chapter examines climate change and variability emergency disaster risks on the small-holder farmers' livelihoods and food security of the local communities' agricultural communities around the south-western Ugandan Lake Mburo National Park. Ugandan women contribute up to 75 percent % of domestic food production and yet due to fragile ecosystems and climate change impacts like agricultural droughts or water stress. To ensure sustainable food security in the face of climate change vulnerability risks, the role of women is vital. Communication strategy to promote local climate information service (CIS) delivery system has been developed by the local government district planners in the park areas but there is a lack of capacity to raise public awareness of the gender equality for the empowerment of women and girls for sustainable food security through agriculture production in Uganda for enhanced livelihood assets.

Chapter 11 explores the effects of climate change and anthropogenic activities on algivorous cichlid fish in Lake Tanganyika. Over the last century, water temperatures in Lake Tanganyika have risen due to climate change which increased thermal stratification and reduced the magnitude of nutrient availability. A rise in temperature increases the C:N:P ratio resulting in a poor algal diet. In addition, lake littoral habitat is experiencing increased sediment load due to deforestation of the watershed caused by anthropogenic activities. Sediments cover benthic algae and reduce its nutritional value, consequently affecting the foraging behavior, distribution, and growth performance of algivorous fish. Algae and algivorous fish are an important link in the lake food chain, therefore, if the rise in temperature will continue as predicted, then, this may have a cascading effect for the rest of the community in the food chain including human being. This, in turn, may contribute to food insecurity at local and regional levels. To counteract this adaptation and mitigation measures such as environmental monitoring systems, and creating new opportunities should be considered.

In summary, it is the purpose of this text, and of the authors who contributed their scholarly work, that peoples in all corners of the world will take to heart the importance of climate change in relation to agriculture and aquaculture survival and productivity. A hungry world awaits increased food security and sustainable socio-economic development.

Acknowledgment

In this collective edited book, an integrated assessment of climate change impacts on agriculture and aquaculture was performed. The book includes 11 studies using different approaches such as case studies, reviews, and modeling methods. Statistical model using machine learning algorithm, transformed Soil Adjusted Vegetation Index, statistical downscaling model, stress model, and framework for agrisector considering Disaster and climatic change are examples used in this collection.

This book is the result of collaboration between four editors and several academics and researchers from Morocco, USA, Saudi Arabia, Spain, India, Kenya, Namibia, Uganda, and Tanzania, with the continuous support of the IGI global publisher over a period of 10 months. I would like to thank all the members of the IGI Global team that helped us in the book production. I would like to especially thank Jordan Tepper, Josh Christ, and Carlee Nilphai for their support and patience throughout the development of this book.

Information in this book was reviewed by an international scientific committee. Without the experiences and support from the reviewers and the co-editors, this book would not exist. They have given me the opportunity to lead a great group of academics.

My time in the research work wouldn't have been made possible without the support of my family. I'm eternally grateful to my wife Siham, my two children, Yassin and Amine, who supported me in the production steps. To my father Mimoun for being the person who believes in me.

Finally, a special thanks to all those who have been a part of our research projects: Ahmed Amghar and Drisse El Bouzidi.

Section 1
Climate Change and Agriculture

Chapter 1
Downscaling of Open Coarse Precipitation Data Using a Machine Learning Algorithm

Ismail Elhassnaoui
Mohammadia School of Engineers, Rabat, Morocco

Zineb Moumen
University Politecnica de Catalunya, Spain

Hicham Ezzine
Mohammadia School of Engineers, Rabat, Morocco

Marwane Bel-lahcen
Advanced Digital Entreprise Modeling and Information Retrieval, Morocco

Ahmed Bouziane
Mohammadia School of Engineers, Rabat, Morocco

Driss Ouazar
Mohammadia School of Engineers, Rabat, Morocco

Moulay Driss Hasnaoui
Ministry of Equipment, Transport, Logistics, and Water, Rabat, Morocco

ABSTRACT

In this chapter, the authors propose a novel statistical model with a residual correction of downscaling coarse precipitation TRMM 3B43 product. The presented study was carried out over Morocco, and the objective is to improve statistical downscaling for TRMM 3B43 products using a machine learning algorithm. Indeed,

DOI: 10.4018/978-1-7998-3343-7.ch001

the statistical model is based on the Transformed Soil Adjusted Vegetation Index (TSAVI), elevation, and distance from the sea. TSAVI was retrieved using the quantile regression method. Stepwise regression was implemented with the minimization of the Akaike information criterion and Mallows' Cp indicator. The model validation is performed using ten in-situ measurements from rain gauge stations (the most available data). The result shows that the model presents the best fit of the TRMM 3B43 product and good accuracy on estimating precipitation at 1km according to R2, RMSE, bias, and MAE. In addition, TSAVI improved the model accuracy in the humid bioclimatic stage and in the Saharan region to some extent due to its capacity to reduce soil brightness.

INTRODUCTION

Precipitation is the most significant component in the hydrologic cycle (Elhassnaoui et al., 2019). Indeed, precipitation data is a fundamental requirement for full features of meteorology, hydrology, groundwater, streamflow, flood, drought, agriculture, and economics (Maidment, 1993). In a context of climate change, providing high precise precipitation data as a vital variable that describe a variety of features and phenomenon is very significant (Chen & li, 2016; l. Tang et al., 2015; Zhao et al., 2017).

From ancient times, rain gauge stations have been an essential tool for precipitation observation in a hydro-meteorological perspective(Schneider et al., 2014; Schwaller & Morris, 2011). However, due to the topographical characteristic of the catchments, the rain gauge network suffers from sparse spatial distribution (Guofeng et al., 2016; Maggioni et al., 2016). The sparse rain gauge network cannot provide a significant statistical distribution rainfall using interpolation (Kro & Law, 2005). Furthermore, rain gauge stations face many impediments to record and monitor precipitation data namely the weather modification, wind speed, type of precipitation, which lead to significant errors in rainfall observation (Essery & Wblcock, 1991; peck, 1974; Rodda & Smith, 1986; John Rodda & Dixon, 2011; Spring & Peck, 1980).

To overcome rain gauge errors, and provide a quantitative spatial measurement of precipitation, remotes sensing techniques have been developed (Gregg & Casey, 2004; Karaska et al., 2004). Besides, passive satellite sensors can provide accurate global coverage of precipitation data without data interpolation (Duan & Bastiaanssen, 2013; Kidd & Levizzani, 2011; Zhang & Li, 2018). Indeed, satellite sensors with a set of algorithms that embody microwaves radiation have a high potential for measuring precipitation, because, microwaves are directly related to the raindrop through emission, absorption and scattering techniques (Ezzine et al., 2017; Liu et al., 2018; Maidment, 1993). The temperature threshold method is the standard method for estimating rainfall by remote sensing (Arkin et al., 1980.; Arkin et al., 1979).

Satellite sensors programs have become an advanced and efficient tool for providing accurate precipitation data (Silva & Lopes, 2017), covering a broad spatial distribution (Ezzine et al., 2017; Irvem & Ozbuldu, 2019; Kidd, 2001).

Straight away, many derived data from satellite sensors are freely available to the public, namely the Climate Precipitation Center morphing technique (COMRPH), the Tropical Rainfall Measure Mission (TRMM) Multi-satellite Precipitation Analysis (TMPA), and the Integrated Multi-Satellite Retrievals for Global Precipitation Measurement (GPM) mission, ERA-Interim (ERAI), Global Precipitation Climatology Centre (GPCC) and Multi-Source Weighted Ensemble Precipitation (MSWEP) (Belay et al., 2020; Ezine et al., 2017; Mondal et al., 2014; Nichol & Abbas, 2015; Xue et al., 2015; Zheng, 2015).

Among these satellite products, the Tropical Rainfall Measure Mission (TRMM) is a passive microwave sensor that provides precipitation data (Najja et al., 2018). It is considered as a convenient sensor to estimate precipitation at a spatial resolution of 0.25°×0.25°(Chen & Li, 2016; Milewski et al., 2015). TRMM product has been used TRMM was developed jointly by the National Aeronautics and Space Administration (NASA) and the Japanese Aerospace Exploration Agency (JAXA) in 1997. Moreover, TRMM was retired on April 8, 2015 after more than 17 years of rich data collection (Zhou et al., 2020).

The Tropical Rainfall Measure Mission (TRMM) data promise great potential for hydrologic, meteorological and agriculture application (Darand et al., 2017; Naumann et al., 2012; Yong et al., 2008; Immerzeel et al., 2009; Mehta et al., 2008; Sciences et al., 2007; Tang et al., 2016; Wang et al., 2016; Yuan et al., 2017) since it has been used in different studies worldwide (Ciabatta et al., 2016; Kim et al., 2016; Paredes-trejo et al., 2019.; Thiemig et al., 2013; Tuo et al., 2016). However, it exhibits marked coarse spatial distribution. Corpse precipitation data cannot identify accurate local phenomenon like weather, topography, winds, maritime, and atmospheric tidal phenomenon fin scale area (Chokkavarapu & Ravibabu, 2019; Immerzeel et al., 2009). Also, Precipitation data is highly sensitive to spatial resolution (Brienen et al., 2010; Fowler & Ekstr, 2009; Gutjahr, 2013).

Assessing the hydrological impact of present and future climate change at a fine-scale spatial distribution requires accurate precipitation data at a small-scale taking into account the local environmental factors (Brienen et al., 2010). Indeed, the spatial resolution at the fine-scale is crucial in providing long-term precipitation projection, prediction of precipitation variability, flood modeling, or rainfall-runoff modeling (Segond et al., 2007; Wheater, 2017). In this sense precipitation downscaling plays a significant role in reducing the gaps between global precipitation data and local characteristics, in a way to capture climate change at fine-scale spatial distribution (Brienen et al., 2010).

To overcome satellite corpse data issues, researchers opt for downscaling methods. The main downscaling methods are delta/ratio method, statistical downscaling, and dynamic downscaling (Chokkavarapu & Ravibabu, 2019). Besides, relevant studies have displayed that precipitation is correlated to topography vegetation and distance from the sea. In this context, multiple statistical downscaling models, based on an environmental factor have been developed(Chen et al., 2014; Duan et al., 2017; Ezzine et al., 2017; Liu et al., 2018; Park, 2013; Chen et al., 2019; Zhang et al., 2018).

LITERATURE REVIEW

To meet the requirement of precipitation resolution data, many researchers have improved downscaling models. Jia et al., (2011) developed a downscaling model based on the relation between precipitation, Normalized Difference Vegetation Index (NDVI) and elevation, in the Qaidam Basin of Chin. The significant results allow the downscaling of TRMM 3B43 at a fine-scale of 1km using multivariate regression. Fang et al., (2013) improved downscaling accuracy at 1km resolution by the introduction of DEM, humidity, and temperature as environmental factors. Ezzine et al., (2017) enhanced the accuracy of the model based on the Normalized Difference Vegetation Index (NDVI) and Digital Elevation Model (DEM) by adding the distance from the sea to integrate the maritime phenomena. This study enabled to downscale TRMM 3B43 products to a resolution of 1km using stepwise regression. Tsanis et al, (2016) made a great effort to enhance the downscaling accuracy by using a machine learning algorithm based on artificial neural networks (ANN). This study aims to downscale the TRMM 3B43 product to 1km. Chen et al., (2014) developed a model-based Normalized Difference Vegetation Index (NDVI) and Digital Elevation Model (DEM) based on geographically weighted regression (GWR) to downscale TRMM 3B43 precipitation product at 1km. Similar study was used Normalized Difference Vegetation Index (NDVI) and Digital Elevation Model (DEM) to downscale TRMM 3B43 product at 1 km using a regression model with a residual correction (Zhang & Li, 2018). The downscaling accuracy has been enhanced by (Jing et al, 2017), through developing a model based on Normalized Difference Vegetation Index (NDVI), land surface temperature (LST), and digital elevation model (DEM). This study provides TRRMM product downscaling at 1km using two machine learning algorithms, support vector machine (SVM) and random forests (RF).

This chapter proposes a novel downscaling model, which best accurate precipitation data at a fine scale of 1km. This model is based on Transformed Soil Adjusted

Vegetation Index (TSAVI), Digital Elevation Model (DEM), and Distance from the Sea, using a machine learning algorithm.

Precipitation is highly correlated with environmental factors (Jing et al, 2017; Zhang & Li, 2018). The major downscaling models are based mainly on Normalized Difference Vegetation Index (NDVI) due to its capacity to better accurate precipitation data (Wang et al., 2003.).Indeed, precipitation is a fundamental component of the hydrological cycle. Therefore precipitation is a limiting factor of soil moisture and vegetation growth. Soil moisture is considered as a significant node that relates precipitation and NDVI (Wang et al., 2003.). In some extent, many studies have demonstrated that NDVI performs a good correlation with precipitation(Chen et al., 2014; l. Duan et al., 2017; Jia et al., 2011; J. Liu et al., 2018; Jing et al, 2017; Y. Zhang & li, 2018). However, NDVI values are influenced by soil (Baret et al., 1991.; Rondeaux et al., 1996).

Soil brightness induces noise into red and near-infrared bands (Rondeaux et al., 1996). Indeed, red and near-infrared are linearly correlated in a way that the intercept and the slope depending on the soil types (Rondeaux et al., 1996). Furthermore, the soil is properties are characterized by its inorganic solids, organic matter, and soil moisture (Rondeaux et al., 1996). The development of remote sensing provides a proper assessment of soil properties dues to its reflectance in the visible and near-infrared spectra (Rondeaux et al., 1996). Therefore, soil line parameters influence the vegetation information in visible and near-infrared bands (Baret et al., 1996; Rondeaux et al., 1996). In order to sort-out, the soil line parameters the extracting the bottom in red and near-infrared bands is essential. In this context, the Transformed Soil Adjusted Vegetation Index as an improved vegetation index was developed to overcome soil background effects (Baret et al., 1996).

This chapter aims to improve downscaling models by reducing soil brightness noise by using the Transformed Soil Adjusted Vegetation Index. In this sense, TSAVI, DEM, and distance from the sea were considered as a predictor for desegregating precipitation, using stepwise regression, AIC, and Cp criterion, to evaluate the model error and performance The MSE, MAE, RMSE, and R-Square statistical metrics are used. Besides, this chapter provides scientific advances in assessing accurate precipitation at a fine-scale spatial resolution to enhance the accuracy of estimating long-term climate change projection, prediction of climate change variability.

MATERIALS AND METHODS

Study Area

This study was conducted in Morocco (Figure 1). Morocco is located in northwestern Africa, between [1.5° -17°] W longitudes and [20.5°-36°] N latitudes. The Kingdom is bathed in the North by the Mediterranean Sea, in the West by the Atlantic Ocean (Ezzine et al., 2017). The country is bordered to the east by Algeria, and to the south and south-east by Mauritania. Morocco covers an area of 710,850 km^2, included fourteen watersheds, with 25% of agricultural land (Moumen et al., 2019). Besides, Morocco is surmounted in the West by the Atlas Mountains, in the North by the Rif, and in the south by a large Sahara. The climate ranges from the Humid to Saharan

Figure 1. Moroccan map highlighting watershed and bioclimatic stages
(Source: the authors)

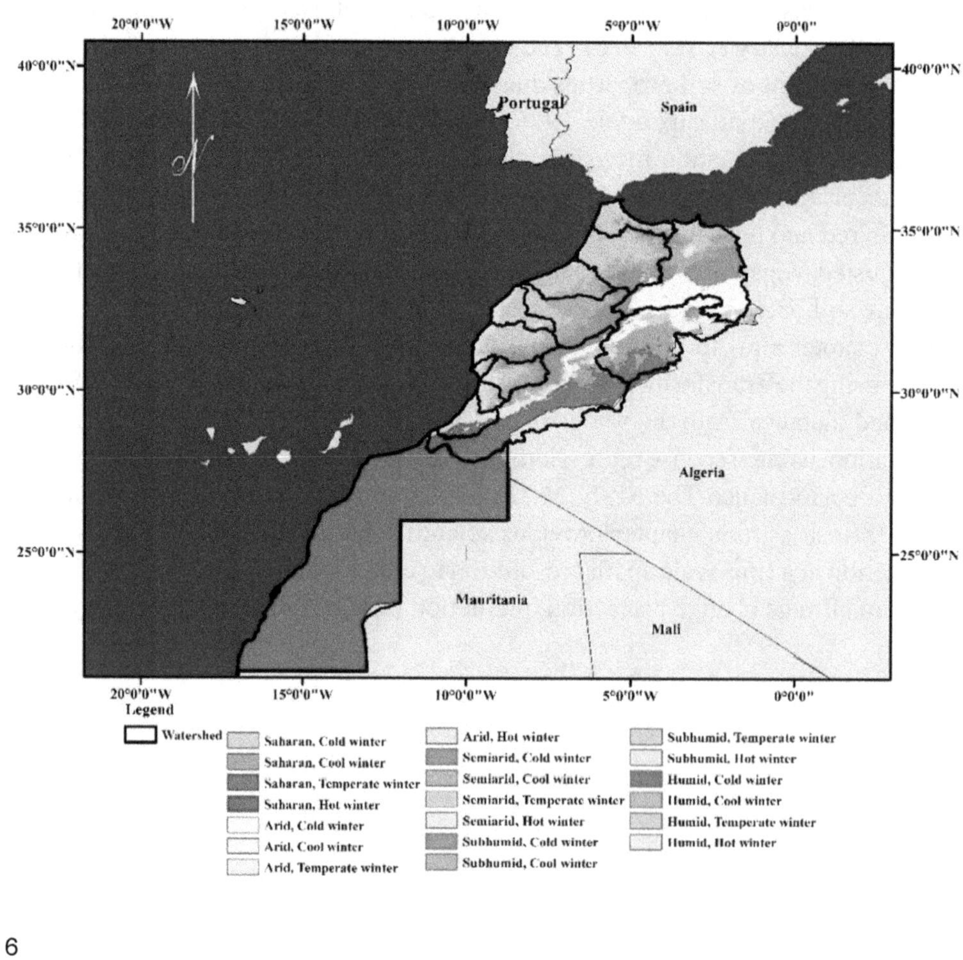

bioclimatic stage (Ezzine et al., 2017). Furthermore, Morocco is a highly water-stressed country with erratic rainfall and frequent droughts(Africa et al., 2014). Indeed, Morocco receives an amount of rainfall ranging from 100mm to 1200 mm, with an evaporation rate of 80% and an interception rate of 0.6%(Schilling et al., 2020).

Data Collection

Satellite Sensor Precipitation Data

In this chapter, the TRMM 3B43 Version 7 product is used as a free precipitation data with a spatial resolution of 0.25°×0.25°. TRMM 3B34 precipitation data over the period 1999 to 2014 are considered.

Climatic Data

Moroccan bioclimatic map was generated by using Emberger's quotient (Condés & García-Robredo, 2012; Mokhtari et al., 2013). Emberger's quotient was calculated using precipitation and temperature data for a period ranged from 1970 to 2000. Weather data was retrieved freely from worldclim portal https://worldclim.org/

Emberger's quotient formula:

$$Q = \frac{2000 \times P}{(M + m + 546.4) \times (M - m)} \tag{1}$$

Where: P : The annual rainfall (mm), M : The average maxima of the hottest month, m the average minima of the coldest month

Hydroclimatic Data

The hydroclimatic data were provided over Morocco. The most available monthly rainfall data were obtained from the recording of the rain stations for a period of 12 years (1999-2012).

Gis Data

- The digital elevation model (DEM) has been derived from the Shuttle Radar Topography Mission (SRTM), Version 4.1, with a spatial resolution of 90m. DEM data is available on the portal http://srtm.csi.cgiar.org/.

- Remote sensing imagery has been retrieved from the most recent generation of Landsat satellite Landsat 8 (OLI) (Mancino et al., 2020). Landsat 8 (OLI) is the 9 Operational Land Imager (OLI) and Thermal Infrared Sensor (TIRS), this satellite sensor is composed of nine bands: band four is red, and band five is near-infrared (https://www.usgs.gov/). Landsat 8 imagery is used in the Mediterranean region especially when it concern providing vegetation indices (Mancino et al., 2020). Moreover, Landsat imagery is freely available on earth explorer portal https://earthexplorer.usgs.gov/, with a spatial resolution of 30 m. Landsat 8 images from 2010 to 2019, was used in this chapter to retrieve bare soil lines and the Transformed Soil Adjusted Vegetation Index (TSAVI). The rasters over the period 2010-2019 concern the month's May, June, June, July, July, and August.
- Distance from the sea map was generated about the Atlantic and the Mediterranean coastline.

Table 1. Data sets

Data	Characteristics		
	Period	**Spatial Resolution**	**Temporal Resolution (Synthesis)**
TRMM 3B43	1999–2012	0.25°	Monthly
Landsat 8 (OLI)	2010-2019	30 m	—
Hydroclimatic data	1999–2012	1 km	Monthly
DEM (SRTM V4.1)	2008	90 m	—
Distance from sea	2019	1 km	—
Bioclimatic map	2019	1 km	—

(Source: the authors)

Methods

The methodology adopted aims to build a robust model able to best downscale precipitation data using machine learning. The diagram below presents a schematic illustration of the structure and the working of the downscaling model (Figure 2).

Figure 2. Downscaling algorithm used in this chapter
(Source: the authors)

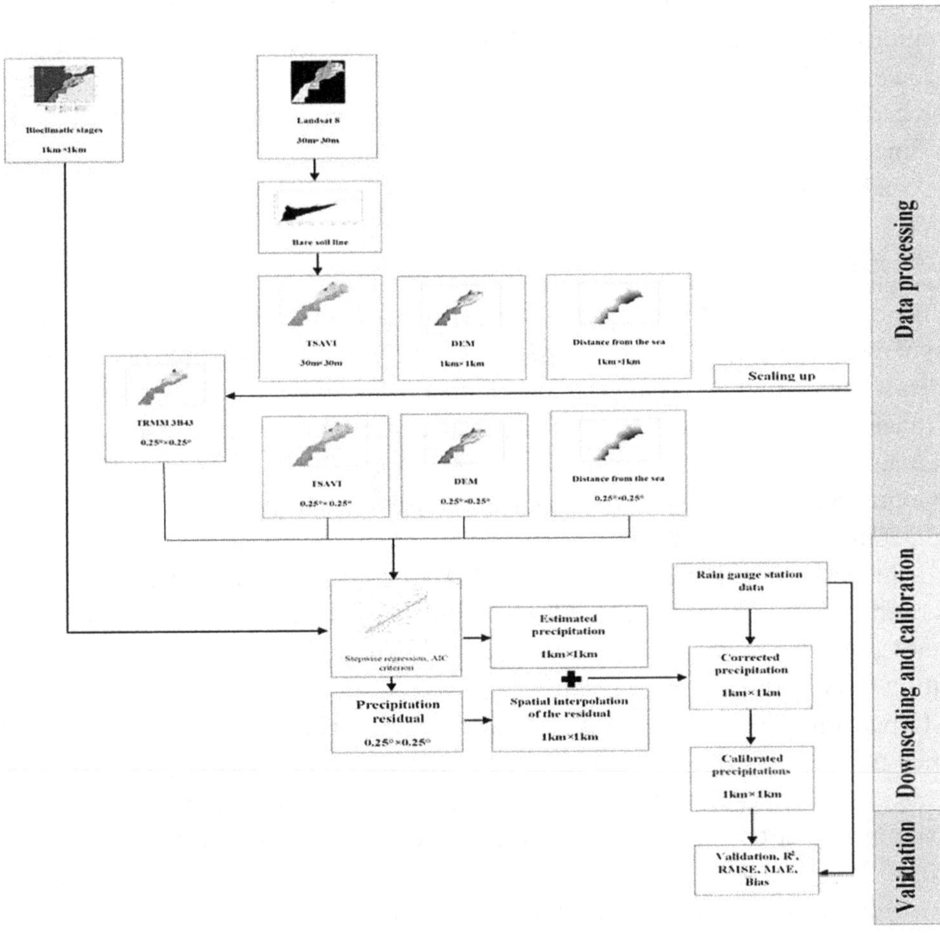

Data Preparation

The first step consists of the calculation of the annual precipitation, from the monthly TRMM 3B43 product over the period 1999-2012. After, the annual precipitation will be upscaled to a spatial resolution of 0.25°x0.25°. Besides, the distance from the sea and the DEM rasters will also be scaled up to a spatial resolution of 0.25°x0.25°.

Transformed Soil Adjusted Vegetation Index Generation

Accurate data through satellite sensors has reached a mature level. However, handling this amount of data is not evident because of its characteristics of bid data (P. Liu et al., 2018). In this sense, collecting and processing Landsat 8 imagery for Morocco needs the integration of many tools and methods that deals with big data. The imagery reassembling was conducted through GIs, besides statistical process was conducted through R software

Bare Line Soil

By the development of data science, quantile regression performs best than least square regression, which has been used extensively as a statistical method. Indeed, the least square regression is a parametric model based on assumptions; however, the quantile regression does not forge any assumptions between the distributions of residuals (Foulds, 1984.; Koenker et al., 2007). Quantile regression is efficient in extracting boundary lines in two-dimensional scatter plots (Mills et al., 2006; Mills et al., 2009; Xu & Guo, 2013)

In this sense, quantile regression is used to retrieve the bare soil line. The method has been coded in R software and is available in a software package called raster quantile, which can be downloaded from the R Archive Network (CRAN) mirror at http://www.r-project.org/.

To perform the quantile regression, the quantile tau is set to 0.00001 in order to extract the vary soil line in Near-Infrared (NIR) versus Red scatter plots. We vary the tau is step by step, to extract the best bottom line with the minimum p-value and maximum t-value (Ahmadian et al., 2016; Xu & Guo, 2013). We gain the quantile by:

$$\frac{the\,number\,of\,points\,below\,or\,on\,the\,bottom\,line}{the\,total\,number\,of\,points}$$

The bottom line extraction is processed for the fourteen watersheds of Morocco.

TSAVI Generation

According to the algorithm shown in the table (2), the Transformed Soil Adjusted Vegetation Index is generated for the fourteen watersheds of Morocco.

Table 2. Transformed Soil Adjusted Vegetation Index

Vegetation Index	Algorithm	Description	Citation
TSAVI	$\frac{a\times\left(NIR-a\times RED-b\right)}{RED+a\times NIR-a\times b+X\times\left(1+a^2\right)}$	$a=$ slope of the Soil Line, $b=$ intercept of the SL, $X=0.08$	(Baret & Guyot, 1991; Rondeaux et al., 1996)

(Source: the authors)

Correlation Between TRMM 3B43 Products and TSAVI

Least square regression is used to assess the correlation between the annual precipitation of the TRMM 3B43 product as a dependent variable and TSAVI as an independent variable, over the period 1999-2004. The performance of the correlation is checked based on the R squareR^2.

Correlation Between TRMM 3B43 and Predictors Using Machine Learning

Machine learning techniques are a significant method used widespread among data scientists. There are two types of machine learning techniques: supervised and unsupervised learning. In this chapter, we will apply supervised learning, namely regression. The downscaling model is carried out using stepwise regression based on the Akaike information criterion (AIC) and Mallows' Cp.

Stepwise regression is a multiple regression method aiming to identify the best fit model predictors. This method tries to optimize the choice of the explanatory variables that predict the dependent variable. Stepwise regression automatically consists of iterative model construction. The choice of the best explanatory variables is based on different criteria like F-tests, t-tests, Adjusted R square, Akaike information criterion, Bayesian information criterion, and Mallows' Cp indicator (Ormann et al., 2018). This criterion seeks to identify a set of independent variable that explains the dependent variable significantly (Ormann et al., 2018).

Stepwise regression involves forwards and backward selection. In one hand forward selection consist on the introduction of the variables one by one and keep the most significant ones with the less p-value, until the result is optimal, in the other hand Backward selection consists on setting all the explanatory variables and discard those who are statistically not significant, with the less p-value. Stepwise regression aims to best fit a linear regression between the dependent and the independent variables, according to the following equation.

Linear regression equation:

$$y = a_0 + a_1 x_1 + a_2 x_2 + \ldots\ldots + a_p x_p + \varepsilon \tag{2}$$

Where: y is a dependent variable, $x_{1\leq j\leq p}$ are the independent variables, $a_{1\leq j\leq p}$ are the parameters to be estimated, and ε is the intercept or the residual.

In order to enhance the model selection in stepwise regression, we use the Akaike information criterion and Mallows's Cp as a powerful technique used in the weighted likelihood methodology for model selection (Agostinelli et al., 2002).

On the one hand, the Akaike information criterion (AIC) (Akaike et al, 1974) seeks to estimate the likelihood of the independent variable to explain the dependent. A considered as a good predictor of the variable that gas the minimum AIC among the other variable. The AIC equation is as follow:

The Akaike information criterion formula:

$$AIC = -2 \times \ln(L) + 2 \times k \tag{3}$$

Where: L is the value of the likelihood and k is the number of estimated parameters.

On the other hand, Mallows's Cp (Taylor & Mallows, 2012)seeks to assess the precision and bias of the predictors. It is considered unbiased and precise estimators of the predictors that have a small value near to the number of predictors. (Agostinelli et al., 2002; Lance et al., 2014)

The Mallows's Cp formula:

$$C_p = \frac{RSS_p}{MSE_k} + 2 \times p - n \tag{4}$$

Where: RSS_p is the residual sum of squares for the p-variable, MSE_k is the mean squared error for the k-variable, and n is the sample size

Stepwise regression is widely used in data mining, with high use in processing big data (Smith et al., 2018). In this chapter, we use Stepwise regression, with the combination of the forward and backward selection techniques, based on Akaike information criterion and Mallows's Cp indicator, in order to identify the best variable combination among TSAVI, DEM and distance from the sea, that explain TRMM 3B43 product in the best way.

The stepwise regression with AIC and Mallows's Cpcriteria was performed over the period 1999-2004 in order to identify the best model after the model is generalized to all the study period (1999-2014).

Downscaling and Calibration of Precipitation

The group model identified using stepwise regression, were used to downscale precipitation data, with a residual correction. First of all, we estimate precipitation at a spatial resolution of 0.25°×0.25°. Then we estimate the residual component by generating the difference between the original TRMM 3B43 data and the estimated precipitation. After that, the residual component was downscaled at a spatial resolution of 1km using spatial interpolation. Downscaling precipitation was conducted by using the regression model variables at 1km, which were corrected by adding the residual component.

The calibration of the downscaled data was carried out by using the Geographical Differential Analysis(GDA) calibration method developed by Jehanzeb & Cheema, (2012). GDA method consists of minimizing the difference between the TRMM 3B43 precipitation products and the in-situ rain gauge measurements. Over the period 1999-2014, and for every station gauge, the difference between downscaled precipitation and in situ measurement was generated. Second, the difference was interpolated based on the Distance Weighting algorithm. Finally, we add the geographical difference to the downscaled precipitation at a spatial resolution of 1vkm.

Validation

The validation of the downscaled and calibrated precipitation data was carried out using the common statistical metrics, widely used in GIS, mathematic, environment and hydrology, especially the root mean square error (RMSE), the bias, and the mean absolute error (MAE) (Duan et al., 2017; Ezzine, 2017; Liu et al., 2018; Jing et al., 2017; Zhang & Li, 2018).

The mean absolute error formula:

$$MAE = \frac{\sum_{t=1}^{t=n} \left| \widehat{Y} - Y \right|}{n} \tag{5}$$

The root mean square error formula:

$$RMSE = \sqrt{\frac{\sum_{t=1}^{t=n}(\widehat{Y} - Y)^2}{n}} \tag{6}$$

The bias formula:

$$Bias = \frac{\sum_{t=1}^{t=n}\widehat{Y}}{\sum_{t=1}^{t=n}Y} - 1 \tag{7}$$

Where: $\hat{Y}$: is the dependent variable, Y is the independent variable, and n is the sample size.

Sensitivity to the Bioclimatic Stages

When assessing the sensitivity of the downscaling model to the bioclimatic stages (Ezzine, 2017) found that the model based on NDVI, DEM, and distance from the sea is weak in explaining precipitation in arid and Saharan bioclimatic stages. In this chapter, we aim to enhance the performance of the downscaling method in these bioclimatic stages. Indeed, explaining the precipitation in the denuded area from vegetation suffer from soil blitheness, which weakened the NDVI based model. The proposed model in this chapter, based on TSAVI, overcomes this problem. The resulting of the stepwise regression model was performed in every bioclimatic stage and asses the sensitivity of the model to the climatic zoning based on the p-value test.

RESULTS AND DISCUSSION

Bare Soil Line

Figure (3) present the NIR versus RED scatterplots with the extracted soil line, for all Moroccan watershed. The NIR versus Red scatters plots present some scattered points in low and high Red values, which may impede extracting bares soil line. Using quantile regression, we tried to fit the line with the bottom boundary line, instead of the existence of some sparse points. Especially the case of Tami, which present a high rate of the sparse point below the bottom boundary line. Quantile regression is used in many studies to extracting the bare sol line (Ahmadian et al.,

2016; Xu & Guo, 2013). According to (Ahmadian et al., 2016; Xu & Guo, 2013), quantile regression shows a better capacity in extracting bare soil line. Furthermore, it shows that the best period of extracting bare soil line is the early green-up season starting from late March to late April, and late senescence season starting from early August to late October (Ahmadian et al., 2016; Xu & Guo, 2013). Table (3) turns –up the slope and the intercept of the extracted bare soil line, based on near-infrared and red scatterplot; using the quantile regression method. The mean bare soil intercept and slope over Morocco are respectively -771.838 and 1.006.

Figure 3. Bare soil line extracted for all Moroccan watershed, from Landsat 8 OLI images, based on NIR versus Red scatterplot; using the quantile regression method *(Source: the authors)*

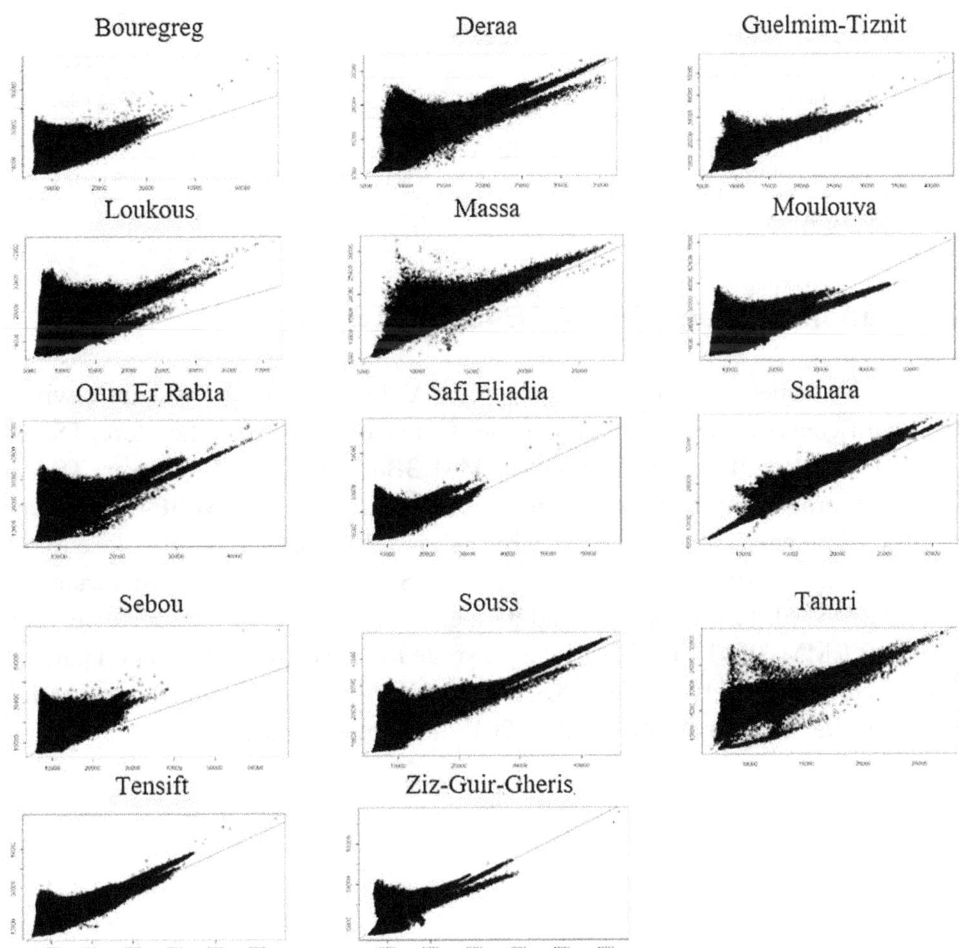

Table 3. Slope and intercept of extracted bare soil line from all watersheds over Morocco, using the quantile regression method

Method	Quantile Regression (NIR versus RED)	
Moroccan Watershed	Slope	Intercept
Bouregreg	0.816	-399.329
Deraa	1.053	-546.630
Guelmim-Tiznit	1.215	-1999.385
Loukous	0.645	831.088
Massa	1.102	-767.073
Moulouya	1.048	-488.026
Oum Er Rabia	1.094	-1476.695
Safi Eljadia	0.942	-1010.796
Sahara	1.103	-629.042
Sebou	0.684	735.890
Souss	1.013	-44.562
Tamri	1.313	-3369.862
Tensift	0.971	-1059.236
Ziz-Guir-Gheris	1.092	-582.077

(Source: the authors)

Correlation Between TRMM 3B43 and TSAVI

The correlation between TRMM 3B43 and TSAVI shows a high R-squareR^2 with a very significant p-value. The p-value varies from 0 to 5.3×10^{-268}; therefore, TSAVI is statistically significant to explain TRMM 3B43 products. Moreover, TSAVI has a high probability of obtaining an accurate result as TRMM 3B43 products. Besides, R^2 ranges from 0.7 to 0.88, which shows that the data is very close to the regression line. Also, these results show that TSAVI conveys the most variability of the TRMM 3B43 products around its mean (Figure 4). Testing the correlation between TRMM 3B43 and TSAVI demonstrate the best agreement in comparison with the correlation between TRMM 3B43 and NDVI or NDWI, which was carried out by several studies namely (Ezzine, 2017; Nichol & abbas, 2015)

Figure 4. Correlation between TRMM 3B43 precipitation and TSAVI over the period 1999-2004
(Source: the authors)

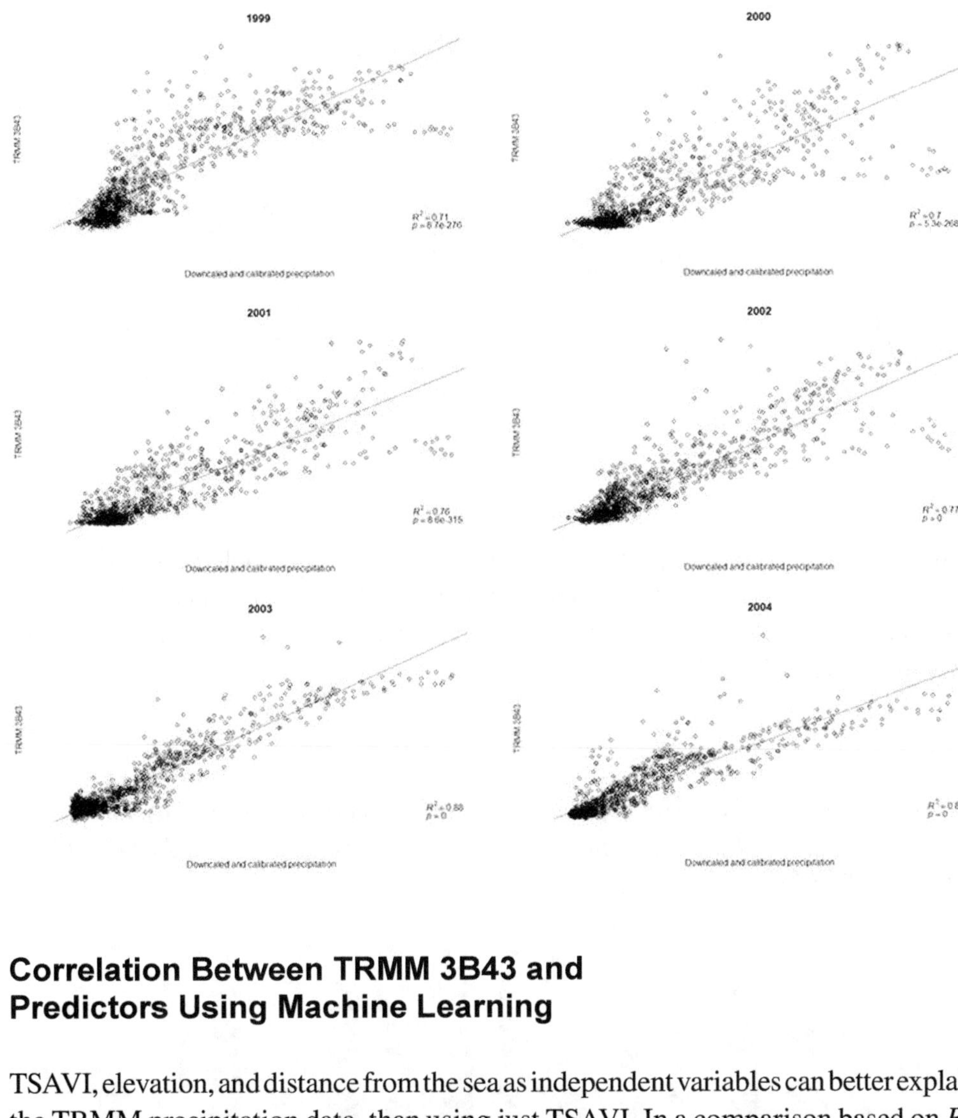

Correlation Between TRMM 3B43 and Predictors Using Machine Learning

TSAVI, elevation, and distance from the sea as independent variables can better explain the TRMM precipitation data, than using just TSAVI. In a comparison based on R^2, Figure 5 shows that TSAVI, elevation, and distance from the sea model enhanced the TSAVI based model by an average of 0.17 points. Indeed, the lower R^2increased from 0.7 to 0.73, and the high R^2increased from 0.88 to 0.89. Figure 5 shows that the average R^2using TSAVI, elevation, and distance from the sea model, is equal to 0.77, which explain that the model can significantly predict TRMM precipitations. Moreover, figure 6 shows that the range of R^2over the period 1999-2014 ranges

Figure 5. R square of the comparison between TRMM versus TSAVI and TRMM versus TSAVI, DEM and distance from the sea, over the period 1999-2014
(Source: the authors)

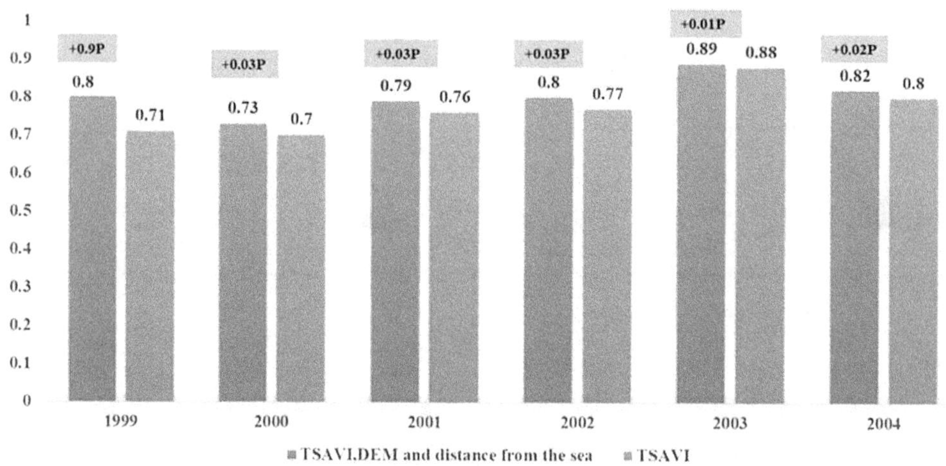

from 0.62 to 0.89. In comparison with the downscaling models based on NDVI and proposed by (Ezzine, 2017; Nichol & Abbas, 2015), the independent variable based on TSAVI, elevation, and distance from the sea shows the best agreement in explaining TRMM precipitation data.

Figure 6. R square of the correlation between TRMM precipitation and the independent variables (TSAVI, DEM, and distance from the sea; over the period 1999-2014
(Source: the authors)

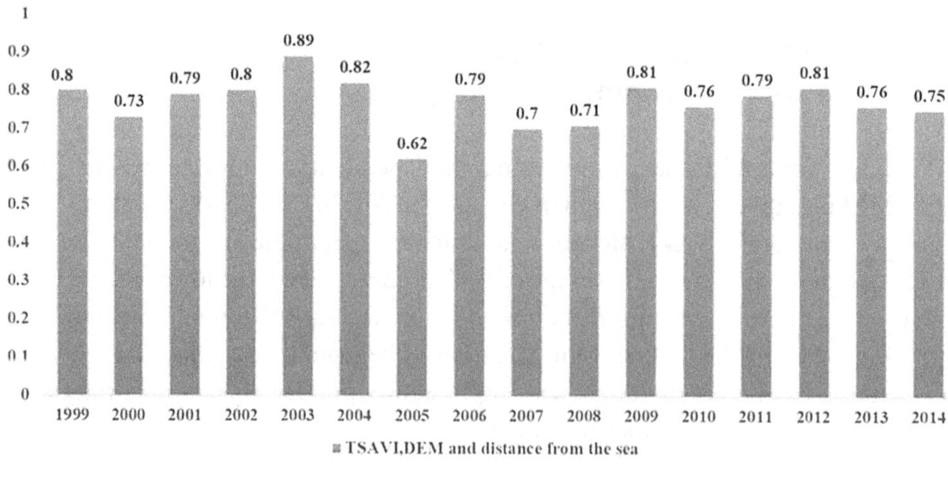

Figure 7. Statistical model Accuracy Metrics: R square, Adjusted R square, Cp, and AIC; for the TRMM vs. TSAVI, DEM, and distance from the sea.
(Source: the authors)

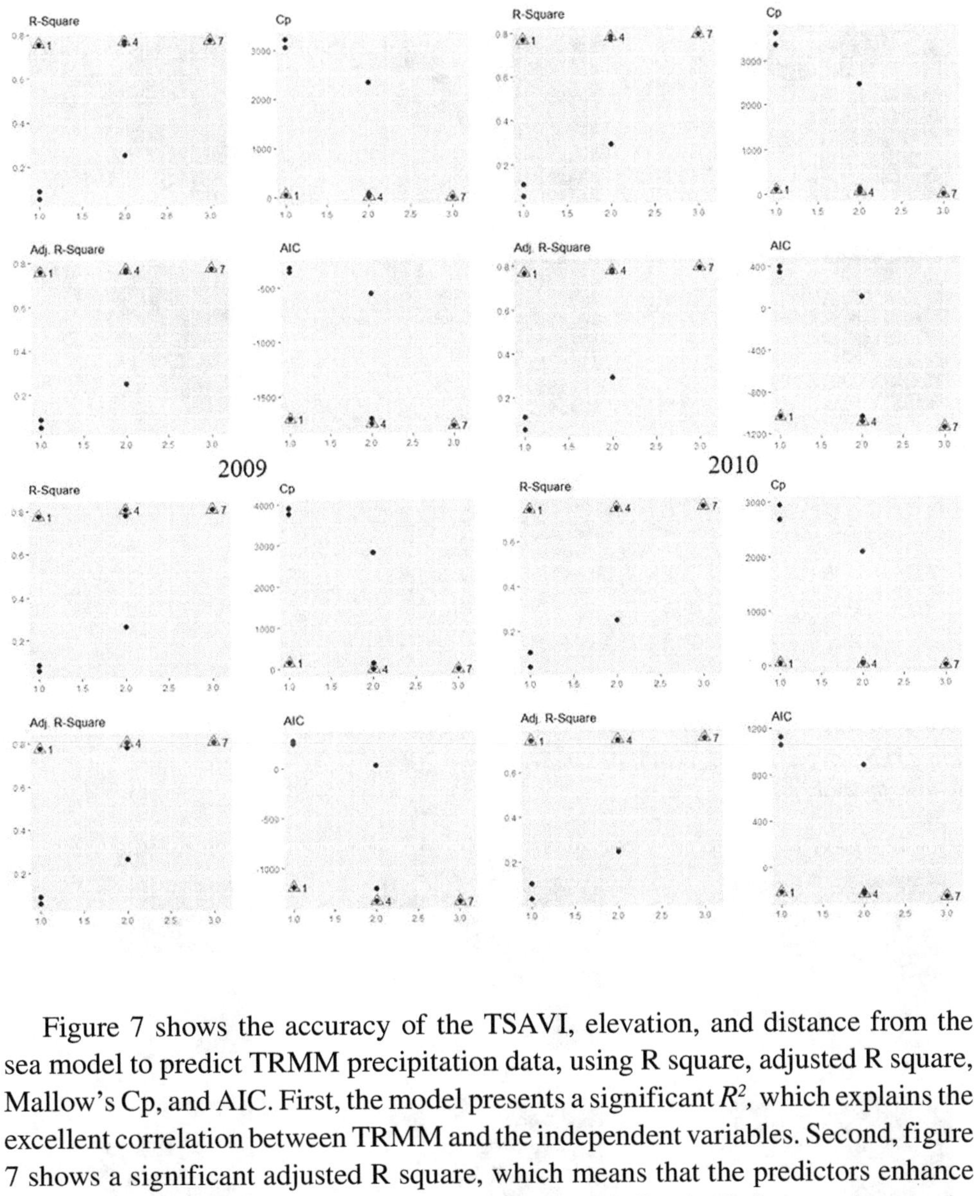

Figure 7 shows the accuracy of the TSAVI, elevation, and distance from the sea model to predict TRMM precipitation data, using R square, adjusted R square, Mallow's Cp, and AIC. First, the model presents a significant R^2, which explains the excellent correlation between TRMM and the independent variables. Second, figure 7 shows a significant adjusted R square, which means that the predictors enhance the model correlation significantly. Furthermore, Mallow's Cp is very close to the number of predictors; there for the model best fits with high precision and low bias. Besides, the low AIC values explain the goodness of the model quality in a way that the predictors explain TRMM precipitation data with less loss of information.

Figure 8. Estimated precipitation using TSAVI, DEM and distance from the see in a spatial resolution of 0.25°×0.25°, over the reference period of 1999-2004
(Source: the authors)

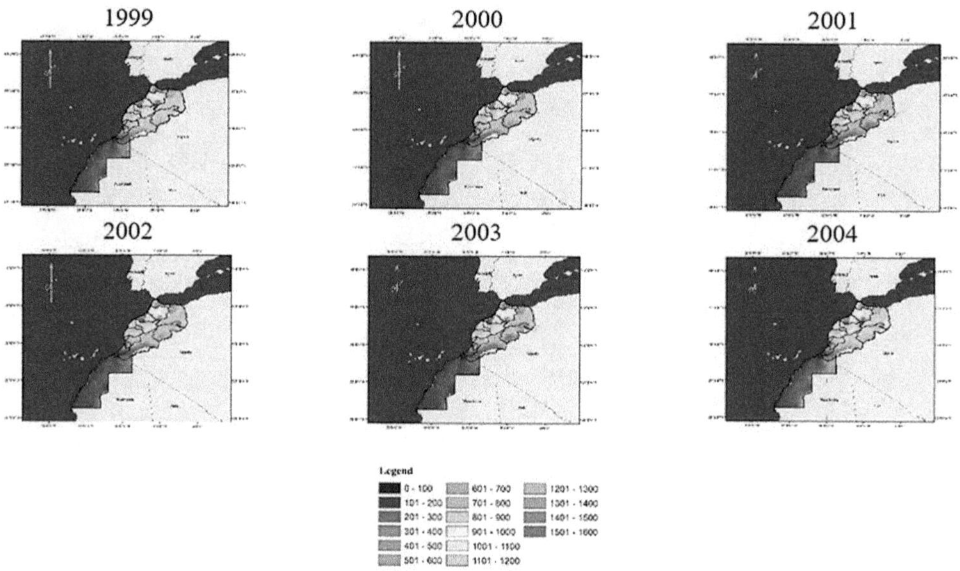

Figure 9. Spatial distribution of the regression residual at a spatial resolution of 1km×1km, over the reference period of 1999-2004
(Source: the authors)

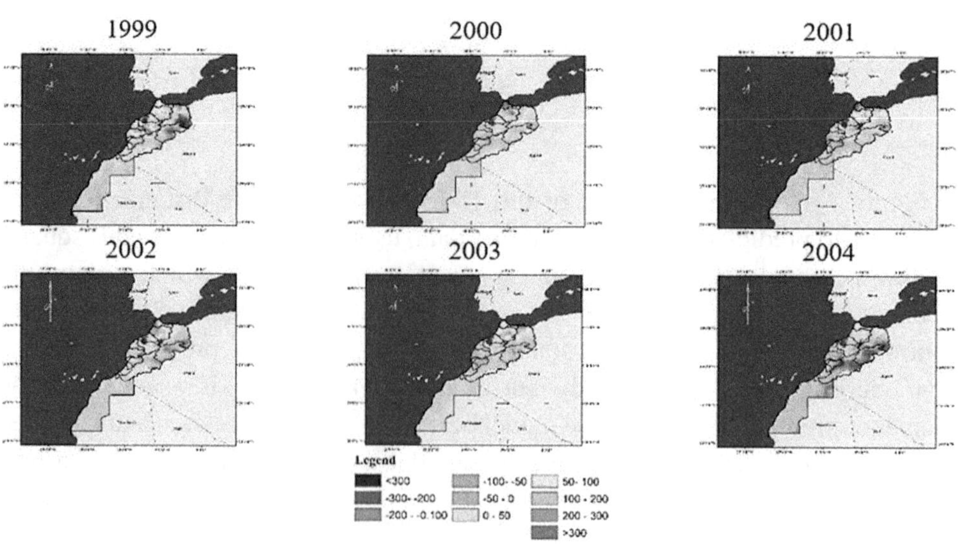

Downscaling and Calibration of Precipitation

Figure 8 shows the spatial distribution pf estimated precipitation using TSAVI, elevation, and distance from the sea predictors over the reference period of 1999-2004. The figure shows that precipitation is abundant in the North in Morocco and very scarce in the south, which explains the fact that Morocco suffers from water scarcity(Moumen et al., 2019).

The spatial distribution of the model error is highlighted in figure 9. The residual of precipitation that the model could not estimate is used parameter of correction to enhance the model performance. Indeed, positive values indicate the model underestimated TRMM product data. However, negative values indicate that the model overestimates TRMM product data.

*Figure 10. Downscaled and calibrated precipitation at a spatial resolution of 1km*1km, over the reference period 1999-2004*
(Source: the authors)

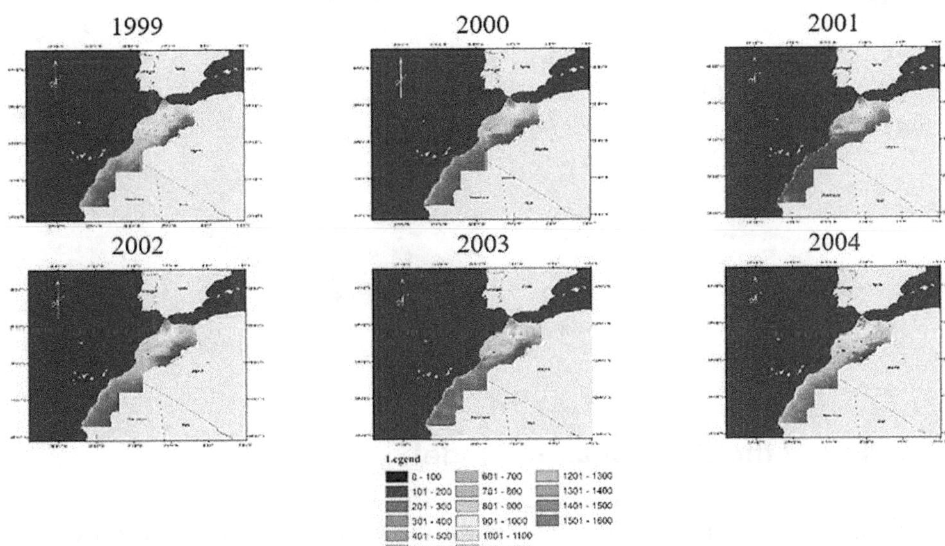

Downscaled and calibrated precipitation using TSAVI, elevation, and distance from the sea model shows a significant correlation with TRMM 3B43 precipitation data. For the reference period, the correlation between TRMM vs. downscaled and calibrated precipitation presents a high R^2, which ranges from 0.85 to 0.91. Furthermore, the p-value of the correlation is equal to zero, which explains the significance of the predictors (Figure 10 and Figure 11).

Figure 11. TRMM vs. Downscaled and calibrated precipitation over the reference period 1999-2004
(Source: the authors)

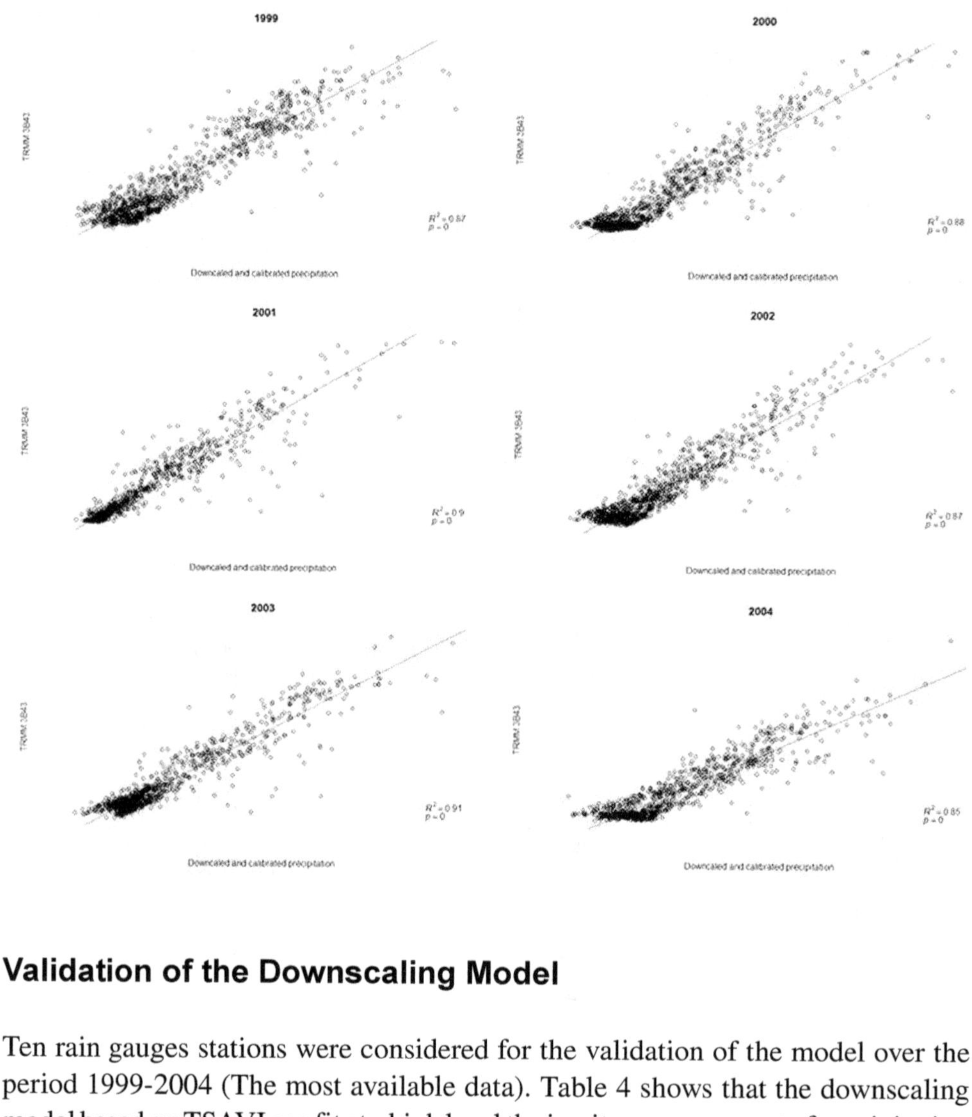

Validation of the Downscaling Model

Ten rain gauges stations were considered for the validation of the model over the period 1999-2004 (The most available data). Table 4 shows that the downscaling model based on TSAVI can fit at a high level the in-situ measurement of precipitation at a spatial resolution of 1km, in comparison to the NDVI based models (Ezzine, 2017; Nichol & Abbas, 2015). Over the period 1999-2004, R^2 ranges from 0.83 to 0.99, and bias varied between -0.012 and -0.04. Besides, during the same period, RMSE (root mean squared error), and MAE (mean absolute error) shows the model performs with a minimum error. Indeed, RMSE ranges from 15.73 to 47.62, and MAE ranges from 12.65 to 38.94, which indicates that the model has high accuracy in

Table 4. Spatial statistical metrics calculated the downscaled and validated precipitation versus in-situ measure precipitation from rain gauge stations

Statistical Metrics	1999	2000	2001	2002	2003	2004
R^2	0.83	0.96	0.95	0.97	0.99	0.99
RMSE	47.62	46.93	20.05	42.32	15.73	15.92
Bias	-0.12	-0.09	-0.04	-0.07	-0.02	-0.04
MAE	38.94	44.05	15.59	30.49	12.65	13.82

(Source: the authors)

predicting precipitation at a spatial resolution of 1km. The current model was applied over the period 2005-2014, to assess the spatial distribution of the precipitation at 1km over Morocco.

Sensitivity to Bioclimatic Stages

There is a high correlation between TRMM 3B43 product and TSAVI for the humid bioclimatic stage. Indeed, Table 5 shows a high R^2 value and a significant p-value for the humid bioclimatic stages. For the humid bioclimatic stage, R^2 ranges from 0.76 to 0.88, and the p-value is in range of 7.1×10^{-9}-1.7×10^{-6}. Arid and semiarid bioclimatic stages present an average R^2 of 0.30. Furthermore, semiarid and subhumid bioclimatic stages present the lower R^2average.

The methodology shows its capacity to better explain TRMM 3B43 product in the Humid bioclimatic stage by comparison to the NDVI based methodology, namely (Ezzine, 2017). Moreover, the TSAVI based model could slightly enhance the assessment of TRMM 3B43 product in the Saharan bioclimatic stage. However, the model is weak in explaining TRMM 3B43 precipitation data in both arid and semiarid stages.

CONCLUSION

This chapter provides scientific advances in assessing accurate precipitation at a fine-scale spatial resolution to enhance the accuracy of estimating long-term climate change projection, prediction of climate change variability. Furthermore, the chapter seeks to propose a novel model for downscaling corpse precipitation, taking into account the bioclimatic zoning. The downscaling model is based on TSAVI, DEM, and distance from the sea. The study was carried out over Morocco. In this chapter, the TRMM 3B43 product was considered as a free satellite sensor data over the

Table 5. Correlation between TRMM 3B43 product and TSAVI, for the five bioclimatic stages over the period 1999-2004

Year	Bioclimatic Stage	R^2	*p*-Value
1999	Humid	0.76	1.7×10^{-6}
	Sub humid	0.36	0.015
	Semiarid	0.11	0.011
	Arid	0.40	9.5×10^{-14}
	Saharan	0.30	1.8×10^{-11}
2000	Humid	0.85	3.2×10^{-8}
	Sub humid	0.16	0.00072
	Semiarid	0.2	7×10^{-4}
	Arid	0.20	3.6×10^{-6}
	Saharan	0.21	4.5×10^{-8}
2001	Humid	0.80	3×10^{-7}
	Sub humid	0.26	1.2×10^{-5}
	Semiarid	0.11	0.012
	Arid	0.32	8.4×10^{-11}
	Saharan	0.41	1.2×10^{-8}
2002	Humid	0.88	7.1×10^{-9}
	Sub humid	0.36	1.7×10^{-7}
	Semiarid	0.15	0.0084
	Arid	0.27	3.9×10^{-9}
	Saharan	0.22	5.8×10^{-23}
2003	Humid	0.82	1.8×10^{-7}
	Sub humid	0.10	0.017
	Semiarid	0.10	0.012
	Arid	0.22	9.9×10^{-8}
	Saharan	0.43	1.4×10^{-10}
2004	Humid	0.83	10^{-7}
	Sub humid	0.12	0.0052
	Semiarid	0.2	0.00069
	Arid	0.37	1.1×10^{-11}
	Saharan	0.27	3.2×10^{-10}

(Source: the authors)

period 1999-2014. The assessment of the relationship between the model component and TRMM 3B43 product is conducted using stepwise regression based on the AIC criterion and Mallows' Cp indicator, as a machine learning technique. The results show a significant correlation between TRMM 3B43 products and TSAVI with an R^2 ranges from 0.7 to 0.88 and a *p*-value ranges from 0 to5.3× 10^{-268}. The other environmental factors, namely DEM and distance from the sea, enhanced the model performance. The correlation between TRMM vs. downscaled and calibrated precipitation present a high R^2, which ranges from 0.85 to 0.91

Furthermore, the Transformed Soil Adjusted Vegetation Index (TSAVI) enhanced the downscaling model over the Humid and Saharan bioclimatic stages due to its capacity to reduce the soil brightness with an average R^2 respectively of 0.82 and 0.30.

The statistical metrics RMSE, MAE, and bias show that downscaled and calibrated precipitations best fit with the in-situ measured precipitation data.

REFERENCES

Agostinelli, C. (2002). Robust model selection in regression via weighted likelihood methodology. *Statistics & Probability Letters*, *56*(3), 289–300. doi:10.1016/S0167-7152(01)00193-6

Ahmadian, N., Demattê, J. A. M., Xu, D., Borg, E., & Zölitz, R. (2016). *A New Concept of Soil Line Retrieval from Landsat 8 Images for Estimating Plant Biophysical Parameters*. Remote Sensing Article. doi:10.3390/rs8090738

Akaike, H. (1974). ANew Look at the Statistical Model Identification. *Institute of Electrical and Electronics Engineers*, *19*, 716–723.

Arkin, F. R. P. (1980). *On the Relationship between Satellite-Observed Cloud Cover and Precipitation*. Academic Press.

Arkin, P. A. (1979). *The Relationship between Fractional Coverage of High Cloud and Rainfall Accumulations during GATE over the B-Scale Array*. doi:10.1175/1520-0493(1979)107<1382:TRBFCO>2.0.CO;2

Baret, F., & Guyot, G. (1991). Potentials and Limits of Vegetation Indices for LAI and APAR Assessment. *INRA Bioclimatologie. Montfavet, France*, *173*, 161–173.

Baret, F., Guyot, G., & Major, D. J. (1989). Tsavi: A vegetation index which minimizes soil b r i g w e s s effects on w and. *INRA Bioclimatologie. Montfavet, France*, *2*, 1355–1358.

Belay, H., Ayalew, S., & Hailu, D. (2020). Hydrological performance evaluation of multiple satellite precipitation products in the upper Blue Nile basin, Ethiopia. *Journal of Hydrology: Regional Studies, 27*(August), 100664. doi:10.1016/j.ejrh.2020.100664

Chen, F., & Li, X. (2016). Evaluation of IMERG and TRMM 3B43 Monthly Precipitation Products over Mainland China. *Remote Sensing*, *8*(6), 1–18. doi:10.3390/rs8060472

Chen, F., Liu, Y., Liu, Q., & Li, X. (2014). International Journal of Remote Spatial downscaling of TRMM 3B43 precipitation considering spatial heterogeneity. *International Journal of Remote Sensing*, (August), 37–41. doi:10.1080/01431161.2014.902550

Chen, S., Zhang, L., She, D., & Chen, J. (2019). Spatial Downscaling of Tropical Rainfall Measuring Mission (TRMM) Annual and Monthly Precipitation Data over the Middle and Lower Reaches of the Yangtze River Basin, China. *Water (Basel)*, *11*(3), 1–17. doi:10.3390/w11030568

Chokkavarapu, N., & Ravibabu, V. (2019). Comparative study of GCMs, RCMs, downscaling and hydrological models : A review toward future climate change impact estimation. *SN Applied Sciences*, *1*(12), 1–15. doi:10.100742452-019-1764-x

Ciabatta, L., Brocca, L., Massari, C., Moramarco, T., Gabellani, S., Puca, S., & Wagner, W. (2016). International Journal of Applied Earth Observation and Geoinformation Rainfall-runoff modelling by using SM2RAIN-derived and state-of-the-art satellite rainfall products over Italy. *International Journal of Applied Earth Observation and Geoinformation*, *48*, 163–173. doi:10.1016/j.jag.2015.10.004

Condés, S., & García-robredo, F. (2012). Forest Ecology and Management An empirical mixed model to quantify climate influence on the growth of Pinus halepensis Mill. stands in South-Eastern Spain. *Forest Ecology and Management Journal*, *284*, 59–68. doi:10.1016/j.foreco.2012.07.030

Darand, M., Amanollahi, J., & Zandkarimi, S. (2017). Evaluation of the performance of TRMM Multi-satellite Precipitation Analysis (TMPA) estimation over Iran. *Atmospheric Research*, *190*, 121–127. Advance online publication. doi:10.1016/j.atmosres.2017.02.011

Duan, L., Fan, K., Li, W., & Liu, T. (2017). Spatial downscaling algorithm of TRMM precipitation based on multiple high-resolution satellite data for Inner Mongolia, China. *Theoretical and Applied Climatology*, *2001*, 15.

Duan, Z., & Bastiaanssen, W. G. M. (2013). Remote Sensing of Environment First results from Version 7 TRMM 3B43 precipitation product in combination with a new downscaling – calibration procedure. *Remote Sensing of Environment, 131*, 1–13. doi:10.1016/j.rse.2012.12.002

Elhassnaoui, I., Moumen, Z., Serrari, I., Bouziane, A., Ouazar, D., & Hasnaoui, M. I. (2019). Generation of synthetic design storm hyetograph and hydrologic modeling under HEC HMS for Ziz watershed. *International Journal of Innovative Technology and Exploring Engineering*, *8*(10), 3308–3319. doi:10.35940/ijitee.J1214.0881019

Essery, C. I., & Wilcock, D. N. (2009). The variation in rainfall catch from standard UK Meteorological Office raingauges : A twelve year case study. *Hydrological Sciences Journal, 6667*. Advance online publication. doi:10.1080/02626669109492482

Ezzine, H., Bouziane, A., Ouazar, D., & Hasnaoui, M. D. (2017). Downscaling of Open Coarse Precipitation Data through Spatial and Statistical Analysis, Integrating NDVI, NDWI, Elevation, and Distance from Sea. *Hindawi, 2017*, 20. doi:10.1155/2017/8124962

Fang, J., Du, J., Xu, W., Shi, P., Li, M., & Ming, X. (2013). Advances in Water Resources Spatial downscaling of TRMM precipitation data based on the orographical effect and meteorological conditions in a mountainous area. *Advances in Water Resources*, *61*, 42–50. doi:10.1016/j.advwatres.2013.08.011

Foulds, L. R. (1984). Linear Programming and Extensions 1.1. *Combinatorial Optimization for Undergraduates*, 67.

Fowler, H. J., & Ekstr, M. (2009). Multi-model ensemble estimates of climate change impacts on UK seasonal precipitation extremes. *International Journal of Climatology*, *416*(January), 385–416. doi:10.1002/joc.1827

Gregg, W. W., & Casey, N. W. (2004). Global and regional evaluation of the SeaWiFS chlorophyll data set. *Remote Sensing of Environment*, *93*(4), 463–479. doi:10.1016/j.rse.2003.12.012

Guofeng, Z., Dahe, Q., Yuanfeng, L., Fenli, C., & Pengfei, H. (2016). Accuracy of TRMM precipitation data in the southwest monsoon region of China. *Theoretical and Applied Climatology*, *10*. Advance online publication. doi:10.100700704-016-1791-0

Gutjahr, O. (2013). Comparing precipitation bias correction methods for high-resolution regional climate simulations using COSMO-CLM Effects on extreme values and climate change signal. *Theoretical and Applied Climatology, 2012*. Advance online publication. doi:10.100700704-013-0834-z

Immerzeel, W. W., Rutten, M. M., & Droogers, P. (2009). Remote Sensing of Environment Spatial downscaling of TRMM precipitation using vegetative response on the Iberian Peninsula. *Remote Sensing of Environment, 113*(2), 362–370. doi:10.1016/j.rse.2008.10.004

Irvem, A., & Ozbuldu, M. (2019). Evaluation of Satellite and Reanalysis Precipitation Products Using GIS for All Basins in Turkey. *Hindawi, 12*, 1–11. Advance online publication. doi:10.1155/2019/4820136

Jehanzeb, M., & Cheema, M. (2012). International Journal of Remote Local calibration of remotely sensed rainfall from the TRMM satellite for different periods and spatial scales in the Indus Basin. *International Journal of Remote Sensing*, (November), 37–41. doi:10.1080/01431161.2011.617397

Jia, S., Zhu, W., Aifeng, L., & Yan, T. (2011). Remote Sensing of Environment A statistical spatial downscaling algorithm of TRMM precipitation based on NDVI and DEM in the Qaidam Basin of China. *Remote Sensing of Environment, 115*(12), 3069–3079. doi:10.1016/j.rse.2011.06.009

Jing, W., Zhang, P., Jiang, H., & Zhao, X. (2017). Reconstructing Satellite-Based Monthly Precipitation over Northeast China Using Machine Learning Algorithms. *Remote Sensing*, *9*(8), 1–17. doi:10.3390/rs9080781

Karaska, M. A., Huguenin, R. L., Beacham, J. L., Wang, M., Jensen, J. R., & Kaufmann, R. S. (2004). AVIRIS Measurements of Chlorophyll, Suspended Minerals, Dissolved Organic Carbon, and Turbidity in the Neuse River, North Carolina. *Photogrammetric Engineering and Remote Sensing*, *70*(1), 125–133. doi:10.14358/PERS.70.1.125

Kidd, C. (2001). Satellite Rainfall Climatology : A Review. *International Journal of Climatology*, *1066*(9), 1041–1066. doi:10.1002/joc.635

Kidd, C., & Levizzani, V. (2011). Status of satellite precipitation retrievals. *Hydrology and Earth System Sciences*, *15*(4), 1109–1116. doi:10.5194/hess-15-1109-2011

Kim, J. P., Jung, I. W., Park, K. W., Yoon, S. K., & Lee, D. (2016). Hydrological Utility and Uncertainty of Multi-Satellite Precipitation Products in the Mountainous Region of South Korea. *Remote Sensing*, *19*(7), 608. Advance online publication. doi:10.3390/rs8070608

Koenker, R., Bassett, G., & Jan, N. (1978). Regression Quantiles. *Econometrica*, *46*(1), 33–50. doi:10.2307/1913643

Kro, S., & Law, R. J. (2005). Sensing the sea. *Trends in Biotechnology*, *23*(5). Advance online publication. doi:10.1016/j.tibtech.2005.03.004 PMID:15866003

Lance, B. C. E. (2014). Mallows ' Cp Statistic I. *Wiley StatsRef: Statistics Reference Online*, 1–2.

Liu, J., Zhang, W., & Nie, N. (2018). Spatial Downscaling of TRMM Precipitation Data Using an Optimal Subset Regression Model with NDVI and Terrain Factors in the Yarlung Zangbo River Basin, China. [Research]. *Hindawi, 13*, 1–13. Advance online publication. doi:10.1155/2018/3491960

Liu, P., Di, L., Du, Q., & Wang, L. (2018). Remote Sensing Big Data : Theory, Methods and Applications. *Remote Sensing*, *4*(5), 711. Advance online publication. doi:10.3390/rs10050711

Maggioni, V., & Meyers, P. C. M. D. R. (2016). A Review of Merged High Resolution Satellite Precipitation Product Accuracy During the Tropical Rainfall Measuring Mission (TRMM) - Era. *Journal of Hydrometeorology*. Advance online publication. doi:10.1175/JHM-D-15-0190.1

Maidment, D. R. (1993). Handbook of Hydrology. McGraw-Hill Professional.

Mallows, C. L. (1973). Some Comments on Cp. *Technometrics,* 37–41. doi:10.10 80/00401706.1973.10489103

Mancino, & Ferrara. (2020). Landsat 7 (ETM +) Derived Vegetation Indices in a Mediterranean Environment. *Remote Sensing*, *8*, 20. doi:10.3390/rs12020291

Maraun, Wetterhall, Ireson, Chandler, Kendon, Widmann, Brienen, Rust, Sauter, Themeßl, Venema, Chun, Goodess, Jones, Onof, & Vrac. (2010). Precipitation downscaling under climate change: Recent developments to bridge the gap between dynamical models and the end user. *Rev. Geophys*, *48*(2009), 1–34. doi:10.1029/2009RG000314

Mehta, A. V., Yang, S., Group, M. S., & Springs, C. (2008). Advances in Geosciences Precipitation climatology over Mediterranean Basin from ten years of TRMM measurements. *Advances in Geosciences*, *17*, 87–91. doi:10.5194/adgeo-17-87-2008

Milewski, A., Elkadiri, R., & Durham, M. (2015). Assessment and Comparison of TMPA Satellite Precipitation Products in Varying Climatic and Topographic Regimes in Morocco. *Remote Sensing*, *21*(5), 5697–5717. Advance online publication. doi:10.3390/rs70505697

Mills, A., Fey, M., Donaldson, J., Todd, S., & Theron, L. (2009). *Soil infiltrability as a driver of plant cover and species richness in the semi-arid Karoo*. doi:10.100711104-009-9904-5

Mills, A. J., Fey, M. V., Gröngröft, A., Petersen, A., & Medinski, T. V. (2006). Unravelling the effects of soil properties on water infiltration : Segmented quantile regression on a large data set from arid south-west Africa. *Australian Journal OfSoil Research*, *44*(8), 783–797. doi:10.1071/SR05180

Mokhtari, N., Mrabet, R., Lebailly, P., & Bock, L. (2014). Spatialisation des bioclimats, de l ' aridité et des étages de végétation du Maroc. *Rev. Mar. Sci. Agron. Vét*, 50–66.

Mondal, P., Jain, M., Defries, R. S., Galford, G. L., & Small, C. (2014). Sensitivity of crop cover to climate variability : Insights from two Indian agro-ecoregions. *Journal of Environmental Management*. Advance online publication. doi:10.1016/j.jenvman.2014.02.026 PMID:24680541

Morocco. (2014). Drought Management in Morocco. *Workshop on Developing and Implementing Mitigation and Preparedness Of Water Scarcity and Drought (WS&D) Management Plans Zaragoza*, 38.

Moumen, Z. (2019). *Test the Capacity of SWAT Model to Simulate the Hydrological Behavior of Watershed in Semi-Arid Climate Hydrologic Modeling Using SWAT.* doi:10.4018/978-1-5225-9771-1.ch008

Moumen, Z., El, N., El, A., Tvaronavičienė, M., & Lahrach, A. (2019). Water security and sustainable development To cite this version : HAL Id : hal-02342701. *Insights into Regional Development*, *1*(4), 301–317. doi:10.9770/ird.2019.1.4(2)

Najja, S., Zad, M., Zulkafli, Z., & Muharram, F. M. (2018). Satellite Rainfall (TRMM 3B42-V7) Performance Assessment and Adjustment over Pahang River. *Remote Sensing*, 1–24. doi:10.3390/rs10030388

Naumann, Barbosa, Carrao, & Singleton. (2012). Monitoring Drought Conditions and Their Uncertainties in Africa Using TRMM Data. *American Meteorological Society*, 1867–1874. doi:10.1175/JAMC-D-12-0113.1

Nichol, J. E., & Abbas, S. (2015). Science of the Total Environment Integration of remote sensing datasets for local scale assessment and prediction of drought. *The Science of the Total Environment*, *505*, 503–507. doi:10.1016/j.scitotenv.2014.09.099 PMID:25461052

Ormann, C. A. F. D., Alabrese, J. U. M. C., Rroita, G. U. G. U., Atechou, E. L. M., & Ahn, V. O. B. (2018). Model averaging in ecology : A review of Bayesian, information-theoretic, and tactical approaches for predictive inference. *Ecological Monographs*, *0*(0), 1–20. doi:10.1002/ecm.1309

Paredes-trejo, F., Barbosa, H., & Santos, C. A. C. (2019). Evaluation of the Performance of SM2RAIN-Derived Rainfall Products over Brazil. *Remote Sensing*, *11*(9), 1–28. doi:10.3390/rs11091113

Park, N. (2013). Spatial Downscaling of TRMM Precipitation Using Geostatistics and Fine Scale Environmental Variables. *Hindawi*, *2013*, 1–9. doi:10.1155/2013/237126

Peck, L. (1974). Accuracy of Precipitation Measurements for Hydrologic Modelin. *Water Resources Research*, *I*(4), 857–863.

Rodda, J. C., & Dixon, H. (2011). Rainfall measurement revisited. *Royal Meteorogical Society*, *1842*, 131–136. doi:10.1002/wea.1884

Rodda, J. C., & Smith, S. W. (1986). Technical note the significance of the systematic error in rainfall measurement for assessing wet deposition. *Atmospheric Environment*, *20*(5), 1059–1064. doi:10.1016/0004-6981(86)90293-3

Rondeaux, G., Steven, M., & Baret, F. (1996). Optimization of Soil-Adjusted Vegetation Indices. *Remote Sensing and Environment, 107*(August), 95–107.

Schilling, J., Hertig, E., Tramblay, Y., & Scheffran, J. (2020). Climate change vulnerability, water resources and social implications in North Africa. *Regional Environmental Change*, *20*(1), 1–12. doi:10.100710113-020-01597-7

Schneider, U., Becker, A., Finger, P., Meyer-christoffer, A., Ziese, M., & Rudolf, B. (2014). GPCC ' s new land surface precipitation climatology based on quality-controlled in situ data and its role in quantifying the global water cycle. *Theoretical and Applied Climatology*, *115*(1-2), 15–40. doi:10.100700704-013-0860-x

Schwaller, M. R., & Morris, K. R. (2011). A Ground Validation Network for the Global Precipitation Measurement Mission. *Journal of Atmospheric and Oceanic Technology*, *28*(3), 301–320. doi:10.1175/2010JTECHA1403.1

Segond, M., Wheater, H. S., & Onof, C. (2007). The significance of spatial rainfall representation for flood runoff estimation : A numerical evaluation based on the Lee catchment, UK. *Journal of Hydrology (Amsterdam)*, *347*(1-2), 116–131. doi:10.1016/j.jhydrol.2007.09.040

Silva, E., & Lopes, S. (2017). Validation of Satellite Rainfall Products over a Mountainous Watershed in a Humid Subtropical Climate Region of Brazil. *Remote Sensing*, 1–18. doi:10.3390/rs9121240

Smith, G. (2018). Step away from stepwise. *Journal of Big Data*, 1–12. doi:10.118640537-018-0143-6

Song Yang, E. A. S. (2008). Convective – Stratiform Precipitation Variability at Seasonal Scale from 8 Yr of TRMM Observations : Implications for Multiple Modes of Diurnal Variability. *American Meteorological Society*. doi:10.1175/2008JCLI2096.1

Spring, S., & Peck, E. L. (1980). Robert E. Norton Memorial Lecture. *American Meteorological Society*, *61*(8), 894–902.

Tang, G., Ma, Y., Long, D., Zhong, L., & Hong, Y. (2016). Evaluation of GPM Day-1 IMERG and TMPA Version-7 legacy products over Mainland China at multiple spatiotemporal scales. *Journal of Hydrology*, *533*, 152–167. doi:10.1016/j.jhydrol.2015.12.008

Tang, L., Tian, Y., Yan, F., & Habib, E. (2015). An improved procedure for the validation of satellite-based precipitation estimates. *Atmospheric Research*, *163*, 61–73. Advance online publication. doi:10.1016/j.atmosres.2014.12.016

Thiemig, V., Rojas, R., Zambrano-bigiarini, M., & De Roo, A. (2013). Hydrological evaluation of satellite-based rainfall estimates over the Volta and Baro-Akobo Basin. *Journal of Hydrology (Amsterdam)*, *499*, 324–338. doi:10.1016/j.jhydrol.2013.07.012

Tian, Peters-Lidard, & Choudhury. (2007). Multitemporal Analysis of TRMM-Based Satellite Precipitation Products for Land. *Journal of Hydrometeorology*, 1165–1183. doi:10.1175/2007JHM859.1

Tsanis, D. D. A. I. K. (2016). Comparison of multiple linear regression and artificial neural network models for downscaling TRMM precipitation products using MODIS data. *Environmental Earth Sciences*. Advance online publication. doi:10.100712665-016-5883-z

Tuo, Y., Duan, Z., Disse, M., & Chiogna, G. (2016). Science of the Total Environment Evaluation of precipitation input for SWAT modeling in Alpine catchment : A case study in the Adige river basin (Italy). *The Science of the Total Environment*, *573*, 66–82. doi:10.1016/j.scitotenv.2016.08.034 PMID:27552731

Wang, J. (2013). Temporal responses of NDVI to precipitation and temperature in the central Great Plains, USA. *International Journal of Remote Sensing*, (March), 37–41. doi:10.1080/01431160210154812

Wang, W., Lu, H., Yang, D., Sothea, K., Jiao, Y., Gao, B., Peng, X., & Pang, Z. (2016). Modelling Hydrologic Processes in the Mekong River Basin Using a Distributed Model Driven by Satellite Precipitation and Rain Gauge Observations. *PLoS One*, *11*(3), 1–19. doi:10.1371/journal.ponc.0152229 PMID:27010692

Wheater, H. S. (2017). Modelling hydrological processes in arid and semi-arid areas : an introduction to the workshop. Cambridge University Press. doi:10.1017/CBO9780511535734.002

Xu, D., & Guo, X. (2013). A Study of Soil Line Simulation from Landsat Images in. *Remote Sensing*, *1*(9), 4533–4550. doi:10.3390/rs5094533

Xue, C., Song, W., Qin, L., Dong, Q., & Wen, X. (2015). A spatiotemporal mining framework for abnormal association patterns in marine environments with a time series of remote sensing images. *International Journal of Applied Earth Observation and Geoinformation*, *38*, 105–114. doi:10.1016/j.jag.2014.12.009

Yuan, F., Zhang, L., Wah, K., Win, W., Ren, L., & Zhao, C. (2017). Assessment of GPM and TRMM Multi-Satellite Precipitation Products in Streamflow Simulations in a Data-Sparse Mountainous Watershed in Myanmar. *Remote Sensing*, *9*(3), 1–23. doi:10.3390/rs9030302

Zhang, T., Li, B., Yuan, Y., Gao, X., Sun, Q., Xu, L., & Jiang, Y. (2018). Remote Sensing of Environment Spatial downscaling of TRMM precipitation data considering the impacts of macro-geographical factors and local elevation in the Three-River Headwaters Region. *Remote Sensing of Environment*, *215*(May), 109–127. doi:10.1016/j.rse.2018.06.004

Zhang, Y., Li, Y., Ji, X., Luo, X., & Li, X. (2018). Fine-Resolution Precipitation Mapping in a Mountainous Watershed : Geostatistical Downscaling of TRMM Products Based on Environmental Variables. *Remote Sensing*, *10*(1), 1–27. doi:10.3390/rs10010119

Zhao, H., Yang, B., Yang, S., & Huang, Y. (2017). NU SC Institute of Environment and Sustainable Development in Agriculture, Chinese Academy of. *Atmospheric Research*. Advance online publication. doi:10.1016/j.atmosres.2017.11.005

Zheng, X., & Zhu, J. (2015). A methodological approach for spatial downscaling of TRMM precipitation data in North China. *International Journal of Remote Sensing*, *36*(January), 37–41. doi:10.1080/01431161.2014.995275

Zhou, Z., Guo, B., Su, Y., Chen, Z., & Wang, J. (2020). Multidimensional evaluation of the TRMM 3B43V7 satellite-based precipitation product in mainland China from 1998 – 2016. *PeerJ*, *8*, 1–24. doi:10.7717/peerj.8615 PMID:32117637

KEY TERMS AND DEFINITIONS

Akaike Information Criterion (AIC): Is a statistical criterion that aims to choose the best independent variables that can explain the dependent variable. A considered as a good predictor of the variable that gas the minimum AIC among the other variable.

Machine Learning: Is an algorithmic-based technique known as training data. Machine learning is an algorithmic-based on a mathematical model that analyzes and computes data in a way to make predictions or decisions. This concept, a significant method used widespread among data scientists. There are two types of machine learning techniques: supervised and unsupervised learning.

Mallows' Cp: Is a statistical indicator, aiming to assess the precision and bias of the predictors. It is considered unbiased and precise estimators of the predictors that have a small value near to the number of predictors.

Quantile Regression: Is a regression technique used in data science. It is an advanced version of linear regression, which does not forge any assumptions between the distributions of residuals.

Remote-Sensing: Is the science of retrieving information about an object, a landscape, or a phenomenon without being in contact with them. Remote sensing techniques are based on sensors that retrieve information through the analysis of the electromagnetic spectrum reflected from an object, landscape, or a phenomenon. There are two types of sensors: passive sensors that respond to the amount of light naturally reflected from an object, a landscape or a phenomenon, and active sensors, which are radars that measure the transmitted light by the sensor that was reflected.

Spatial Downscaling: Is a technique that aims to reduce the gap between global or regional data with local data. Spatial downscaling mainly seeks to assess global or regional at fine-scale, taking into account the environmental factors in the local area.

Stepwise Regression: Is a multiple regression method aiming to identify the best fit model predictors. This method tries to optimize the choice of the explanatory variables that predict the dependent variable. Stepwise regression automatically consists of iterative model construction.

Transformed Soil Adjusted Vegetation Index (TSAVI): Is a vegetation index that seeks to minimize the soil brightness that noise to the electromagnetic wavelengths reflected from vegetation.

Chapter 2
Designing a Framework for Agri Sector Considering Disaster and Climatic Change:
A Case Study of Odisha

Suchismita Satapathy
https://orcid.org/0000-0002-4805-1793
KIIT University, India

ABSTRACT

Change of technology, utilization of genetically modified crops, and organic farming can be used to improve the fertility of land and to increase production in farming, but the effect of climate change is a big problem for the Indian farmers. Farmers have to face extreme weather conditions, the heavy workload during the fieldwork like weeding, harvesting, etc. The conventional method of farming and lack of advanced technology makes farming too difficult. Due to climate change, high heat, heavy rain, and frost, productivity decreases and lands become barren, and farmers also suffer from mental and physical disorders. Hence, an effort is taken to design a framework for the agriculture sector keeping climate change factors in view. To improve the agricultural system in Indian agrisector, some implications must be done to avoid wastage of rainwater. By creating an efficient drainage system in the crop field, the extra water can be reused. To avoid loss due to heavy rain, greenhouse or poly house can act as an effective way during natural calamities.

DOI: 10.4018/978-1-7998-3343-7.ch002

INTRODUCTION

Agricultural sector is a very challenging sector. It is dominated by nature and man made disasters. Rain is essential for increasing production of vegetables, pulses and agriproducts but sometimes heavy rain fall acts like curse and due to heavy rain flood washes away everything. Similarly sunlight is like a blessings for farming but hot climate dries fertile farming land and creates draught situation .There are many causes for Climate change like deforestation, pollution, population rise, waste disposal system, industrialization etc. The rise of air and water temperature rises sea levels, so supercharged storms and higher wind speeds, more intense and prolonged droughts, heavier precipitation and flooding takes place. Climate hazards are natural events in weather cycles. hurricanes and droughts, flooding and high winds always witnessing a scale of damage and destruction that is new and terrifying. It always makes millions of people homeless and indigent. Flooding and strong winds from an extraordinary succession of powerful hurricanes have destroyed homes and crops barrens fertile lands. Extreme weather disasters affect all countries, rich and poor. Climate change is forcing people from their homes, bringing poverty on top of poverty and increasing hunger.

Agriculture is the back bone of India and it acts as the skeleton of India's economy. More than 50% of the rural households manage their bread and butter on agricultural commodities by supplying food and necessary products to all over the world . It is found that the agricultural sector is the highest contributors of Gross Domestic Product (GDP) of India. Although the demand of agricultural sector is very high and the govt. Provides loans and subsidies for the development of Agricultural workers/ farmers, still the problems of farmers remains un resolved.

LITERATURE REVIEW

The nominal farmers of agricultural sector of India has to work in high noise, vibration and pollution . Modification of equipments have solved the problems of lengthy working procedure some how able to reduce heavy work load, safety related problems but created problems due to pollution and sustainability. Sustainability has turned out to be a principal concern for the country, researchers and international businesses (Drexhage and Murphy, 2010). The sustainable agricultural development was primarily concerned on the environmental factors followed by economic, social as well as political factors (DFID, 2003). Therefore the economic, social, and environmental dimensions need to be integrated to achieve sustainable development, because a subsystem's development on long-term basis is dependent on the others to some extent (Voinov and Smith, 1998). Political decisions at the local, regional or national levels are involved for sustainable decision-making aiming at impartial growth of socio-environmental systems. Further, defining as well as measurement of the sustainable development is a critical issue in sustainable decision-making. Moreover, as there is no exclusive sustainable pathway, and therefore, different criteria and approaches should be chosen by policy makers for efficient sustainable decisions for every country (Andriantiatsaholiniaina et al., 2004). An extensive and integrated way for dealing the social, monetary and natural processes for development was found to be fundamentally included for sustainability (Sathaye et al., 2007; Tracey and Anne, 2008) and a shared activity plan seemed of requiring differing partners interest and view points to build up for the development (Kates et al., 2005). Agricultural sustainability has been a fundamental concern because of its major impacts on production of foods, continual use of natural resources and environmental outcomes (Bell and Morse, 2008). The economical, social and environmental concerns were reported to be included for the sustainability of agriculture (Jackson-Smith, 2010; Nedea, 2012). Economic sustainability comprised of the ability of farmers for producing an adequate amount of food for feeding them as well as their society, and also maintaining the economic agricultural feasibility (Van Calker et al., 2008; Jackson-Smith, 2010). While, social sustainability comprised of the fairness along with the quality of life of farm workers, consumers as well as the community members. However, the environmental sustainability included the enrichment of the quality of the landscape environmental quality and proper use of natural resources (Sydorovych and Wossink, 2007; Jackson-Smith, 2010). It was reported that by the sustainable agriculture the food demands can be fulfilled by diverse initiatives for the present as well as future (DFID, 2004). Many of the assessment tools for sustainability have been developed over the past decade for the farm level assessment of sustainability performance (Marchand et al., 2014; Wustenberghs et al., 2015). For instance, Sharghi et al. (2010) have used a series of questionnaires communicated to 56 researchers in the research centers of Yazd and Tehran provinces via mail, email and fax, and found the effective factors such as Infrastructure, economy, policy-making, society, research, participation, extension and education, to achieve sustainable agriculture in Iran. Moreover, in case of sustainability assessment and performance on multiple themes, a set of indicators were used such as energy, soil and animal health within environmental, economical and social dimensions (Gasso et al., 2015; Van Cauwenbergh et al., 2007). Gaviglio et al. (2017) have considered the 'South Milan Agricultural Park' in Italy, and developed a framework for the assessment of farm sustainability based on three main steps: (i) collection of data through interviews with farmers and institutions, (ii) elaboration of data through an aggregative structure, and (iii) analysis of scores. Sulewski et al. (2018) have assessed the inter-dependencies between different dimensions for sustainability of farms, which was carried out by considering 601 farms participated in the "Polish Farm Accountancy Data Network". These data was supported by supplementary information from interviews, and on the basis of a number of variables, economical, social, environmental, and composite indices were collected for sustainability. Irfan et al. (2018) have used survey-based method and partial least squares structural equation modeling (PLSSEM) in their study by considering 425 respondents from public-sector organizations in Pakistan. They found a positive relationship between economic, environmental, as well as social sustainability and the corporate reputation. Sami et al. (2013) have proposed a comprehensive index considering two large maize and wheat production farms in Iran, to measure the sustainability in agricultural production system taking advantage of fuzzy logic combining all selected six indexes as the representatives of three dimensions of sustainability. Lechenet et al. (2014) have made an assessment for sustainability of cropping system by using a set of environmental, economic and social indicators. They have started with initiating diversified crop rotation by re-designing in the existing cropping system and pest management by combining a wide range of available techniques. Moreover, a modified Likert scale based on fuzzy set theory was developed in which in the data collection process, the respondents can be assigned a degree of membership associated with the agreement level or disagreement level (Li, 2010; Li, 2013). Mahmood et al. (2014) have used fuzzy logic approach to assess the sustainability of hollow fiber membrane intended for designers and manufacturers for improvement of sustainability in future. Mirsaeidi et al. (2016) have developed a framework considering large gas refineries based on fuzzy for prioritizing the health, safety and environmental associated risks for employees, working conditions, and process equipments. An elusiveness or vagueness is described by fuzzy logic for decision-making and controlling, which involves uncertainty. Fuzzy logic has been successfully applied in a number of applications, such as in evaluating pain intensity rating scales (Araujo and Miyahira, 2009), assessment of soil (Gruijter et al., 2011), assessment of life cycle impact (Feri and Hong-Chao, 2014), for classification of crops and agricultural land fitness (Murmu and Biswas, 2015), soil quality index & landfill sitting in agriculture (Rodrígueza et al., 2016), etc. Especially the uncertainties in the measurement of sustainability are usually managed by the use of fuzzy logic (Lamastra et al., 2016). However, uncertainties or errors may be contained in the assessment of sustainability based on the used attributes and methods (An et al., 2017; Arodudu et al., 2017). Amiri and Khadivar (2017) have developed a fuzzy expert system by using MATLAB software, for diagnosis & treatment of musculoskeletal disorders in wrist. Houshyar et al. (2017) have used a combined analytical hierarchy process along with fuzzy modeling with the geographical information system (GIS) for the development of a reference base for evaluating the sustainability of winter wheat in a study in Iran. Bottani et al. (2017) have embodied the architecture of the fuzzy inference tool in an ad hoc software developed in Microsoft Excel TM (Microsoft Corporation, Albuquerque, NM, USA), for a case study in a food machinery company to evaluate the sustainability at the company's level. Kozarevića and Puška (2018) have considered the food industry companies to measure the influence of an independent variable of supply chain practice, on the dependent variable of supply chain performance by the use of fuzzy logic. Therefore, an initiative movement for a sustainable agricultural system which is economically reasonable, socially acceptable and environmentally sound would help in the formulation of current policies of the agricultural practices.

RESEARCH METHODOLOGY

A total of 145 farmers from five villages were selected randomly from Odisha state in India. Prior consent was obtained from the respective village heads or leaders. The author and co-author visited the farmers personally and collected the information and data collected using likert scale and by standardized Nordic questionnaires as shown in Table 1 in likert scale (i.e.,1=agree, 2=totally agree, 3=no opinion, 4=totally disagree, 5=disagree). Then, seven experts from different areas of expertise were consulted for the subsequent decision analysis. The House of Quality (HOQ) as shown in figure 1 is a 'voice of customer' analysis tool & the key element of the Quality Functional Deployment (QFD) technique. First of all customer's requirement is converted in terms of engineering design by developing relationship matrix. These techniques help to understand & interpret the customer needs and help to provide a product of superior value. Quality Function Deployment (QFD) is a planned approach to define the customer needs/ requirements and to translate them into specific plans to achieve the needs. The customers value was collected by survey, then house of quality is designed for the customer requirement. The customer requirement is called as What in House of quality and design requirement is called as HOW's in HOQ.

This method will help the stakeholders and policy makers to know the essentials or requirement's of consumers/customers and help them design and frame product or service to fulfill the consumer's need. Thus, QFD is considered as an effective communicating and quality planning tool. Satapathy (2014) has used the ANN, QFD and ISM approach for the electric utilities services in India for the consumer's satisfaction and discussed the interrelationship between various design requirements. Lombardi et al. (2018) have prioritize the risks and hazards for agricultural equipments like trailers, ploughs, harrows, etc.

RESULT AND DISCUSSION

Based on the interaction with the farmers through questionnaires, some important reasons for agricultural sustainability, noted on Table 2 found from expert analysis and literature review.

Assigned numbers and symbols for QFD: 8 for Strong [O], 6 for Moderate [Θ], 4 for Weak [Δ], 2 for Very weak [●].

In Figure 1 the HOQ matrix is designed, to find relation between What's and How's.

- Customer needs (What's): A structured list of requirements derived from expert's feedback.
- Design requirement (How's): To fulfill customer's demand (What's) the relevent set of system design requirements are structured .
- Planning Matrix: With customer requirement and the relative importance of customer need planning matrix is made.
- Interrelationship matrix (centre matrix): Gives the expert's perceptions of interrelationships between design requirements and customer needs. Scale of .2,.4,.6 and .8 is applied with symbols and figure.
- Design correlation(top) matrix: The correlation between system design requirements are made and matrix is structured.

The individual rating for each design requirement is obtained from the centre matrix using the following relation as in equation (1):

Customer Rating= Z_i +1/n-1(åB_{ij} Z_j) (1)

Where, B_{ij} = i^{th} initial customer rating and Z_j denotes the relationship between customer need and customer rating respectively.

And, n = number of customer needs,

B_{ij} = [B_{1ij}, $B_{2ij....}$ B_{nij}].

Design requirement = 1/nå A_{ij} X_j (2)

Where A_{ij} and X_j denote the relative importance of the ith characteristics with respect to the jth customer need in the relationship matrix and the importance of jth customer needs (customer ratings), respectively.

n= Number of customer needs.

Table 1 shows the Questionnaire.

Table 1. Questionnaire

Questionnaire	1	2	3	4	5
1. Reduction of productivity due to temperature rise 2. Quality degradation due to temperature rise 3. Increase of weeds, blights and pest 4. Increase of agricultural disasters such as moisture stress and drought 5. Increase of soil erosion 6. too much or too little rainfall can be harmful, even devastating to crops 7. Drought can kill crops and increase erosion 8. wet weather can cause harmful fungus growth. 9. extreme temperature and precipitation can prevent crops from growing 10. Extreme events, especially floods and droughts, can harm crops and reduce yields. 11. Low areas hold cold pockets that can lower temperatures significantly. These locations also gather moisture which will freeze and cause frost heaves, damaging roots. 12. Plants on higher locations become victim to cold winds and sunscald caused by exposure to winter sun. 13. Disasters destroy critical agricultural assets and infrastructure 14. Floods damage all immovable properties that get submerged in flood waters.					

Table 2 shows the Design Requirement.

Table 2. Design Requirement

Design Requirement
1. Creating an efficient drainage system in the crop field 2. The space between crops can be covered with a 2–3 inch layer of dry crop residues (natural mulch) to conserve soil moisture. 3. create water bodies to store water during seasonal rain fall and use it in summer 4. planning of crops in a year must be done in line with the meteorology advisory devised by the government 5. protecting the crops by covering them or setting up the farm in a protected agriculture system 6. Using plastic sheets to cover the land between the plants to prevent evaporation. 7. Green house or poly house can prevent losses due to unconventional rains 8. dig deep trenches along the field length and fill it with large boulders or stones. This makes way for the water to flow out of the farm and reduce crop damage. 9. Adopting crops which will not be affected by such natural calamities 10. Insurance is one way of recouping losses due to events beyond human control 11. bamboo border many problems solve by this eg wind, cold waves, wild animals, insect attacks and maintain 10 degree Celsius temperature 12. rain protection cover made of polypropylene are in use in the farming industry globally to prevent direct damage of crops from rains. 13. pruning avoid damage of fruit trees in Strom and drought 14. Don't sow unseasonably seasonal crops or short term crops are mostly weather dependent and intercrop reduce some risk 15. Organic farming and fertilizers can help

Table 3 shows HOQ (house of Quality) Quality Function Deployment (QFD).

Table 3. Quality Function Deployment (QFD)

Sr. No.	Ratings	1	2	3	4	5	6	7	8	9	10	11	12	13	14	15	Costumer Needs
1	7	O	●	O	⊙	●	O	O	Δ	Δ	⊙	Δ	Δ	O	●	⊙	10.932
2	9	Δ	⊙	Δ	●	O	Δ	O	●	⊙	●	●	⊙	Δ	⊙	●	13.434
3	5	●	Δ	⊙	Δ	O	●	●	⊙	Δ	O	⊙	⊙	●	⊙	⊙	7.538
4	9	●	Δ	⊙	O	●	⊙	Δ	O	●	●	O	Δ	⊙	O	O	13.846
5	6	Δ	⊙	Δ	O	⊙	●	O	●	Δ	⊙	●	O	●	●	Δ	8.954
6	8	Δ	⊙	O	●	O	Δ	⊙	O	Δ	⊙	⊙	Δ	●	⊙	⊙	12.553
7	4	Δ	⊙	●	Δ	O	Δ	⊙	⊙	⊙	Δ	●	O	●	O	●	6.276
8	5	●	Δ	⊙	O	●	O	⊙	Δ	●	Δ	O	Δ	⊙	●	Δ	7.538
9	4	⊙	O	⊙	●	Δ	O	⊙	●	O	⊙	Δ	●	⊙	⊙	⊙	6.276
10	3	●	Δ	O	Δ	●	⊙	Δ	⊙	Δ	●	Δ	Δ	O	●	Δ	4.384
11	5	⊙	Δ	●	O	●	O	●	●	Δ	⊙	O	⊙	●	⊙	●	7.538
12	5	Δ	●	⊙	O	⊙	Δ	⊙	Δ	●	Δ	O	●	⊙	Δ	O	7.538
13	3	⊙	Δ	●	⊙	O	Δ	⊙	●	⊙	Δ	O	⊙	Δ	●	⊙	4.523
14	4	⊙	O	⊙	Δ	●		Δ	O	⊙	Δ	⊙	●	⊙	O	Δ	6.786
Initial rating design requirement		44.40	55.0	60.55	59.33	56.71	57.79	63.76	50.58	50.87	49.46	93.70	54.20	49.12	55.65	54.25	
Revised rating		35.52	11.0	48.44	35.59	11.34	46.23	51.0	40.46	40.69	29.67	74.96	43.36	39.29	11.13	12.52	
Normalized rating		0.47	0.14	0.64	0.48	0.15	0.08	0.68	0.53	0.54	0.39	1	0.57	0.52	0.13	0.07	
Rank		8	10	3	7	9	12	2	5	4	9	1	3	6	11	13	

In these table of quality function deployment shows that the customer requirements "HOW" Creating an efficient drainage system in the crop field is ranked first. Green house or poly house can prevent losses due to unconventional rains is ranked second create water bodies to store water during seasonal rain fall and use it in summer is ranked third. Adopting crops which will not be affected by such natural calamities is ranked fourth. Dig deep trenches along the field length and fill it with large boulders or stones. This makes way for the water to flow out of the farm and reduce crop damage is ranked fifth.

CONCLUSION

Indian Economy totally depends on agrisector of India. Although Government of India frames many policies and spends crores of money for the development of agrisector. Still the growth of agrisector is very slow. Poor and nominal farmers of India unable to purchase costly equipment and uses conventional method of farming in high heat and heavy rain and suffers with physical and mental problems. Extreme weather condition, heavy work load during their working procedure gives them early old age, bone and muscle problems. So more and more researches must

Figure 1. House of Quality (HOQ)

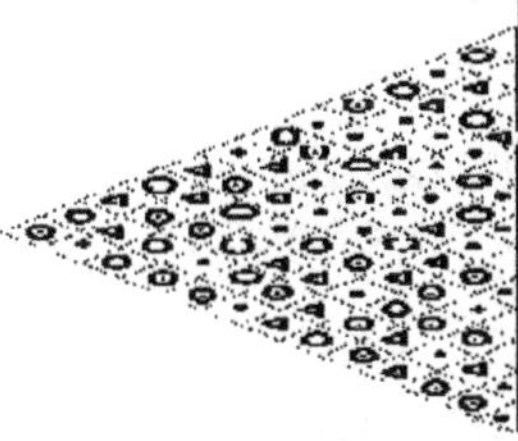

Sr.no	Ratings	1	2	3	4	5	6	7	8	9	10	11	12	13	14	15	Costumer needs
1	7	O	•	O	Θ	•	O	O	Δ	Δ	Θ	Δ	Δ	O	•	Θ	10.932
2	9	Δ	Θ	Δ	•	O	Δ	O	•	Θ	•	•	Θ	Δ	Θ	•	13.434
3	5	•	Δ	Θ	Δ	O	•	•	Θ	Δ	O	Θ	Θ	•	Θ	Θ	7.538
4	9	•	Δ	Θ	O	•	Θ	Δ	O	•	•	O	Δ	Θ	O	O	13.846
5	6	Δ	Θ	Δ	O	Θ	•	O	•	Δ	Θ	•	O	•	•	Δ	8.954
6	8	Δ	Θ	O	•	O	Δ	Θ	O	Δ	Θ	Θ	Δ	•	Θ	Θ	12.553
7	4	Δ	Θ	•	Δ	O	Δ	Θ	Θ	Θ	Δ	•	O	•	O	•	6.276
8	5	•	Δ	Θ	O	•	O	Θ	Δ	•	Δ	O	Δ	Θ	•	Δ	7.538
9	4	Θ	O	Θ	•	Δ	O	Θ	•	O	Θ	Δ	•	Θ	Θ	Θ	6.276
10	3	•	Δ	O	Δ	•	Θ	Δ	Θ	Δ	•	Δ	Δ	O	•	Δ	4.384
11	5	Θ	Δ	•	O	•	O	•	•	Δ	Θ	O	Θ	•	Θ	•	7.538
12	5	Δ	•	Θ	O	Θ	Δ	Θ	Δ	•	Δ	O	•	Θ	Δ	O	7.538
13	3	Θ	Δ	•	Θ	O	Δ	Θ	•	Θ	Δ	O	Θ	Δ	•	Θ	4.523
14	4	Θ	O	Θ	Δ	•		Δ	O	Θ	Δ	Θ	•	Θ	O	Δ	6.786
Initial rating design requirement		44.4 0	55.0	60.5 5	59.3 3	56.7 1	57.7 9	63.7 6	50.5 8	50.8 7	49.4 6	93.7 0	54.20	49.12	55.65	54.2 5	
Revised rating		35.5 2	11.0	48.4 4	35.5 9	11.3 4	46.2 3	51.0	40.4 6	40.6 9	29.6 7	74.9 6	43.36	39.29	11.13	12.5 2	
Normalized rating		0.47	0.14	0.64	0.48	0.15	0.08	0.68	0.53	0.54	0.39	1	0.57	0.52	0.13	0.07	
Rank		8	10	3	7	9	12	2	5	4	9	1	3	6	11	13	

be done to find solution to weather and environmental problems. Similarly the traditional farming system must be modified in such a manner that is suitable to fight extreme and varying weather condition. Particularly modification of drainage system, closed greenhouse, less water consumed farming will definitely change the scenario of farming.

REFERENCES

Akao, Y. & Mazur, G. H. (2003). The leading edge in QFD: Past, present, and future. *International Journal of Quality and Reliability Management, 20*(1), 20-35.

Amiri, F. M., & Khadivar, A. (2017). A fuzzy expert system for diagnosis and treatment of musculoskeletal disorders in wrist. *Technical Gazette, 24*(Suppl.1), 147–155.

An, D., Xi, B., Ren, J., Wang, Y., Jia, X., He, C., & Li, Z. (2017). Sustainability assessment of groundwater remediation technologies based on multi-criteria decision making method. *Resources, Conservation and Recycling, 119*, 36–46. doi:10.1016/j.resconrec.2016.08.002

Andriantiatsaholiniaina, L. A., Kouikoglou, V. S., & Phillis, Y. A. (2004). Evaluating strategies for sustainable development: Fuzzy logic reasoning and sensitivity analysis. *Ecological Economics, 48*(2), 149–172. doi:10.1016/j.ecolecon.2003.08.009

Araujo, E., & Miyahira, S. A. (2009). Unidimensional Fuzzy Pain Intensity Scale. *IEEE International Conferences on Fuzzy Systems*, 185-190. 10.1109/FUZZY.2009.5277285

Arodudu, O., Helming, K., Wiggering, H., & Voinov, A. (2017). Towards a more holistic sustainability assessment framework for agro-bioenergy systems-A review. *Environmental Impact Assessment Review, 62*, 61–75. doi:10.1016/j.eiar.2016.07.008

Aung, N. N., Myo, P. P., & Moe, H. Z. (2014). Field performance evaluation of a power reaper for rice Harvesting. *International Journal of Scientific Engineering and Technology Research, 12*(3), 2631–2636.

Bell, S., & Morse, S. (2008). *Sustainability indicators: measuring the immeasurable?* Earthscan.

Bottani, E., Gentilotti, M. C., & Rinaldi, M. (2017). A Fuzzy Logic-Based Tool for the Assessment of Corporate Sustainability: A Case Study in the Food Machinery Industry. *Sustainability, 9*(4), 583. doi:10.3390u9040583

Brandl, C., Mertens, A., & Schlick, C. M. (2017a). Ergonomic analysis of working postures using OWAS in semi-trailer assembly, applying an individual sampling strategy. *Int. J. Occup. Saf. Ergon, JOSE*, 1-8. doi:10.1080/10803548.2016.1191224

Brandl, C., Mertens, A., & Schlick, C. M. (2017b). Effect of sampling interval on the reliability of ergonomic analysis using the Ovako working posture analysing system (OWAS). *International Journal of Industrial Ergonomics, 57*, 68–73. doi:10.1016/j.ergon.2016.11.013

Bull, J. W., Jobstvogt, N., Böhnke-Henrichs, A., Mascarenhas, A., Sitas, N., Baulcomb, C., Lambini, C. K., Rawlins, M., Baral, H., Zähringer, J., Carter-Silk, E., Balzan, M. V., Kenter, J. O., Häyhä, T., Petz, K., & Koss, R. (2016). Strengths, Weaknesses, Opportunities and Threats: A SWOT analysis of the ecosystem services framework. *Ecosystem Services*, *17*, 99–111. doi:10.1016/j.ecoser.2015.11.012

Chan, L. K. & Wu, M. L. (2002a). Quality function deployment: A comprehensive review +of its concepts and methods. *Quality Engineering, 15*(1), 23-35.

Chavan, P. B., Patil, D. K., & Dhondge, D. S. (2015). Design and development of manually operated reaper. *IOSR Journal of Mechanical and Civil Engineering*, *12*(3), 15–22.

Chieh-Tsung, C. (2012). A new electric brush cutter. *WSEAS Transactions on Systems and Control*, *3*(7), 2224–2856.

DFID. (2004). *Agricultural sustainability: Working / Supporting Paper*. Retrieved on May 10 2018 from http://dfid-agricultureconsultation.nri.org/summaries/wp12.pdf

DFID, Department for International Development-UK. (2003). *Sustainable agriculture. Key sheet*. Retrieved on March 16 2018 from http://www.odi.org.uk/sites/odi.org.uk/files/odiassets/publications-opinion-files/3143.pdf

Drexhage, J., & Murphy, D. (2010). *Sustainable development: from Brundtland to Rio 2012*. Background Paper for the High Level Panel on Global Sustainability, United Nations. Retrieved on April 10 2018 from https://www.un.org/wcm/webdav/site/climatechange/shared/gsp/docs/GSP1-6_Background%20on%20Sustainable%20Devt.pdf

Feri, A., & Hong-Chao, Z. (2014). A fuzzy logic based aggregation method for LCIA. *Journal of Cleaner Production*, *67*, 159–172. doi:10.1016/j.jclepro.2013.12.010

Gandhewar, V. R., Moghe, S. M., Kapratwar, S., & Bobate, N. G. (2015). A review on human powered mini paddy harvester. *International Journal for Engineering Applications and Technology*.

Garg, I. K., & Sharma, V. K. (1991). Tractor-operated vertical-conveyer reaper windrower. *Indian Farming*, *40*(11), 32–36.

Gasso, V., Oudshoorn, F. W., De Olde, E., & Sørensen, C. A. G. (2015). Generic sustainability assessment themes and the role of context: The case of Danish maize for German biogas. *Ecological Indicators*, *49*, 143–153. doi:10.1016/j.ecolind.2014.10.008

Gaviglio, A., Bertocchi, M., & Demartini, E. (2017). A Tool for the Sustainability Assessment of Farms: Selection, Adaptation and Use of Indicators for an Italian Case Study. *Resources*, *6*(4), 60. doi:10.3390/resources6040060

Gruijter, J., Walvoort, D. J., & Bragato, G. (2011). Application of fuzzy logic to boolean models for digital soil assessment. *Geoderma*, *166*(1), 15–33. doi:10.1016/j.geoderma.2011.06.003

Hatefi, S. M. (2018). Strategic planning of urban transportation system based on sustainable development dimensions using an integrated SWOT and fuzzy COPRAS approach. *Global J. Environ. Sci. Manage.*, *4*(1), 99–112. doi:10.22034/gjesm.2018.04.01.010

Hellig, T., Mertens, A., & Brandl, C. (2018). The interaction effect of working postures on muscle activity and subjective discomfort during static working postures and its correlation with OWAS. *International Journal of Industrial Ergonomics*, *68*, 25–33. doi:10.1016/j.ergon.2018.06.006

Hossain, M. A., Hoque, M. A., Wohab, M. A., Miah, M. A. M., and Hassan, M. S. (2015). Technical and economic performance of combined harvester in farmers' field. *Bangladesh J. Agril. Res.*, *40*(2), 291-304.

Hossain, M. D. (2003). *Fabrication of self-propelled reaper by locally available materials* (MS thesis). Department of Farm Power and Machinery, BAU, Mymensingh, Bangladesh.

Houshyar, E., Smith, P., Mahmoodi-Eshkaftaki, M., & Azadi, H. (2017). Sustainability of wheat production in Southwest Iran: A fuzzy-GIS based evaluation by ANFIS. *Cogent Food & Agriculture*, *3*(1), 1327682. doi:10.1080/23311932.2017.1327682

Hunt, D. (1995). *Farm power and machinery management, Cost determination* (9th ed.). Iowa State University Press.

Irfan, M., Hassan, M., & Hassan, N. (2018). Unravelling the Fuzzy Effect of Economic, Social and Environmental Sustainability on the Corporate Reputation of Public-Sector Organizations: A Case Study of Pakistan. *Sustainability*, *10*(3), 769. doi:10.3390u10030769

Jackson-Smith, D. (2010). *Toward Sustainable Agricultural Systems in the 21st Century*. National Academies Press.

Karhu, O., Harkonen, R., Sorvali, P., & Vepsalainen, P. (1981). Observing working postures in industry: Examples of OWAS application. *Applied Ergonomics*, *12*(1), 13-17. doi:10.1016/0003-6870(81)90088-0

Karhu, O., Kansi, P., & Kuorinka, I. (1977). Correcting working postures in industry: A practical method for analysis. *Applied Ergonomics*, *8*(4), 199–201. doi:10.1016/0003-6870(77)90164-8 PMID:15677243

Kates, R. W., Parris, T. M., & Leiserowitz, A. A. (2005). What is sustainable development? Goals, indicators, values, and practice. *Environment, 47*(3), 8-21.

Kozarevića, S., & Puška, A. (2018). Use of fuzzy logic for measuring practices and performances of supply chain. *Operations Research Perspectives*, *5*, 150–160. doi:10.1016/j.orp.2018.07.001

Lamastra, L., Balderacchi, M., Di Guardo, A., Monchiero, M., & Trevisa, M. (2016). A novel fuzzy expert system to assess the sustainability of the viticulture at the wine-estate scale. *The Science of the Total Environment*, *572*, 724–733. doi:10.1016/j.scitotenv.2016.07.043 PMID:27601053

Lechenet, M., Bretagnolle, V., Bockstaller, C., Boissinot, F., Petit, M.-S., Petit, S., & Munier-Jolain, N. M. (2014). Reconciling Pesticide Reduction with Economic and Environmental Sustainability in Arable Farming. *PLoS One*, *9*(6), e97922. doi:10.1371/journal.pone.0097922 PMID:24887494

Li, Q. (2010). *A new Likert Scale Based on Fuzzy Sets Theory* (Ph.D. dissertation). University of Connecticut.

Li, Q. (2013). A novel Likert scale based on fuzzy sets theory. *Expert Systems with Applications*, *40*(5), 1609–1618. doi:10.1016/j.eswa.2012.09.015

Mahmood, S., Hemdi, A. R., Saman, M. Z. M., & Noordin, M. Y. (2014). Fuzzy Logic Approach for Assessing Sustainability: Methodology Development for Hollow Fiber Membrane Module. *Advanced Materials Research*, *845*, 579–583. doi:10.4028/www.scientific.net/AMR.845.579

Marchand, F., Debruyne, L., Triste, L., Gerrard, C., Padel, S., & Lauwers, L. (2014). Key characteristics for tool choice in indicator based sustainability assessment at farm level. *Ecology and Society, 19*(3), 46.

Meisner, C. A., Petter, H., Badruddin, M., Razzaque, M. A., Giri, G. S., & Scott, J. (1997). Mechanical revolution among small landholders of South Asia: The growing use of Chinese hand tractors. *Proc. of the Joint Intl. Conf. on Agril. Eng. & Tech. Exhibition Vol.'97, 3*, 781-787.

Mirsaeidi, A., Moradi, H., & Zekri, M. (2016). Fuzzy-based Prioritization of Health, Safety, and Environmental Risks: The Case of a Large Gas Refinery. *Iranian Journal of Health, Safety & Environment, 4*(1), 641-653.

Mishra, R., Singh, Y. P., Mishra, Y. D., Singh, S., & Singh, H. (2013). Dissemination of improved sickles for female agriculture workers for crop harvesting. *A Journal of Multidisciplinary Advance Research, 2*(1), 118-123.

Murmu, S., & Biswas, S. (2015). Application of fuzzy logic and neural network in crop classification: A review. *Aquatic Procedia, 4*, 1203–1210. doi:10.1016/j.aqpro.2015.02.153

Nag, P. K., Goswami, A., Ashtekar, S. P., & Pradhan, C. K. (1988). Ergonomics in sickle operation. *Applied Ergonomics, 19*(3), 233-239.

Nalawade, S. M., Turbatmath, P. A., & Gajakos, A. V. (2009). Design and development of tractor operated Jowar reaper windrower. *New Agriculturist, 20*(12), 75-81.

Nedea, P. S. (2012). The Evolving Concept of Sustainable Agriculture. *Anale. Seria Stiinte Economice. Timisoara, 18*, 64–69.

Noby, M., Hasan, M. K., Ali, M. R., Saha, C. K., Alam, M. M., & Hossain, M. M. (2018). Performance evaluation of modified BAU self-propelled reaper for paddy. *J Bangladesh Agril Univ, 16*(2), 171–177. doi:10.3329/jbau.v16i2.37956

Rahman, S., Rashid, M. A., & Sanajpati, J. N. (1980). Design and development of hand-operated grain harvester. *AMA, 11*(1), 79-80.

Rodrígueza, E., Pechea, R., Garbisub, C., Gorostizac, I., Epeldeb, L., Artetxed, U., Irizare, A., Sotoe, M., & Mabecerrild, J. (2016). Dynamic quality index for agricultural soils based on fuzzy logic. *Ecological Indicators, 60*, 678–692. doi:10.1016/j.ecolind.2015.08.016

Roman-Liu, D. (2014). Comparison of concepts in easy-to-use methods for MSD risk assessment. *Applied Ergonomics, 45*(3), 420–427. doi:10.1016/j.apergo.2013.05.010 PMID:23849897

Sami, M., Shiekhdavoodi, M. J., Almassi, M., & Marzban, A. (2013). Assessing the sustainability of agricultural production systems using fuzzy logic. *Journal of Central European Agriculture, 14*(3), 318–330. doi:10.5513/JCEA01/14.3.1323

Sathaye, J., Najam, A., Cocklin, C., Heller, T., Lecocq, F., Llanes-Regueiro, J., Pan, J., … Winkler. (2007). Sustainable development and mitigation. In *Climate change 2007: Mitigation. Contribution of working group iii to the fourth assessment report of the intergovernmental panel on climate change*. Cambridge University Press. Retrieved on April 29 2018 from https://www.ipcc.ch/pdf/assessmentreport/ar4/wg3/ar4-wg3-chapter12.pdf

Shalini, P. S., Wankhade, A. M., & Syed, M. N. (2015). Design and fabrication of crop reaper. *International Journal for Engineering Applications and Technology*.

Singh, R., Khakha, J., & Singh, G. (2009). *Development and testing of manually operated vertical conveyor reaper* (Unpublished, B-Tech. thesis). Indira Gandhi Agricultural University, Raipur.

Singh, S., Vatsa, D. K., & Verma, M. K. (2007). Feasibility and Performance Evaluation of Power Tiller Operated Reaper in Hills of Himachal Pradesh. *Agricultural Engineering Today, 31*(2), 6-10.

Singh, S. P. (2012). Physiological workload of farm women while evaluating sickles for paddy harvesting. *Agric Eng. Int: CIGR Journal, 14*(1), 1540.

Singh, T. P., & Singh, B. (1995). Design and development of animal-drawn, engine-operated reaper. *AMA, 26*(2), 29-34.

Vinodh S. & Chintha S. K. (2011). Application of fuzzy QFD for enabling leanness in a manufacturing organization. *International Journal of Production Research, 49*(6), 1627–1644.

KEY TERMS AND DEFINITIONS

Climate Change: Environmental change with rise of temp or decrease in temp due to heavy or low rain.

HOQ: House of quality. The “house of quality” is a main design tool known as quality function deployment (QFD).

QFD: Quality function deployment. QFD is used to find customer requirements and convert it to design requirement.

Sustainability: Sustainability consists of three pillars: economic, environmental, and social.

Chapter 3
Mitigation of Climate Change Through Carbon Sequestration

Alkesha Naik
Department of Environmental Science and Engineering, Marwadi University, India

Koushik Singha Roy
Department of Environmental Science, SRM University, Delhi - NCR Campus, India

ABSTRACT

Climate change is considered one of the burning issues and threats to the globe. One of the most important drivers of climate change is the increasing concentration of atmospheric greenhouse gases that trigger the altered climate patterns. Therefore, researchers are now tackling the adverse impact of climate change as well in minimizing the concentration and emissions of atmospheric greenhouse gases (i.e., carbon sequestration seems to be one of the most attractive choices for the researchers that could be useful for mitigating climate change conditions). The process deals with the capturing of atmospheric carbon dioxide (CO_2) in various segments of natural ecosystems like soil, biological systems, oceans, thereby offsetting the adverse impact of climate change. In this review, the author examines the current scenario of carbon sequestration in the biosphere.

DOI: 10.4018/978-1-7998-3343-7.ch003

LIST OF ABBREVIATIONS

C: Carbon
CO_2: Carbon Dioxide
SIC: Soil inorganic carbon
SOC: Soil organic carbon
GHGs: Greenhouse gases
MBC: Microbial biomass carbon
WSC: Water soluble carbohydrate
DOC: Dissolved organic carbon
RMC: Readily mineralized carbon
POM-C: Particulate organic matter carbon
SOM: Soil organic matter
HA: Humic acid
FA: Fulvic acid
Mg: Magnesium
GRSP: Glomalin-related soil protein
AM: Arbuscularmycrrohizal
REDD+: Reducing Emissions from Deforestation and Forest Degradation
UNFCCC: United Nation's Framework Convention on Climate Change
PNNL: Pacific Northwest National Laboratory
CCS: Carbon capture and storage
RLD: Root length density
$KMnO_4$-C: $KMnO_4$ oxidizable organic carbon
HRPs: High rate ponds
CSTRs: Continuously stirred tank reactors

INTRODUCTION

Global climate change is a significant long – term weather pattern all over the world (IPCC, 2006). The changed climate pattern has aggravated in the recent times due to various anthropogenic man – made activities such as like urbanization, deforestation, and altered land use scenarios (IPCC, 2006). All these resulted into increased emission of greenhouse gases (GHGs) and lead to the alarming condition like climate change (Barbara, 2008). There is the sense of urgency that the altered pattern of climate change must be handled judiciously. Carbon (C) sequestration is considered as one of the attractive choices of the scientists and to manage and minimize the harsh impact of climate change (Paustian et al. 2019). Carbon sequestration is the process that captures atmospheric CO_2 followed by storing it into natural ecosystems. The

Carbon present in the natural soil system a shows significant association with the atmospheric emissions of GHGs. Moreover, the Carbon pool is inherently connected to atmospheric CO_2 level through Carbon sequestration, photosynthesis process, respiration and decomposition of organic matter (Davidson and Jassens, 2006). Carbon sink is the way to remove greenhouse gas emissions and to sequester atmospheric CO_2 in these view forests are the major sinks of Carbon and accounts half of its dry weight and therewith globally forests are the net Carbon sinks in which tropical forests contribute to 2.8 billion tons Carbon year^{-1}. Carbon sinks through forestation and reforestation; because soil sequesters about 20 Pg Carbon in 25 years.

Figure 1. Carbon dynamics in terrestrial system

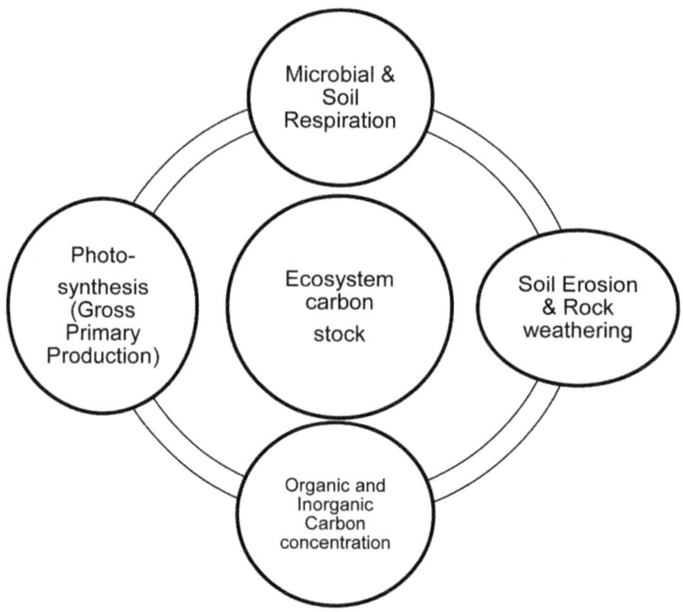

Globally Carbon sequestration processes take place in soil, ocean, geologic formation and other terrestrial ecosystems (Table 1, Figure 1). Physical, chemical and biological processes are carried out to sequester Carbon in the soil of various ecosystems. Majorly, the Carbon sequestration process is divided into chemical, geological and biological sequestration (Figure 2). Furthermore, chemical Carbon sequestration is categorized into two parts; these are (1) Carbon sequestration by activated charcoal and (2) Carbon sequestration by nano-particles. The geological C sequestration includes Carbon sequestration through pressurized CO_2 in rock system (Dooley et al., 2006). The advantage over terrestrial sequestration is that geological

Table 1. Case studies on carbon sequestration in diverse systems

Types of Systems Place	Outcome	References (Authors, Year)
Agriculture	Straw return was effective in increasing Carbon sequestration in maize-wheat system. The Carbon stock was found 10.5 Mg ha^{-1} after four years of maize-wheat cropping system.	Li et al. (2016)
Horticulture	The orchard and horticulture cropping system lead to increase of 0.44 Mg C ha^{-1} yr^{-1} during the first 20 years (range 0.41-0.52 Mg C ha^{-1} yr^{-1}) and led to a total SOC accumulation of about 30 Tg C after 100 years.	Pardo et al. (2016)
Agroforestry	The rate of Carbon sequestration varied 5-10 kg C ha^{-1} in 25 years of extensive tree-intercropping systems of arid and semiarid lands to 100-250 kg C ha^{-1} in about 10 years in species-intensive multi-strata shaded perennial systems and home-gardens of humid tropics.	Nair et al. (2009)
Forest	The 20-year-old plantations of three forest species (*Pinus patula, Cupressus lusitanica, Eucalyptus grandis*) lead to Carbon storage at 50 cm depth was in the range of 101.2-180.4 Mg ha^{-1}.	Lemma et al. (2006)
Geological system (rocks)	The rocks like dunite, harzburgite and olivine basalt upon pulverization followed by milling were reacted in seawater under batch reactor systems showed 10% sequestration in the form of precipitated aragonite.	Rigopoulos et al. (2018a)
Mining system	Biosolids were used for the reclamation of copper and molybdenum tailings. Carbon pools increased with increasing biosolids application and ranged from 23-155 Mg C ha^{-1} after 13 years of reclamation. The net SOC sequestration rates (i.e. the Carbon sequestration potential) ranged from 0.72-6.3 Mg C ha^{-1} yr^{-1}.	Antonelli et al. (2018)
Microalgal system	Carbon sequestered at the range of 34-3107 g m^{-2} and stored as algal biomass.	Gorain et al. (2018)
Wetlands	The rice straw and urea in 1:1 combination proved highly potential for Carbon sequestration. This technology could conserve N at the rate of 30 kg ha^{-1} yr^{-1}, sequester 0.35 t ha^{-1} $year^{-1}$ carbon (C) and reduce 9% GHGs emission with a sustainable yield of 5.57 t ha^{-1} (c.v *Gayatri*) as compared to conventional practice.	Bhattacharyya et al. (2012)

formation is having permanence capturing mechanism which can keep CO_2 for thousands to millions of years. The biological Carbon sequestration is divided into 3 different parts; i.e. (1) forest based Carbon sequestration (i.e. mangrove forest, tropical forest, temperate forest, grassland) (2) algae – based Carbon sequestration (i.e. microalgae, macroalgae, bacteria) and (3) Carbon sequestration in agro-ecosystem (rice, wheat, maize, cropping system) (Figure 2).

Climate change and Carbon cycling has significant interrelationship as evident from various research studies (Lal, 2001, 2003). The understanding of soil Carbon stock and its dynamics seemed to be complex owing to various factors like climate

Figure 2. Categories of carbon sequestration

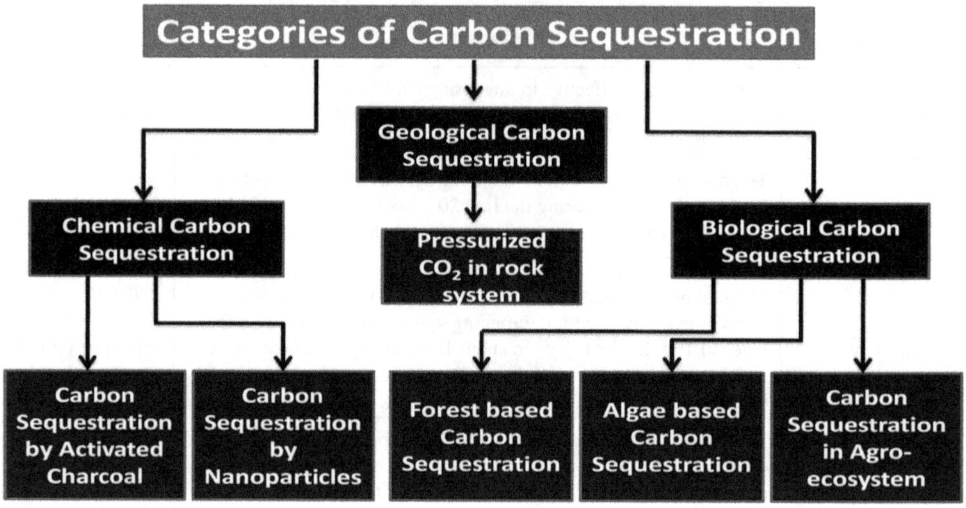

gradients, texture and mineralogical properties of soil, elevation and geological variables. There are still uncertainty prevails of soil Carbon stock in different ecosystems owing to natural and anthropogenic factors (Gren&Zeleke, 2016). It is also proposed that less rainfall could lead to decrease in Carbon stock as significant magnitude (Dintwe et al., 2014, 2018). Cao et al., (2001) concluded that warming could reduce significantly plant production and soil Carbon stock but the situation would be ameliorated owing to more CO_2 level in the atmosphere.

CLIMATE CHANGE FEEDBACK AND CARBON SEQUESTRATION

Human activities like burning fossil fuels, oil pumping results in elevated CO_2 concentration in the atmosphere causes high impact when compared to its removal by natural process. Owing to elevated CO_2 concentration the global temperature has increased significantly over the last few decades (IPCC, 2018). The climate change feedback helps in determining the climate state and the climate sensitivity over a certain region. Feedbacks are the processes which amplify or diminish the effect of each climate forcing. So, there are positive and negative feedbacks in which the positive amplifies the change in the first quantity while the negative diminishes it. Positive feedbacks lead to global warming, so IPCC reported in their reports about the anthropogenic effects and how carbon sequestration is being implemented therewith (IPCC, 2007).

CARBON SEQUESTRATION MECHANISM

The Carbon sequestration mechanism needs to be evaluated in order to understand the process of removal of different forms of Carbon (Figure 3). The mechanism show that how soil Carbon changes from one form to another. The unprotected soil Carbon is either adsorbed into silt and clay particles in soil or readily degraded by the microbial actions. This form of easily degradable Carbon is labile Carbon and measured as soil microbial biomass Carbon (MBC), water soluble carbohydrate Carbon (WSC) and dissolved organic Carbon (DOC).

Figure 3. Mechanism of carbon sequestration

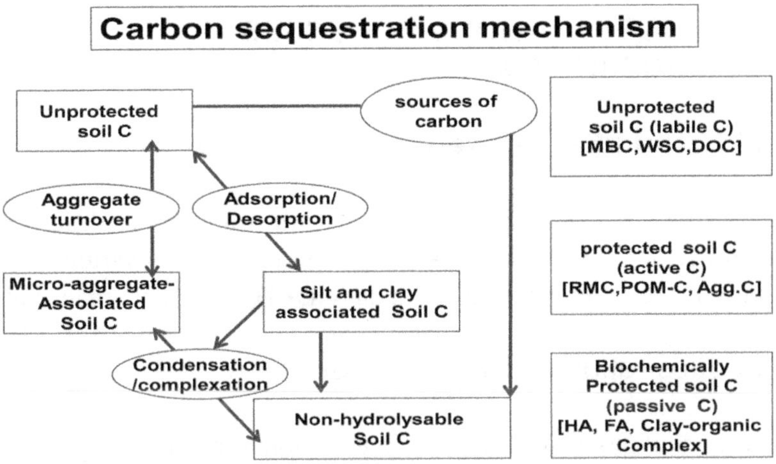

These silt and clay associated soil Carbon is the phase in which the Carbon got attached in between the pore sizes of the particles and bounded together to form an associated soil Carbon or protected soil Carbon (active Carbon) that is measured through readily mineralized Carbon (RMC), particulate organic matter Carbon (POM-C) and aggregated Carbon. The active Carbon afterwards gets converted into non-hydrolysable soil Carbon which is biochemically protected soil Carbon (passive C) and is measured through humic acid (HA), fulvic acid (FA) and clay organic complex content. The non-hydrolysable soil Carbon upon condensation/ complexation process gets converted into a micro-aggregate associated soil Carbon. Therefore, researchers are focusing upon two phases of Carbon sequestration i.e. silt and clay associated with soil Carbon to non-hydrolysable soil Carbon and micro-aggregate associated soil Carbon to unprotected soil Carbon. Environmental changes that influence soil Carbon dynamics could slow atmospheric CO_2 rise and associated warming by promoting soil Carbon storage (Cramer et al., 2001) or they could exacerbate warming by causing soil Carbon to decline (Knorr et al., 2005). Yet, the very properties that make soil Carbon a key reservoir in the global Carbon cycle- namely its large size and slow turnover- also make it difficult to study, empirically. Small changes in the size or turnover of soil Carbon are biochemically significant but difficult to detect in experiments (Smith, 2004). Rising atmospheric CO_2 concentration generally increase the plant growth and Carbon input to soil, suggesting that soil might help mitigating increasing atmospheric CO_2 and global warming. The large size of the soil Carbon pool not only makes it a potential buffer against rising atmospheric CO_2, but also makes it difficult to measure changes amid the existing background (Hungate et al. 2009).

CARBON SEQUESTRATION PROCESSES & ITS SCENARIO IN ASIA

Chemical Carbon Sequestration

Carbon sequestration is one of the best promising solutions to deal with global warming. Furthermore, chemical Carbon sequestration includes various processes to capture and store CO_2 into the different zonations of earth (Bouwman et al., 2010).

Carbon Sequestration by Nanoparticles

Major contributor of global warming is CO_2 emission from anthropogenic activities, i.e.,power plants, land use, etc. (Khdary et al., 2016). Simultaneously, Carbon needs to be stored in various parts of ecosystems. Many studies have been reported on nano technology based application for Carbon sequestration. Recent studies focus on the wide application of nano-technology to capture or absorb CO_2. Naturally occurring nanoparticles in soil are abundantly present but its functional mechanisms are poorly understood. The adsorption by these nanoparticles affects organic Carbon sequestration. The application of biochar nanoparticles over the land field is a powerful strategy for the Carbon sequestration and soil remediation. The catalytic activity of nickel nanoparticles results into CO_2 hydration and towards Carbon sequestration.

Geological Carbon Sequestration

Geological Carbon sequestration is a process in which the CO_2 is captured from the exhaust of fossil fuel power plant and other major sources to prevent its release into the atmosphere that contributes global warming (Singh et al., 2016). Compared to terrestrial CO_2 sequestration, geological CO_2 sequestration is less over the period of time. Due to potential permanence and capacity of geologic storage, larger rates of sequestration are envisaged to take the advantage. There are several geological CO_2 trapping mechanisms are generally used but the effectiveness of permanence depends feasibility.

The physical trapping mechanism can retain CO_2 for thousands to millions of years beneath the impermeable barrier which is similar to that of natural geology trapping of oil and gas (Manitoba Eco-network, 2018). When $CO_2$is injected, some of them will react chemically to form carbonate minerals with the surrounding rocks. The factors which affect the geological Carbon sequestration are the volume and distribution of potential storage sites. The US department of Pacific Northwest National Laboratory (PNNL) carried out the technique of injecting pressurized CO_2 with basalt rock to turn it into solid rock and letting the natural chemical reactions trigger the transformation (Mcgrail et al., 2017).

Furthermore, the increased rate of weathering of mafic and ultramafic rocks is found to be a promising means of Carbon sequestration (Brown et al., 2009). Rigopoulos et al. (2018a) observed that predates and basalts upon enhanced weathering could sequester CO_2 in seawater. The rocks like dunite, harzburgite and olivine basalt upon pulverization followed by milling were reacted in seawater under batch reactor systems showed 10% sequestration in the form of precipitated aragonite (Rigopoulos et al. 2018a).

From this study it concludes that the mineral carbonation is one of the promising technologies for C capture and storage (CCS) strategy to mitigate climate change. It has been observed that partially serpentinized harzburgite rocks have undergone different degrees of ball milling process to produce nanoscale ultramafic powders having increased CO_2 storage potential (Rigopoulos et al., 2018b). Rosenbauer et al. (2012) found that, basaltic rocks showed potential repositories for sequestering CO_2 owing to their capacity for trapping CO_2 in carbonate minerals. The amount of uptake at 100°C, 300 bar ranged from 8-26% for tholeite and picrite respectively. The actual amount of CO_2 uptake was depending on the magnesium (Mg) content and overall reaction period. Therefore, the porous rock (i.e. the basaltic rocks which have a high sequestration rate of 8-26%) injection is suitable for the sequestration but economically it is not feasible.

Biological Carbon Sequestration

Biological C sequestration is a natural phenomenon or a process that balance the global Carbon cycle based on mainly the plant species and the forest systems. Harris et al., (2018) explained the mechanisms of plant based CO_2 capture followed by allocation of stored C used in respiration and the rest stored as biomass and sequestering Carbon for long term basis. The bio-energy crops have been found as potential means for GHG reductions and maintaining biogenic Carbon cycle (Thornley and Adams, 2018). Khadary et al., 2016 explained that the genetically engineered trees can tackle climate change condition more efficiently than normal plant species. Atmospheric CO_2 can be temporarily stored in vegetation through photosynthesis and sequester relatively large volumes of Carbon at low cost, protecting and improving the soil structure, water resources, biodiversity, habitat and can mitigate climate change in an easier way.

Forest Based Carbon Sequestration

Major Carbon is sequestered in above-ground plant biomass and below-ground soil as well (IPCC,2014). The global Carbon stock was 4.1 Pg in these two forest Carbon pools. It was estimated that Carbon sequestration in forest system results into 30% reduction in emission from fossil fuels (IPCC, 2014).

Mangrove Forest Based Carbon Sequestration

Mangroves are the tropical and subtropical latitude sustaining trees and grow in low oxygen soil under saline or brackish environment (Kristensen et al., 2008). The mangroves are having high ability to fix and store CO_2 due to high productivity (218 Tg C y^{-1}) and unique conditions (Bouillon et al., 2008). Mangroves account approximately 10-15% Carbon sequestration (Alongi 2009, 2014) in ~0.5% of global coastal ocean area (Giri et al., 2011). Carbon is assimilated both in aboveground and belowground biomass including stems, leaves, roots and branches. Majorly 90% of C stocks are present as soil organic matter (Donato et al., 2011; Stringer et al., 2015). Hien et al. (2018) carried out experiments to determine the Carbon stocks and Carbon burial rates in 18 years old mangrove forest soil. The concentration of Carbon in 100 cm soil depth and in roots of *Kandeliaobovata* were found 146.78 Mg C ha^{-1} and 12.67 Mg C ha^{-1} respectively. The mean Carbon burial rate was also found higher (6.94 Mg C ha^{-1}) than many mangrove forests. This study showed the capacity of mangroves to store Carbon belowground at high potential rate to mitigate climate change.

The glomalin-related soil protein (GRSP) that contribute to soil Carbon sequestration was isolated in mangrove forest soil predominated by *Avicennia marina*, *Kandeliaobovata* and *Aegicerascorniculatum* mangrove species along with Arbuscularmycrrohizal (AM) fungi (Table 2). Wang et al. (2015) noted that the presence of AM fungi in aerenchymatous cortex provides oxygen for aiding. AM fungi works significantly to increase the absorption rate of N, P and K as well as enhance the mangrove species growth.

Table 2. Glomalin related soil protein (GRSP) concentration in mangrove sediment

GRSP Concentration Types	GRSP Range (mg g^{-1})
Easily extractable GRSP (EE-GRSP)	1.20-2.22
Total-GRSP (T-GRSP)	1.38-2.61
Particulate organic matter GRSP (POM-GRSP)	1.45-10.78
GRSP in pore water (PW-GRSP)	10.35-39.65

(Wang et al., 2015)

Ray et al. (2017) studied Sundarbans mangrove ecosystem and estimated the rate of sequestration of CO_2 (2.79 Tg C) derived from coal based power plant. Bala Krishnan Prasad and Ramanathan (2008) observed the Carbon cycling in Pichavaram mangrove ecosystem. The levels of organic Carbon at off shore and experimental sites ranged from 0.02-6.9% and 0.1-6.5%, respectively. The sites showed differences in Carbon accumulation in pre and post monsoon period. Highest accumulation of Carbon was found in the post monsoon period (3.4%). Mitra et al. (2011) studied the Carbon stock in various plan components of three mangrove species viz. *Sonneratiaapetala, Avicennia alba* and *Excoecariaagallocha*. Among the three species, the carbon storage was in the order of *Sonneratiaapetala>Avicennia alba>Excoecariaagallocha*. The variation in C stock was due to high siltation rate and nutrient turnover at the concerned location. Mangrove ecosystem captures 55% of CO_2 including blue carbon habitats.

Tropical Forest Based Carbon Sequestration

The concept of "Reducing emissions from deforestation and forest degradation (REDD+)" was introduced by the United Nation's Framework Convention on Climate Change (UNFCCC) in order to mitigate climate change and increasing rate of Carbon sequestration. The sustainability of Carbon sinks potential over tropical forests located between 23° North and South of the Equator. Mohanty and Panda (2011) studied soil respiration rate in relation to microbial diversity in tropical forest soil of Orissa, India. The mean CO_2 flux was found 237 mg m^{-2} h^{-1} and 134 mg CO_2 m^{-2} h^{-1} in forest and deforested site respectively (Mohanty and Panda, 2011). The fungal population showed positive relationship with the rate of soil respiration. According to the survey, tropical forest stores 95% of the CO_2 concentration on the earth and the average storing capacity of tropical forest is 22.6 kg C $year^{-1}$.

Temperate Forest Based Carbon Sequestration

Temperate forests are diverse in plant species, soil and ecosystem (25° and 55° N and S of the equator). Majorly they are found in North-eastern Asia, North America, Central and Western Europe, southern Chile, Mediterranean and New Zealand. In soil profile, 100 Mg C ha^{-1} or often more is generally derived in temperate forest areas. Compared to tropical rainforest ecosystem, temperate forests have high Carbon sink capacity depending upon particular seasons and climatic characteristics (Lal et al., 2012). The afforestation and revegetation projects are implemented for biodiversity conservation in order to meet the goals of Kyoto Protocol under UNFCCC. Heij et al. (2001) observed that temperate trees are preserved through this project and new forests are established over degraded land to increase the terrestrial Carbon sequestration. There are more methods and processes which are needed to develop to understand the microbial mechanism in temperate forests.

Grassland Based Carbon Sequestration

Grasslands are the major contributor of soil organic matter, in terms of extensive fibrous root mats. Grassland also contributes to productivity, policies and income as agricultural and forestry land. Overgrazing and livestock enhancement had turned out to be a destructive basis for the CO_2 sequestration process. Grassland can sequester 0.01-0.30 Gt C yr^{-1} through techniques such as grazing management, fertilizer usage, irrigation applications, etc (Ghosh et al., 2014). Grasslands cover 26% world land area i.e. 3.5 billion ha area, governing 20% of world's soil Carbon stock (O'Maara, 2012). Soil organic matter (SOM) is twice that of Carbon in atmosphere and any changes in SOM results into CO_2 emission into the atmosphere (Heij et al.2001). As per the report of the World resource institute (2000) the grasslands store 34% global terrestrial Carbon stock while forest stores 39% and agro-ecosystem stores 17%. It was also studied that grasslands are having 200-300 Pg of Carbon pool. It is estimated that globally 0.2-0.8 Gt Carbon can be sequestered through grassland soils by 2030. On contrary, grasslands have limitations including less biomass contents which affect the Carbon stock. Further research work needs to be carried out for Carbon sequestration process in grassland ecosystem.

Agro-ecosystem Based Carbon Sequestration

The agro-ecosystem includes cropland, grazing and rangelands major terrestrial pool of organic Carbon in soil and soil organic matter. Original soil organic Carbon levels have depleted by 25-75%, recording a loss of 10-50 C ha^{-1} (Table 3). The decreasing trend in low level of soil organic Carbon results in increased concentration of CO_2 flux in the atmosphere. Reduction in tillage, managing fertility and irrigation can enhance the growth and diversification of soil microbial communities. The fungi form soil aggregates and protects soil organic matter. Soil microbes play a crucial role to restore soil health. Restoring soil health results in Carbon sequestration reducing Carbon concentration entering into the atmosphere. Increasing soil organic matter through techniques contributes to Carbon sequestration also improves nutrient cycling biologically and stimulating biodiversity. Yearly around 39-49 Tg Carbon is sequestered through agro-ecosystems.

- Rice: Lowland rice ecosystem behaves as net Carbon sink (Bhattacharyya et al., 2014) apart from this the submerged soil field decomposes organic matter at a slower rate than well drained soil field and turn into high Carbon stock. Rice fields are the great potential for the sequestration process (Liu et al., 2006). It was reported that 401 kg C ha^{-1} is sequestered in rice ecosystem.
- Maize: 10,497 mg of Carbon is sequestered in maize ecosystem during autumn 2012 by using mean values of above and below ground biomass carbon (Table 4) (Zahra et al., 2016).
- Cropping system: In Rice (*Oryza Savita L.*)- Wheat (*Triticumaestivum L.*) cropping system, 1.94 Mg C ha^{-1} is sequestered without any fertilization control (Brar et al., 2013).

Table 3. Soil C sequestration in agro-ecosystem of major countries

Countries	Rate of Soil C Sequestration in Agro-ecosystem (Tg C $year^{-1}$)
India	39-49
European Union	90-120
USA	75-208
China	105-198
Canada	24

(Wang et al., 2010)

Table 4. Above and below ground biomass concentration

Place	Above Ground Biomass Carbon (mg ha^{-1})	Below Ground Biomass Carbon (mg ha^{-1})
NIAB- I	9.60	1.5
NIAB- II	8.21	1.3
Gojra	9.61	1.4

(Zahra et al., 2016)

Table 5. Total land distributions worldwide

Land Use	Area (Million ha)	Percentage (%)
Permanent crops	132	0.9
Arable land	1369	9.7
Permanent pasture	3460	24.5
Forest and woodland	4172	29.6
Agricultural land	4961	35.2
Total area	14094	100.0

(adopted from FAO, 2001)

Algae Based Carbon Sequestration

Algae is considered as one of the primary producers in aquatic ecosystems that fix CO_2 through the process of photosynthesis and help to maintain ecosystem functioning. According to the theories, Carbon uptake through algae is still unidentified. Only physical and chemical processes are understandable, stimulating natural environment will provide a better understanding for algae in Carbon cycle. There were various studies carried out to establish the potential of algae to capture CO_2 (table 6).

Some of the studies investigated that the freshwater algae (*Chlorella, Oscillatoria, Oedogonium, Anabaena, Microspora and Lyngbya*) is having CO_2 uptake of 159 mg L^{-1} in continuously stirred tank reactors (CSTRs) that is almost twice that of high rate ponds (HRPs) (Tsai et al. 2016). In another study, Ramaraj et al. (2014a,b; 2015) concluded that CSTR was having 178 mg L^{-1} day $^{-1}$ more algal biomass production (96% efficiency) as compared to HRP. The study was unique as it dealt with fresh water algae and estimated that 4000 Gt of Carbon was fixed as bio fixation.

- Microscopic algae: Microalgae are more prominent because of high protein content, carbohydrates and lipids compared to terrestrial plants. Prospective microalgae such as *Chlorella vulgaris*, *Scenedesmus* sp., *Chroococcus* sp. and *Chlamydomonas* sp. are eminent and actively used widely for CO_2 sequestration. 90% capturing efficiency is reported in open ponds with CO_2 or bicarbonate. Algal growth is having more biomass production than terrestrial systems (Hien et al., 2018). Algae are having much efficiency over land system because algal growth is having high photosynthetic efficiency which converts light energy faster into utilizable chemical energy. It was observed every gram of algae biomass captured 1.6-2.0 gram of CO_2 (Figure

Table 6. Some recent case studies on carbon capture by algal species

References	Algal Species	Outcome
Upender et al. (2018)	*Synechococcus* sp (NIT18)	Carbon absorption rate by 71% under CO_2 concentration of 5-20%, pH 7-11 and inoculums size of 5-12.5%
Gorain et al. (2018)	Species of Spirogyra, Rhizoclonium, Ulva, Cladophora, Pithophora, Gracilaria, Lyngbya	Carbon sequestered at the range of 34-3107 g m^{-2} and stored as algal biomass.
Munoz-Rojas et al. (2018)	*Nostoc commune, Tolypothrix distorta* and *Scytonema hyalinum*	Soil Carbon sequestered at the rate of 2-12 g kg^{-1} in mine spoil system and the algal strains were used as ecosystem restoration process.
Pavlik et al. (2017)	*Chlorella vulgaris*	The mass balance analysis concluded that a one metric-ton unit can generate 0.4 kg of dry algae biomass with 51% protein content and sequester about 800g CO_2 day^{-1}.
Eloka-Eboka et al. (2017)	*Chlorella vulgaris, Dunaliella* sp, *Scenedesmus quadricauda* and *Synechococcus* spp.	The rate of removal of CO_2 was found 11.73 mg L^{-1} min^{-1} to 18.84 mg L^{-1} min^{-1}. The species of *Dunaliella* was found as more effective in CO_2 sequestration.
Bhakta et al. (2015)	*Chlorella* sp., *Scenedesmus* sp., *Sphaerocystis* sp. and *Spirulina* sp.	CO_2 sequestration (53-100%; 150-291 mg g^{-1}) with strong growth performance in wastewater was observed.
Zhao et al. (2015)	*Chlorella* sp., *Isochrysis* sp. and *Amphidinium carterae.*	The study was conducted in 1L bubble column photo-bioreactors with an aeration of 15% CO_2. *Chlorella* sp. was selected as the dominant species due to higher growth rate of 0.328 d^{-1}, a biomass production rate of 0.192 gL^{-1} d^{-1} and CO_2 fixation rate of 0.353 gL^{-1} d^{-1}.
Yoo et al. (2010)	*Chlorella vulgaris, Botryococcus braunii,* and *Scenedesmus* sp.	The algal species were grown under high level of CO_2 concentration. The microalgae B. braunii was found as most suitable for biodiesel production while, *Scenedesmus* sp. was most suitable for CO_2 mitigation.

4). Recent studies were about the modifications over the microalgae based reactors which needs more development.

Figure 4. Algae based carbon sequestration

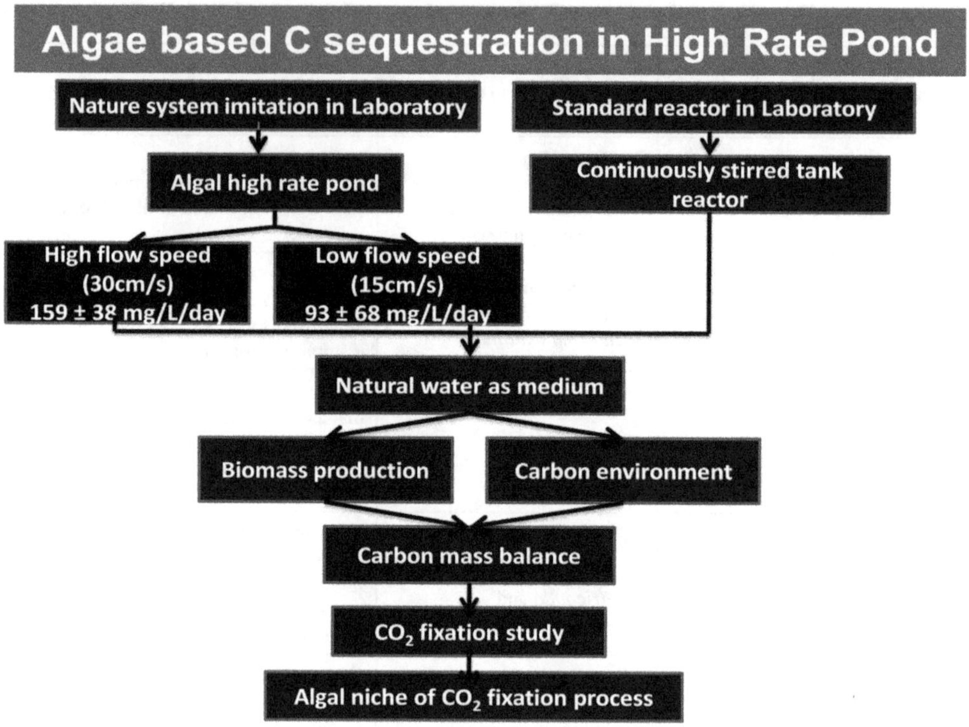

ADVANTAGES OF CARBON SEQUESTRATION

Chemical carbon sequestration:

- 50 tons of Carbon is sequestered through various chemical processes (30% carbon absorption efficiency).
- Recent researches include processes (such as activated charcoal, mineral carbonation, nanoparticles, etc.) which can be used as a pre-cleaner technique in power plants.

Geological carbon sequestration:

- 2 tons of C ha-^{1} year^{-1} is being sequestered through geological processes.
- Earliest technology for Carbon sequestration is geological sequestration because of its permanence capturing capacity.

- Apart from pressurized CO_2 injection in rock system through pipes, current techniques are being focusing upon CO_2 capturing under the rock system of seawater.

Biological carbon sequestration:

- 320 billion metric tons of carbon is sequestered through biological processes including forests, agro ecosystems, algae, etc.
- UNFCCC are implementing many policies in terms of biological sequestration as maximum capturing and storage is reported herewith.
- As per the report of World resource institute (2000) the grasslands store 34% global terrestrial Carbon stock while forest stores 39% and agro-ecosystem stores 17%.

DISADVANTAGES

Chemical carbon sequestration:

- In terms of sequestration through nanoparticles, ongoing researches cannot be implemented on an industrial basis. As these technology needs more time for modifications and economically it cannot be feasible for many industrial stakeholders.

Geological carbon sequestration:

- Injection through pipelines leads to leakage and seepage which contaminates ground water levels.
- As this technique requires large setups which govern huge investment but there are easy options available can be replaceable in terms of physical processes.

Biological carbon sequestration:

- Management practices such as cropping system, algae bioreactor, etc. need to be more focused along with combination as single approach are having drawbacks.

CONCLUSION

The processes those are involved in Carbon sequestration are very much diversified as Carbon could be stored in various segments of ecosystems with varying intensities. The geological Carbon sequestration occurs in very slow space and measures are yet to be implemented for CO_2 entrapment in rock system in field scale at a significant rate. Soil Carbon sequestration is considered as a very much vital process occurring in nature in order to minimize the harsh impact of climate change. In agro-ecosystems, the rate of Carbon sequestration was also found not at a faster rate but cropping systems and specially cultivated rice ecosystems seemed to be promising for reserving Carbon stock in soil system. Soil glomalin protein and microbial exo-polysaccharide could be proved as potential adsorbent of Carbon and further studies must be conducted in order to mass scale production of these products followed by application in agro-ecosystems.

Forest ecosystems are treated as potential Carbon sink and sound policies are further required to manage forest degradation and reforestation programs. The policy, like REDD+ is sound but its implementation strategies need to be refined. Even for mangrove ecosystem which covers 55% of CO_2 emissions, REDD+ needs more modification and implementation. Biological Carbon sequestration using algal species has been found promising of sequestering CO_2 from industrial flue gas. Bioreactor systems have been developed for algae based Carbon sequestration. More modification needs to develop over the microalgae based bioreactors. However, the bioreactor systems are operated not in a robust scale and are limited to industries. Therefore, we need to work out to develop nano-particle based, scrubber based technologies to hasten the rate of Carbon sequestration in terrestrial ecosystems.

As single perspective, it is difficult to manage increasing atmospheric CO_2 levels for that multiple approaches need to be executed. We conclude that genetic engineering can help to sequester CO2 more effectively rather than focusing upon one single policy.

ACKNOWLEDGMENT

Authors are thankful to Karunya Institute of Technology and Sciences Coimbatore and Marwadi University, Rajkot, Gujarat, India for providing us the required funding and support during research work.

Funding Information: There was no funding for the study.

Conflict of Interest Declaration: On behalf of all authors, the corresponding author states that there is no conflict of interest.

REFERENCES

Alongi, D. M. (2014). Carbon sequestration in mangrove forests. *Carbon Management, 3*(3), 313–322. doi:10.4155/cmt.12.20

Antonelli Paul, M., Fraser Lauchlan, H., & Gardner Wendy, C. (2018). Long term carbon sequestration potential of biosolids-amended copper and molybdenum mine tailings following mine site reclamation. *Ecological Engineering, 117*, 38–49. doi:10.1016/j.ecoleng.2018.04.001

Bala Krishna Prasad, M., & Ramanathan, A. L. (2008). Sedimentary nutrient dynamics in a tropical estuarine mangrove ecosystem. *Estuarine, Coastal and Shelf Science, 80*(1), 60–66. doi:10.1016/j.ecss.2008.07.004

Ballav, M. R., & Taranisen, P. (2011). Soil respiration and microbial population in a tropical deciduous forest soil of Orissa, India. *Flora, 206*(12), 1040–1044. doi:10.1016/j.flora.2011.07.009

Barbara, D., Kowalchuk, G., & Van, J. (2008). Climate change goes underground: effects of elevated atmospheric CO2 on microbial community structure and activities in the rhizosphere. Springer-Verlag.

Bhakta, J. N., LahiriSusmita, Pittman J, K., & Jana B. B., 2015. Carbon dioxide sequestration in wastewater by a consortium of elevated carbon dioxide-tolerant microalgae. *Journal of CO_2 Utilization, 10*, 105-112.

Bhattacharyya, P., Neogi, S., Roy, K. S., Dash, P. K., Nayak, A. K., & Mohapatra, T. (2014). Tropical low land rice ecosystem is a net carbon sink. *Agriculture, Ecosystems & Environment, 189*, 127–135. doi:10.1016/j.agee.2014.03.013

Bhattacharyya, P., Roy, K. S., Neogi, S., Adhya, T. K., Rao, K. S., & Manna, M. C. (2012a). Effects of rice straw and nitrogen fertilization on greenhouse gas emissions and carbon storage in tropical flooded soil planted with rice. *Soil & Tillage Research, 124*, 119–130. doi:10.1016/j.still.2012.05.015

Bhattacharyya, P., Roy, K. S., Neogi, S., Adhya, T. K., Rao, K. S., Manna, M. C., & Nayak, A. K. (2013). Influence of elevated carbon dioxide and temperature on belowground carbon allocation and enzyme activities in tropical flooded soil planted with rice. *Environmental Monitoring and Assessment, 185*(10), 8659–8671. doi:10.100710661-013-3202-7 PMID:23612768

Bhattacharyya, P., Roy, K. S., Neogi, S., Chakravorti, S. P., Behera, K. S., Das, K. M., Bardhan, S., & Rao, K. S. (2012). Effect of long-term application of organic amendment on C storage in relation to global warming potential and biological activities in tropical flooded soil planted to rice. *Nutrient Cycling in Agroecosystems*, *94*(2-3), 273–285. doi:10.100710705-012-9540-y

Bhattacharyya, P., Roy, K. S., Neogi, S., Chakravorti, S. P., Behera, K. S., Das, K. M., Bardhan, S., & Rao, K. S. (2012b). Effect of long term application of organic amendment on C storage in relation to global warming potential and biological activities in tropical flooded soil planted to rice. *Nutrient Cycling in Agroecosystems*, *94*(2-3), 273–285. doi:10.100710705-012-9540-y

Boullion, S., Borges, A., Moya, E., Diele, K., Dittmar, T., Duke, N., Kristenesen, E., Lee, S., Marchand, C., Middelburg, J., Monroy, V., Smith, T., & Twilley, R. (2008). Mangrove production and carbon sink: A revision of global budget estimates. *Global Biogeochemical Cycles*, *22*(2), 2. doi:10.1029/2007GB003052

Brar, B. S., Singh, K., Dheri, G. S., & Kumar, B. (2013). Carbon sequestration and soil carbon pools in a rice–wheat cropping system: Effect of long-term use of inorganic fertilizers and organic manure. *Soil & Tillage Research*, *128*, 30–36. doi:10.1016/j.still.2012.10.001

Cao, M., Zhang, Q., & Shugart, H. H. (2001). Dynamic responses of African ecosystem carbon cycling to climate change. *Climate Research*, *17*, 183–193. doi:10.3354/cr017183

Cramer, W., Bondeau, A., Woodward, F. I., Prentice, I. C., Betts, R. A., Brovkin, V., Cox, P. M., Fisher, V., Foley, J. A., Friend, A. D., Kucharik, C., Lomas, M. R., Ramankutty, N., Sitch, S., Smith, B., White, A., & Young-Molling, C. (2001). Global response of terrestrial ecosystem structure and function to CO_2 and climate change: Results from six dynamics global vegetation models. *Global Change Biology*, *7*(4), 357–373. doi:10.1046/j.1365-2486.2001.00383.x

D'Souza, J., & Rodrigues, B. F. (2013). Seasonal Diversity of ArbuscularMycorrhizal Fungi in Mangroves of Goa, India. *International Journal of Biodiversity*, *2013*, 1–7. doi:10.1155/2013/196527

David, P. (2017). Microalgae cultivation for carbon dioxide sequestration and protein production using a high-efficiency photobioreactor system. *Algal Research*, *25*, 413–420. doi:10.1016/j.algal.2017.06.003

Davidson, E., & Janssens, I. (2006). Temperature sensitivity of soil carbon decomposition and feedbacks to climate change. *Nature*, *440*(7081), 165–173. doi:10.1038/nature04514 PMID:16525463

Dintwe, K., & Okin, G. S. (2018). Soil organic carbon in savannas decreases with anthropogenic climate change. *Geoderma*, *309*, 7–16. doi:10.1016/j.geoderma.2017.08.030

Dintwe, K., Okin, G. S., D'Odorico, P., Hrast, T., Mladenov, N., Handorean, A., Bhattachan, A., & Caylor, K. K. (2014). Soil organic C and total N pools in the Kalahari: Potential impacts of climate change on C sequestration in savannas. *Plant and Soil*, *396*(1-2), 27–44. doi:10.100711104-014-2292-5

Donato, D. C., Kauffman, J. B., Murdiyarso, D., Kurnianto, S., Stidham, M., & Kanninen, M. (2011). Mangroves among the most carbon-rich forests in the tropics. *Geoscience.*, *4*(5), 293–297. doi:10.1038/ngeo1123

Drinkwater, L. E., Wagoner, P., & Sarrantonlo, M. (2010). Legume- Based Cropping Systems have Reduced Carbon and Nitrogen Losses. *Nature*, *396*(6708), 262–265. doi:10.1038/24376

Eloka-Eboka, A. C., & Inambao, F. L. (2017). Effects of CO_2 sequestration on lipid and biomass productivity in microalgal biomass production. *Applied Energy*, *195*, 1100–1111. doi:10.1016/j.apenergy.2017.03.071

Fang, H., Yu, G., Cheng, S., Mo, J., Yan, J., & Li, S. (2009). ^{13}C abundance, water-soluble and microbial biomass carbon as potential indicators of soil organic carbon dynamics in subtropical forests at different successional stages and subject to different nitrogen loads. *Plant and Soil*, *320*(1-2), 243–254. doi:10.100711104-009-9890-7

Ghosh, P. K., & Mahanta, S. K. (2014). Carbon Sequestration in grassland systems. *Range Management and Agroforestry*, *35*, 173–181.

Gorain, P. C., Sengupta, S., Satpati, G. G., Paul, I., Tripathi, S., & Pal, R. (2018). Carbon sequestration in macroalgal mats of brackish-water habitats in Indian Sunderbans: Potential as renewable organic resource. *The Science of the Total Environment*, *626*, 689–702. doi:10.1016/j.scitotenv.2018.01.106 PMID:29898555

Gren, I., & Zeleke, A. (2016). Policy design for forest carbon sequestration: A review of the literature. *Forest Policy and Economics*, *70*, 128–136. doi:10.1016/j.forpol.2016.06.008

Harris, Z. M., Milner, S., & Taylor, G. (2018). Biogenic Carbon-Capture and Sequestration. *Greenhouse Gas Balance of Bioenergy Systems*, 55-76.

Hien, H. T., Marchand, C., Aimé, J., Nhon, D. H., Hong, P. N., Tung, N. X., & Cuc, N. T. (2018). Belowground carbon sequestration in a mature planted mangroves (Northern Viet Nam). *Forest Ecology and Management*.

Hungate Bruce, A. (2009). Assessing the effect of elevated carbon dioxide on soil carbon: A comparison of four meta-analyses. *Global Change Biology*, *15*(8), 2020–2034. doi:10.1111/j.1365-2486.2009.01866.x

Ioannis, R. (2018a). Carbon sequestration via enhanced weathering of peridotites and basalts in seawater. *Applied Geochemistry*, *91*, 197–207. doi:10.1016/j.apgeochem.2017.11.001

IPCC. (2007). Summary for Policymakers. In *The physical Science Basis. Contribution of Working Group I to the Fourth Assessment Report of the Intergovernmental Panel on Climate Change*. Cambridge University Press.

Jedli, H., Jbara, A., Hedfi, H., Bouzgarrou, S., & Slimi, K. (2017). Carbon dioxide adsorption isotherm study on various cap rocks in a batch reactor for CO_2 sequestration processes. *Applied Clay Science*, *136*, 199–207. doi:10.1016/j.clay.2016.11.022

Keith, P., Eric, L., Jeffery, K., Ernie, M., & Amy, S. (2019). *Soil C Sequestration as a Biological Negative Emission Strategy*. Frontiers in Climate. doi:10.3389/fclim.2019.00008

Khdary, N. H., Ghanem, M. A., Abdesalam, M. E., & Al-Garadah, M. M. (2016). Sequestration of CO_2 using Cu nanoparticles supported on spherical and rod-shape mesoporous silica. *Journal of Saudi Chemical Society*, *22*(3), 343–351. doi:10.1016/j.jscs.2016.05.004

Knorr, M., Frey, S. D., & Curtis, P. S. (2005). Nitrogen addition and litter decomposition: A meta-analysis. *Ecology*, *8*(12), 3252–3257. doi:10.1890/05-0150

Kothamasi, D., Kothamasi, S., Bhattacharyya, A., & Babu, C. R. (2006). Arbuscular mycorrhizae and phosphate solubilising bacteria of the rhizosphere of the mangrove ecosystem of Great Nicobar island, India. *Biology and Fertility of Soils*, *42*(4), 358–361. doi:10.100700374-005-0035-8

Kristensen, E., Bouillon, S., Dittmar, T., & Marchand, C. (2008). Organic carbon dynamics in mangrove ecosystems: A review. *Aquatic Botany*, *89*(2), 201–219. doi:10.1016/j.aquabot.2007.12.005

Kumar, M., Sundaram, S., Gnansounou, E., Larroche, C., & Thakur, I. S. (2017). Carbon dioxide capture, storage and production of biofuel and biomaterials by bacteria: A review. *Bioresource Technology*. PMID:28951132

Lal, R. (2001). Potential of desertification control to sequester carbon and mitigate the greenhouse effect. *Climatic Change*, *51*(1), 35–72. doi:10.1023/A:1017529816140

Lal, R. (2003). Carbon sequestration in dryland ecosystems. *Environmental Management, 33*, 528–544. PMID:15453406

Lal, R., Lorenz, K., Hüttl, R. F., & Schneider, B. U. (2012). Carbon Sequestration in Temperate Forests, Recarbonization of the Biosphere. Academic Press.

Liu, S., Jiang, Z., Zhang, J., Wu, Y., Huang, X., & Macreadie, P. (2017). Sediment microbes mediate the impact of nutrient loading on blue carbon sequestration by mixed sea grass meadows. *The Science of the Total Environment, 599-600*, 1479–1484. doi:10.1016/j.scitotenv.2017.05.129 PMID:28531956

Marland, G., & Boden, T., A., (1997). *Trends: A Compendium of Data on Global Change*. GHG Concentrations.

Mcgrail, P., Nune, S., Motkuri, R., Glezakou, V., Koech, P., Adint, T., Fifield, L., Fernandaz, C., Liu, J., (2017). *System and process for production of magnesium metal and magnesium hydride from magnesium-containing salts and brines*. Academic Press.

Mitra, A., Sengupta, K., & Banerjee, K. (2011). Standing biomass and carbon storage of above-ground structures in dominant mangrove trees in the Sundarbans. *Forest Ecology and Management, 261*(7), 1325–1335. doi:10.1016/j.foreco.2011.01.012

Muñoz-Rojas, M., Román, J. R., Roncero-Ramos, B., Erickson, T. E., Merritt, D. J., Aguila-Carricondo, P., & Cantón, Y. (2018). Cyanobacteria inoculation enhances carbon sequestration in soil substrates used in dryland restoration. *The Science of the Total Environment, 636*, 1149–1154. doi:10.1016/j.scitotenv.2018.04.265 PMID:29913577

Nair, P. K. R., Nair, V. D., Kumar, B. M., & Haile, S. G. (2009). Soil carbon sequestration in tropical agroforestry systems: A feasibility appraisal. *Environmental Science & Policy, 12*(8), 1099–1111. doi:10.1016/j.envsci.2009.01.010

Neogi, S., Roy, K. S., Panda, B. B., Nayak, A. K., Rao, K. S., Manna, M. C., & Bhattacharyya, P. (2014). Soil respiration, labile carbon pools and enzyme activities as affected by tillage practices in a tropical rice-maize-cowpea cropping system. *Environmental Monitoring and Assessment, 186*(7), 4223–4236. doi:10.100710661-014-3693-x PMID:24609455

Nezar, K., Khdary, H., Ghanem, M. A., Abdesalam, M. E., & Al-Garadah, M. A. (2016). Sequestration of CO_2 using Cu nanoparticles supported on spherical and rod-shape mesoporous silica. *Journal of Saudi Chemical Society, 16*, 1319-6103.

O' Marra, F. P. (2012). The role of grasslands in food security and climate change. *Annals of Botany*, *110*(6), 1263–1270. doi:10.1093/aob/mcs209 PMID:23002270

Pandey, V. C., Sahu, N., Behera, S. K., & Singh, N. (2016). Carbon sequestration in fly ash dumps: Comparative assessment of three plant association. *Ecological Engineering*, *95*, 198–205. doi:10.1016/j.ecoleng.2016.06.010

Pardo, G., del Prado, A., Martínez-Mena, M., Bustamante, M. A., Martín, J. A. R., Álvaro-Fuentes, J., & Moral, R. (2017). Orchard and horticulture systems in Spanish Mediterranean coastal areas: Is there a real possibility to contribute to C sequestration? *Agriculture, Ecosystems & Environment*, *238*, 153–167. doi:10.1016/j.agee.2016.09.034

Patidar, S. K., & Mishra, S. (2017). *Carbon Sequestration by Microalgae: A Green Approach for Climate Change Mitigation*. Earth Systems and Environmental Sciences.

Paudel, M., Sah, S. K., McDonald, A., & Chaudhary, N. K. (2014). Soil Organic Carbon Sequestration in Rice-Wheat System under Conservation and Conventional Agriculture in Western Chitwan, Nepal. *World Journal of Agricultural Research.*, *2*(6A), 1–5. doi:10.12691/wjar-2-6A-1

Pavlik, D., Zhong, Y., Daiek, C., Liao, W., Morgan, B., Clary, W., & Liu, Y. (2017). Microalgae cultivation for carbon dioxide sequestration and protein production using a high-efficiency photo-bioreactor system. *Algal Research*, *25*, 413–420. doi:10.1016/j.algal.2017.06.003

Poussart, J.-N. (2004). Verification of Soil Carbon Sequestration: Sample Requirement. *Environmental Management*, *33*, 416–425. doi:10.100700267-003-9149-7

Power, I. M., Dipple, G. M., & Francis, P. S. (2017). Assessing the carbon sequestration potential of magnesium oxychloride cement building materials. *Cement and Concrete Composites*, *78*, 97–107. doi:10.1016/j.cemconcomp.2017.01.003

Ramaraj, R., Tsai, D. D. W., & Chen, P. H. (2014a). An exploration of the relationships between microalgae biomass growth and related environmental variables. *Journal of Photochemistry and Photobiology. B, Biology*, *135*, 44–47. doi:10.1016/j.jphotobiol.2014.04.001 PMID:24792572

Ramaraj, R., Tsai, D. D. W., & Chen, P. H. (2014b). Freshwater microalgae niche of air carbon dioxide mitigation. *Ecological Engineering*, *68*, 47–52. doi:10.1016/j.ecoleng.2014.03.058

Ramaraj, R., Tsai, D. D. W., & Chen, P. H. (2015). Carbon dioxide fixation of freshwater microalgae growth on natural water medium. *Ecological Engineering*, *75*, 86–92. doi:10.1016/j.ecoleng.2014.11.033

Ray, R., & Jana, T. K. (2017). Carbon sequestration by mangrove forest: One approach for managing carbon dioxide emission from coal-based power plant. *Atmospheric Environment*, *171*, 149–154. doi:10.1016/j.atmosenv.2017.10.019

Rigopoulos, I., Delimitis, A., Ioannou, I., Efstathiou, A. M., & Kyratsi, T. (2018b). Effect of ball milling on the carbon sequestration efficiency of serpentinized peridotites. *Minerals Engineering*, *120*, 66–74. doi:10.1016/j.mineng.2018.02.011

Rosenbauer, R. J., Thomas, B., Bischoff, J. L., & Palandri, J. (2012). Carbon sequestration via reaction with basaltic rocks: Geochemical modeling and experimental results. *Geochimica et Cosmochimica Acta*, *89*, 116–133. doi:10.1016/j.gca.2012.04.042

Roy, K. S., Neogi, S., Nayak, A. K., & Bhattacharyya, P. (2014). Effect of nitrogen fertilization on methane and carbon dioxide production potential in relation to labile carbon pools in tropical flooded rice soils in eastern India. *Archives of Agronomy and Soil Science*, *60*(10), 1329–1344. doi:10.1080/03650340.2013.869673

Sayre, R. (2010). Microalgae: The Potential for Carbon Capture. *Bioscience*, *60*(9), 722–727. doi:10.1525/bio.2010.60.9.9

Sheng, Y., & Zhu, L. (2018). Biochar alters microbial community and carbon sequestration potential across different soil pH. *The Science of the Total Environment*, *622-623*, 1391–1399. doi:10.1016/j.scitotenv.2017.11.337 PMID:29890604

Shukla, S. K., Shee, S., Maity, S. K., Solomon, S., Awasthy, S. K., Gaur, A., Pathak, A. D., & Jaiswal, V. P. (2017). Soil Carbon Sequestration and Crop Yields in Rice-Wheat and Sugarcane-Ratoon-Wheat Cropping Systems through Crop Residue Management and Inoculation of Trichodermaviride in Subtropical India. *Sugar Tech*, *19*(4), 347–358. doi:10.100712355-016-0470-x

Shuo, L., Li, Y., Li, X., Tian, X., Zhao, A., Wang, S., Wang, S., & Shi, J. (2016). Effect of straw management on carbon sequestration and grain production in a maize–wheat cropping system in Anthrosol of the Guanzhong Plain. *Soil & Tillage Research*, *157*, 43–51. doi:10.1016/j.still.2015.11.002

Singh, B. R., & Lal, R. (2005). The potential of soil carbon sequestration through improved management practices in Norway. *Environment, Development and Sustainability*, *7*(1), 161–184. doi:10.100710668-003-6372-6

Smith, P., & Smith, P. (2004). Monitoring and verification of soil C changes under Article 3.4 of the Kyoto Protocol. *Soil Use and Management*, *20*(2), 264–270. doi:10.1079/SUM2004239

Stringer, C. E., Trettin, C. C., Zarnoch, S. J., & Tang, W. (2015). Carbon stocks of mangroves within the Zambezi River Delta, Mozambique. *Forest Ecology and Management*, *354*, 139–148. doi:10.1016/j.foreco.2015.06.027

Trivedi, P., Anderson, I. C., & Singh, B. K. (2013). Microbial modulators of soil carbon storage: Integrating genomic and metabolic knowledge for global prediction. *Trends in Microbiology*, *21*(12), 12. doi:10.1016/j.tim.2013.09.005 PMID:24139848

Tsai, D. D., Chen, P. H., Chou, C. M., Hsu, C., & Ramaraj, R. (2015). Carbon sequestration by alga ecosystems. *Ecological Engineering*, *84*, 386–389. doi:10.1016/j.ecoleng.2015.09.024

Tsai, D. D., Chen, P. H., & Ramaraj, R. (2017). The potential of carbon dioxide capture and sequestration with algae. *Ecological Engineering*, *98*, 17–28. doi:10.1016/j.ecoleng.2016.10.049

Upendar, G., Singh, S., Chakrabarty, J., Chandra Ghanta, K., Dutta, S., & Dutta, A. (2018). Sequestration of carbon dioxide and production of biomolecules using cyanobacteria. *Journal of Environmental Management*, *218*, 234–244. doi:10.1016/j.jenvman.2018.04.031 PMID:29680755

Vass, M. M., (2017). Renewable energies cannot compete with forest carbon sequestration to cost-efficiently meet the EU carbon target for 2050. *Renewable Energy*.

Wang, J., Xiujun Wang, X., Xu, M., Feng, G., Zhang, W., Yang, X., & Huang, S. (2015). Contributions of wheat and maize residues to soil organic carbon under long-term rotation in north China. *Scientific Reports*, *5*(1), 1140. doi:10.1038rep11409 PMID:26100739

Wang, K., Xu, T., Wang, F., & Tian, H. (2016). Experimental study of CO_2–brine–rock interaction during CO_2 sequestration in deep coal seams. *International Journal of Coal Geology*, *154-155*, 265–274. doi:10.1016/j.coal.2016.01.010

Wang, Q., Li, Y., & Alva, A. (2010). Cropping Systems to Improve Carbon Sequestration for Mitigation of Climate Change. *Journal of Environmental Protection*, *1*(03), 207–215. doi:10.4236/jep.2010.13025

Wang, Q., Lu, H., Chen, J., Hong, H., Liu, J., Li, J., & Yan, C. (2018). Spatial distribution of glomalin-related soil protein and its relationship with sediment carbon sequestration across a mangrove forest. *The Science of the Total Environment, 613-614*, 548–556. doi:10.1016/j.scitotenv.2017.09.140 PMID:28926809

Xiaomin Yang, X., Songa, Z., Liu, H., Zwieten, L. V., Song, A., Lie, Z., Hao, A., Zhanga, X., & Wang, H. (2018). Phytolith accumulation in broadleaf and conifer forests of northern China: Implications for phytolith carbon sequestration. *Geoderma*, *312*, 36–44. doi:10.1016/j.geoderma.2017.10.005

Yan, Y. (2018). Integrate carbon dynamic models in analyzing carbon sequestration impact of forest biomass harvest. *The Science of the Total Environment*, *615*, 581–587. doi:10.1016/j.scitotenv.2017.09.326 PMID:28988094

Yana, G., Xinga, Y., Wang, J., Lic, Z., Wang, L., Wanga, Q., Xua, L., Zhanga, Z., Zhange, J., Donga, X., Shana, W., Guoa, L., & Han, S. (2018). Sequestration of atmospheric CO_2 in boreal forest carbon pools in northeastern China: Effects of nitrogen deposition. *Agricultural and Forest Meteorology*, *248*, 70–81. doi:10.1016/j.agrformet.2017.09.015

Yoo, C., Jun, S. Y., Lee, J. Y., Ahn, C. Y., & Oh, H. M. (2010). Selection of microalgae for lipid production under high levels carbon dioxide. *Bioresource Technology*, *101*(1), 71–74. doi:10.1016/j.biortech.2009.03.030 PMID:19362826

Zahra, S. I., Abbas, F., Ishaq, W., Ibrahim, M., Hammad, H. M., Akram, B., & Salik, M. R. (2016). Carbon Sequestration potential of soils under maize production in Irrigated Agriculture of the Punjab province of Pakistan. *The Journal of Animal & Plant Sciences.*, *26*, 706–715.

Zhao, B., Su, Y., Zhang, Y., & Cui, G. (2015). Carbon dioxide fixation and biomass production from combustion flue gas using energy microalgae. *Energy*, *89*, 347–357. doi:10.1016/j.energy.2015.05.123

Zhu, B., Chen, G., Cao, X., & Wei, D. (2017). Molecular characterization of CO_2 sequestration and assimilation in microalgae and its biotechnological applications. *Bioresource Technology*, *244*, 1207–1215. doi:10.1016/j.biortech.2017.05.199 PMID:28606753

Chapter 4
Vulnerability of Oasis Agriculture to Climate Change in Morocco

Mohamed Ait-El-Mokhtar
https://orcid.org/0000-0001-8096-9927
Cadi Ayyad University, Morocco

Abderrahim Boutasknit
https://orcid.org/0000-0002-0311-2441
Cadi Ayyad University, Morocco

Raja Ben-Laouane
https://orcid.org/0000-0002-1656-0462
Cadi Ayyad University, Morocco

Mohamed Anli
https://orcid.org/0000-0002-7028-709X
Cadi Ayyad University, Morocco

Fatima El Amerany
Cadi Ayyad University, Morocco

Salma Toubali
Cadi Ayyad University, Morocco

Soufiane Lahbouki
Cadi Ayyad University, Morocco

Said Wahbi
https://orcid.org/0000-0002-3066-5586
Cadi Ayyad University, Morocco

Abdelilah Meddich
https://orcid.org/0000-0001-9590-4405
Cadi Ayyad University, Morocco

ABSTRACT

Morocco is considered one of the most threatened countries by climate change. Over the last century, oases ecosystems in this country showed a high vulnerability to climate variation, which has led to water scarcity, an increase in land salinity, and therefore, a decrease in agricultural production. Conscious of these issues, several

DOI: 10.4018/978-1-7998-3343-7.ch004

solutions are initiated by the government to cope with climate change adverse effects. Many programs of rehabilitation were launched, and advanced researches are in progress in order to use some biofertilizers to improve tolerance of oasis crops to drought and salinity toward sustainable agriculture. The aim of this chapter is to give an overview of the impacts of climate change on oasis agriculture in Morocco and to provide potentially effective strategies to promote oasis agriculture under climate change. As a conclusion, the authors found that the use of different biofertilizers could be a potential strategy to mitigate climate change effects on oasis agriculture in Morocco.

INTRODUCTION

Arid and semi-arid Mediterranean regions, including Morocco, are among the most affected area by climate change's deleterious effects (Rochdane et *al.*, 2014). It is expected that rainfall will decrease and temperature will increase in the entire Mediterranean region; in particular, in Morocco, which is considered as one of the most vulnerable countries to climate change (Simonneaux et *al.*, 2015). In fact, in recent decades Morocco has experienced precipitation deficits, significant increase in temperatures and devastating flood especially in the south of the country (Driouech, 2010; Ait El Mokhtar et *al.*, 2019b). The oasis is a specific ecological landscape that characterizes the south of Morocco. It is a complex and fragile agro-ecosystem which sustains agriculture under drastic climatic conditions (El Khoumsi et *al.*, 2014). This traditional system proved to be sustainable and productive for centuries as a result of effective management for plants, soil and water (Pegna et *al.*, 2017). Many components of oasis ecosystems are affected by climate change especially water availability and soil quality (Karmaoui et *al.*, 2014; Ait Houssa et *al.*, 2017) which hamper oasian agriculture production. The main factors related to climate variation and affecting agriculture are water scarcity and soil salinization. Furthermore, traditional oasis agriculture is in critical situation because of migration, socio-economic and political mutations leading to entirely abandoned fields and the collapsing of the agricultural infrastructure (De Haas, 2001; Pegna et *al.*, 2017).

To cope with adverse effects of climate change on oasian agriculture, Morocco has initiated several strategies, such as the Green Moroccan Plan and other programs with international organizations to improve the agricultural production under changing climate (Akesbi, 2012, Pegna et *al.*, 2017). Otherwise, Morocco supported a lot of initiatives dealing with climate change at the 22nd COP United Nations Climate

Change Conference (COP22) organized in Marrakech in November 2016. One of the main initiatives being communicated by Morocco for the COP22 is Sustainable Oasis program designed to hinder oasis abandon and degradation, to promote adaptation to the climate change by reducing the vulnerability of natural and human oasis systems and provide support by helping this fragile ecosystem to implement resilient practices climate change mitigation. In addition, the use of agriculture biotechnology such as symbiotic microorganisms, organic amendments and plant tolerant varieties could be a promising way for mitigating the negative impacts of water and salt stress, induced partially by climate change, on agriculture in oasis area.

In this chapter, the authors will give an overview of climate trends in Morocco especially in oasis area and will assess the different adverse effects of climate change on oasian agriculture and related component. Given the importance of adaptation strategies to changing climate, the chapter also presents promising solutions to cope with the negative impacts of climate change on agriculture in oasis ecosystem.

CLIMATE VARIATIONS IN MOROCCAN OASES

Climate Change in Morocco

Morocco is a country located in the North of Africa, opens on both the Atlantic Ocean and the Mediterranean Sea, with about 3,500 km of coastline (Fertah et *al.*, 2017). It covers a territory of 710,000 km^2, from the Strait of Gibraltar to almost the southern part of the great African Sahara (Almeida-Garcia and Chahine, 2015). Its climate is mainly semi-arid but is heavily influenced by the Mediterranean. There are three major climate influences in Morocco: Atlantic, Mediterranean and Saharan (Born et *al.*, 2008).

Semi-arid and Mediterranean regions are expected to be more vulnerable to climate change that is why Morocco is one of the most vulnerable countries to climate variation (GIEC, 2007; Simonneaux et *al.*, 2015). The observations over the past three decades (1976-2006) show signs of likely impacts of climate change: increase in frequency and intensity of droughts, unusual devastating floods, reduced duration of snow cover in the Rif and Atlas peaks, changes in the spatial-temporal distribution of rains and overall a net decrease in the amount of rain, and high summer temperatures. Some of these events have already cost Morocco a great deal in terms of social, economic and environmental components. Morocco has also experienced increasing frequency of weather high-risk events over the last two decades. Torrential rains accompanied by inundations (Figure 1) caused serious damage between 2006 and 2008 in Errachidia (Merzouga, Gourrama and Boudnib regions) and in the north of the country (Tanger, Tetouan, Nador, ...).

Figure 1. Torrential flood in the oasis of Tazarine, Draa, Morocco (August 2018)

Rainfall Trends in Moroccan Oases

The Moroccan oases are characterized by arid to Saharan climate with very irregular rainfall from year to year and with a marked continentalism. Rainfall is often intense and concentrated over time in the form of thunderstorms, causing severe flooding. The annual average rainfall is only 132 mm and the number of rainy days is just twenty (PNUD, 2011a). Climate change research indicates that annual precipitation is expected to decline over the current century (Gibelin and Déqué, 2003; Rowell, 2005). High interannual variability is a characteristic of Mediterranean precipitation. In addition, the coefficient of variability (ratio of the standard deviation to the average) is more than 70% in the southern regions (Knippertz et *al.*, 2003, Driouech et *al.*, 2009).

According to a study by the Moroccan Department of Energy, Mines, Water and Environment, entitled "Adaptations to climate change for resilient oases", even more worrying are Morocco's future climate projections made as part of the Second National Communication on Climate Change, which predict average rainfall declines of -13% and -19% in 2045 and 2075 respectively. Another study was conducted in Ouarzazate on winter rainfall accumulations projected for 2030-2050. The study has shown a rainfall decrease across the entire oasian zone (PNUD, 2011b). This decrease could reach 35% in Ouarzazate over this period, compared to the period 1971-2000 (Figure 2) (Driouech, 2010).

Figure 2. Relative anomalies (in %) of annual average rainfall at the Ouarzazate meteorological station, Morocco (Period 1961-2008)
(Driouech, 2010)

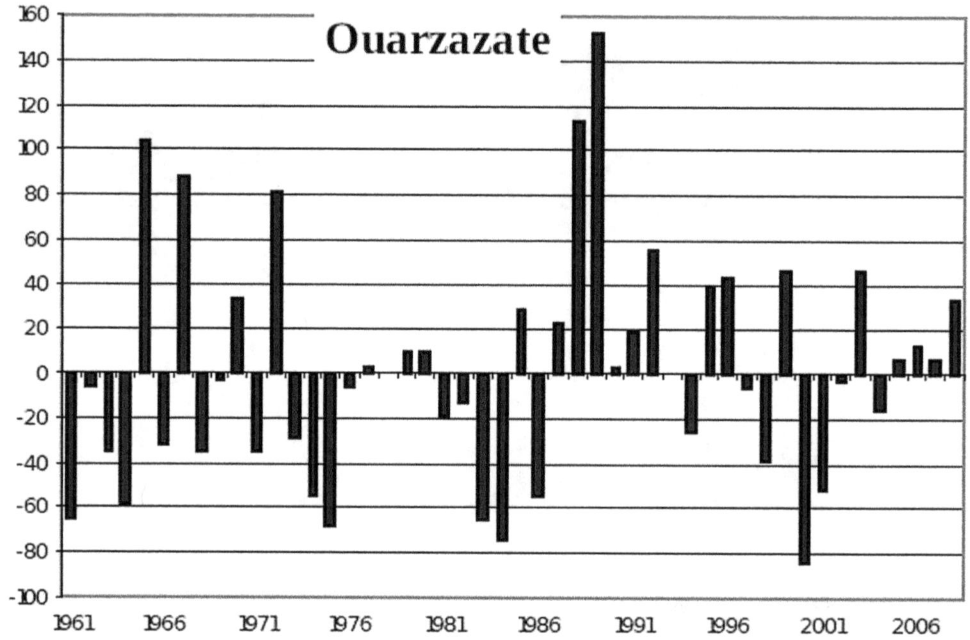

Temperature Trends in Moroccan Oases

In Morocco, and according to the Department of Energy, Mines, Water and Environment, a trend towards temperature average increases of 1.8 °C and 3.2 °C, respectively for the 2045 and 2075 horizons. For example, in the Ouarzazate region and during the period of 1961-2008, the warming of the climate was 0.3 °C per decade (Driouech, 2010). In addition, the projected seasonal and annual warmings for 2030-2050 range from 1 °C to 2.2 °C (Driouech, 2010). The study of temperature variation in the oases of the Tata region, according to the Provincial Directorate of Equipment (DPE) indicates that the average temperature calculated over 13 years since 1985-1986 is of 22 degrees. High summer temperatures are not uncommon, and in August 1998 there were 52 °C.

Overall, climate change is a reality in southern Morocco. It has gone up the air temperature in all seasons over the past 40 years. This increase is the most effective since the 1980s and 1990s (when warming increased with the frequency of droughts). The frequency and severity of heat waves are expected to increase throughout the country (Royaume du Maroc, 2010).

CLIMATE CHANGE IMPACTS ON MOROCCO OASIS AGRICULTURE

Impact of Climate Change on Water Resources

Agriculture sector is closely linked to water resources. In the past thirty years, water resources in oases are suffering from aridity and scarcity (ONEE, 2015). Furthermore, the water requirements are theoretically and practically indispensable for the sustainable development of oasis desert ecosystems (Kuhn et *al.*, 2010). Those resources mainly originate from the precipitation, snow melting in the mountainous areas and groundwater (Fox et *al.*, 2001; Alcalá et *al.*, 2015). Owing to scarce water resources in the oases of South Morocco, ecosystems are vulnerable to climatic changes (Kabiri, 2004; Karmaoui et *al.*, 2014). The impact of climate variation accelerates environmental vulnerability, management and exploitation of water resources in oasis regions (Kabiri, 2005; Karmaoui et *al.*, 2014). Huge amount of studies suggested that the water and ecological degradation in arid regions was largely affected much more by irrational human exploitation than climate change (Karmaoui et *al.*, 2014; Karmaoui, et *al.*, 2019). In addition, the rapid growth of population in oasis regions in Morocco, the increased utilization of surface and groundwater for irrigation caused severe deterioration of water in most ecosystem of South Morocco (De Jong et *al.*, 2006; Diekkrüger et *al.*, 2010). The water resources are increasingly influenced by an annual irregularity and a very inter-annual variability of precipitations, which resulted in a dramatic water scarcity (Ait El Mokhtar et *al.*, 2019b). Thereby, water scarcity has impacts the current ecological imbalance of oasis ecosystems and thereby agricultural system (Kabiri, 2004; Ben Salem et *al.*, 2011).

Impact of Climate Change on Agricultural Soils

Agricultural soils in oasis are characterized by significant physicochemical and biological properties. These properties are influenced by climate change and anthropic activity (Waldrop and Firestone, 2006). However, prolonged climate change is among the major factors that lead to a major agricultural soils degradation in arid and semi-arid areas particularly in oasis regions (Ait El Mokhtar et *al.*, 2019b). On the other hand, climate change in these areas is marked by drought and salinity, which lead to negative influence of the properties of these soils (Gong et *al.*, 2015). In fact, drought and salinity are major constraints that severely change soils characteristics (Tripathi et *al.*, 2006; Geng et *al.*, 2015; Rath and Rousk, 2015). The modification of these characteristics negatively affects ecological function and soil fertility (Chelleri et *al.*, 2014). In affected soils, water and nutrient uptake are inhibited, although there is sufficient quantity of them in soil (Bor et *al.*, 2003). Those effects are negatively correlated with soil properties such as soil microbial activity, soil respiration, and litter and organic matter decomposition (Geng et *al.*, 2015; Mariotte et *al.*, 2015), which can lead to a failure of the supply of important elements in the rehabilitation and have significant negative consequences such as desertification and erosion (Sati, 2015). Furthermore, Ali et *al.* (2017) reviewed the effects of salinity on soil microbial communities and C cycling and concluded that the soil respiration is inhibited. Therefore, drought and salinity is closely linked to other soil degradation issues including erosion of soil, which increased the vulnerability of agricultural soils in oasis ecosystems.

Impact of Climate Change on Oasis Vegetation Biodiversity

In South Morocco, the state of vegetation biodiversity is declining at the level of oasis ecosystems and the pressures upon it are increasing. For example, in M'hamid oasis several changes marked by a great extinction of a large part of the vegetation between the period 1984 and 2016 (Figure 3) (Lamqadem et *al.*, 2017). In addition, it was reported that drought and salinity reduced the date palm area by the half, from 4,575 km^2 to 1,342 km^2 (Benmohammadi et *al.*, 2000). The decline are also due to a range of threats including overexploitation and unsustainable use of natural resources and climate change (Karmaoui and Moumane, 2016). The increased effects of climate change on natural resources, such as the droughts, floods and the reduction of water resources, can negatively affect the resistance of a large number of plant species in oasis ecosystems in South Morocco (ONEE, 2015; Karmaoui et *al.*, 2014). The replacement of these vulnerable species by resistant ones results in a complete reduction in vegetation biodiversity in the oasis (Hannah et *al.*, 2002; Brooks et

Figure 3. Changes of vegetation area in the M'hamid oasis between 1984 and 2016.
(Lamqadem et al., 2017)

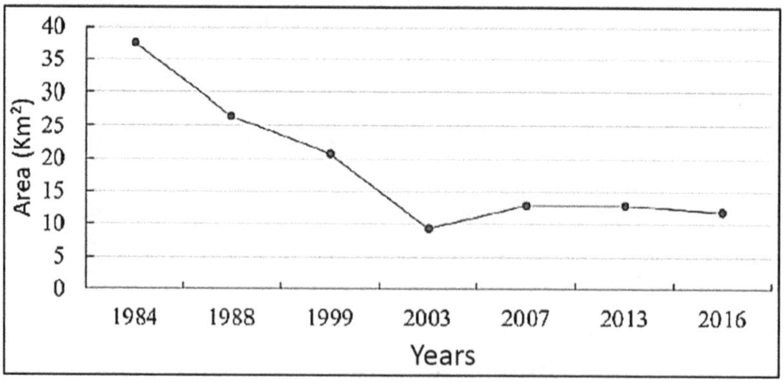

al., 2006). In addition, the vegetation cover degradation is caused by agricultural exploitation of pastures, which has resulted in high degradation of the biodiversity.

Impact of Climate Change on Oasis Agriculture Production

Agriculture and climate change are characterized by a complex cause-effect relationship. Agriculture is among the most vulnerable sectors to the future climate change, and has been the subject of many studies in recent years (Neset et *al.*, 2019). In fact, it is highly exposed to climate variation because its activities depend directly on climatic conditions. However, the effects of these changes on agriculture vary across countries and cultures (Agovino et *al.*, 2019).

Several recent studies on the Mediterranean region showed that Maghreb agriculture will be very affected by the impacts of climate change (Neset et *al.*, 2019), In fact, the expected scarcity of available water (falling rainfall and overexploitation of groundwater) implies a shift and a reduction of growth periods, an acceleration of soil degradation and salinity as well as the loss of productive lands, especially in rural areas (Agovino et *al.*, 2019). In Morocco, agriculture is the first employment position in rural areas. Reduced water availability and increased needs are expected for rainfed and irrigated agriculture, due to changes in the rainfall regime, increased evapotranspiration and sea level rise, thus increasing the risk of drought at crucial periods in crop cycles. This situation directly affects rainfed crops (Ouraich and Tyner, 2014).

Several major cultures are cultivated in the oases, namely wheat, barley, maize, alfalfa, pulses, a vegetable aggregate, henna and date palms as perennial crops (Heidecke et *al.*, 2010). The Moroccan oases are among the most affected ecosystems by climatic variations (Jilali and Zarhoule, 2015). The intense exploitation and the deficit of food, especially during the years of drought cause an intense degradation of the ecological conditions, for all the population and especially for the oasis farmers (Schilling et *al.*, 2015). Indeed, the rivers and the wells are dry, the Khettaras and the grounds are no longer exploited and the socioeconomic infrastructures are threatened by the invasion of the sand (Arrus and Rousset, 2007). This lack of water affects a large part the farmer's income and destabilize the population by encouraging the exodus to neighboring regions, to cities or even now to foreign countries (Agovino et *al.*, 2019).

Arboriculture, mainly the date palm and the olive tree and livestock play a key role in the economic development of Tafilalet (Ait Hmida, 2003). However, the potential productivity of the region is limited by climate changes that affect the agriculture resilience in the oases (Ait El Mokhtar et *al.*, 2019b). Water scarcity, recurring droughts, salinity and bayoud disease (*Fusarium oxyisporum* sp *albedenis*) have contributed to the degradation of the oasis environment. All of these constraints are an obstacle to the optimization of the oases agricultural resources (Ait Hmida, 2003; El Khoumsi et *al.*, 2014).

The date palm is currently exposed to several abiotic threats such as salinity, drought and desertification and biotic including bayoud disease in this region. However, salinity is one of the most prevalent stresses in desert areas which is linked to poor irrigation management and excessive evaporation from rainfall, and causes excessive accumulation of soluble salts. This causes a decrease in productivity and especially the speed of growth and the yield of the date palm (El Khoumsi et *al.*, 2017). In the other hand, even if the palm tree is a drought tolerant species, it will be necessary at least meet its needs to accomplish its function because it can vegetate and produce fruit only after sufficient supply of water needs (Kuhn et *al.*, 2010). The prolonged drought during the 1980s in Morocco resulted in the partial drying of more than 500,000 palm trees and the level of production ranged between 12,000 tons in the extremely dry year 1984 and 120,000 tons in the particularly wet year 1990 (Haddouch, 1996). This indicates that production decreases with increasing drought duration. Then desertification, which is a natural phenomenon that threatens the fertility of arable land and is caused by climate change and / or the consequences of human activities. The aridity of the climate is a factor aggravating the desertification, which reflects the degradation of the grounds and also the degradation of the vegetation which can disappear because of the prolonged drought (Lamqadem et *al.*, 2018).

STRATEGIES TO IMPROVE OASIS AGRICULTURE UNDER CLIMATE CHANGE

Programs of Rehabilitation

Oasis ecosystem in Morocco, like in all countries, is suffering from the effects of climate change. Many studies indicate that, in the absence of adaptation measures, climate change will have devastating impacts on this fragile ecosystem (Pegna et *al.*, 2017; Ait El Mokhtar et *al.*, 2019b). Unfortunately, nowadays oasis agriculture has lost its relevance and this complex environment is in serious danger due not only to climate changes but also to socio-economic and political impacts. For these reasons, the Moroccan country has adopted many programs in collaboration with international organizations and many stakeholders to develop oasis agricultural systems and improve the strength of the sector to climate change (Zucca et *al.*, 2016). The most relevant ones for this study are presented below:

Green Moroccan Plan

The Green Moroccan Plan (GMP) was launched in 2008 by the Ministry of Agriculture, Fisheries, Rural Development and Forests. GMP is the governmental strategy, which aims to develop the agricultural sector in order to be sustainable and competitive at international level. It intends to promote the agricultural incomes for rural inhabitants by sustaining investments, exports and aggregation between the production and the commercial and industrial phases. The implementation of the GMP is based on two pillars and several cross cutting programs. The first pillar is related to the highly productive and intensive agriculture with a direct connection to the market, and the second pillar is related to the consolidation of small holder farmers by improving intensification of crops where appropriate, and the use of more adapted crops with respect to ecological conditions and markets demand (Ouraich and Tyner, 2014). In total, the GMP is made of 1500 projects requiring more than 10 billion USD for implementation until 2020 (Badraoui, 2014).

Date production sector, related to oases area, is one the GPM interests, which considered an important sector for the Moroccan agricultural development strategy, because national consumption is high and date production has a potential to be environmentally sustainable and competitive from an economic point of view, even for exportation. Indeed the largest project in Morocco, which is part of GPM, remains the program to accelerate the rehabilitation of palm groves in the Oriental, Souss-Massa, Guelmim-Oued Noun and Draa-Tafilalet, with the following objectives (Zucca et *al.*, 2016; Pegna et *al.*, 2017):

- Creation of several new date recovery and storage units in Souss-Massa (capacity of 3400 tonnes), Meknès Tafilalt (capacity of 2600 tonnes), Guelmim Es-smara (capacity of 1000), eastern region (capacity of 100 tonnes).
- Establishment of a national laboratory for the cultivation of Errachidia date palm tissue to increase national production to 50,000 budding strains annually.
- Planting 3 million plants by 2020 with the high quality and resistant varieties, thanks to the vitroplant production laboratories.
- Achieving a production of 160,000 tonnes by 2020.

The restoration of a sustainable oasis ecosystem in Morocco is the main goals of others development programs implemented in the last decade. A list of these programs is shown in table 1.

Table 1. Programs of rehabilitation launched by the Moroccan government with different organizations in the last decade.

Programs	Description
Southern Oasis Program (SOP)	SOP was launched by the Agency for the Economic and Social Promotion and Development of the Moroccan Southern Provinces (ADPS) between 2006 and 2015 in collaborations with United Nations Development Program. SOP allows: * To support sustainable regional development * To promote local products and improve people's incomes, * To valorize local heritage * To reduce the loss of biodiversity and protect the environment.
Tafilalet Oasis Program (TOP)	TOP has the same general objectives as SOP, but it concerns Tafilalet oases. TOP was implemented by the Ministry of Energy, Mines, Water and Environment management in collaboration with Directorate of Spatial Planning between 2006 and 2015.
Draa Oasis Program (DOP)	A 2006 – 2015 program aiming to fight desertification, poverty, protection and valorisation of Draa traditional oasis.
ANDZOA Program	The National Agency for the Development of the Oasian and Argan Zones (ANDZOA) is a structure founded in 2010, responsible of the Sustainable Oasis Initiative launched during the COP22 in Marrakesh in 2016. ANDZOA Program aims to mobilize national and international resources to protect and restore the traditional oasis ecosystem, endangered by desertification.
GIAHS program	A program that safeguards the Imilchil-Amellago, Ait Mansour, Akka, Assa and Figuig oases since 2011 and aims to fight the progressive degradation of the oasis area. Implanted by FAO, in collaboration with the Ministry of Environment, the Ministry of Agriculture, ANDZOA, ADPS the National Institute of Agronomic Research (INRA) and Global Environment Facility (GEF).
OASIL	A 2017-2021 project, promoted by FAO in collaboration with ANDZOA and Global Environment Facility (GEF). The project aims: * To revitalize oasis agroecosystems in the Draa-Tafilalet region to achieve productivity, attractiveness and health, and to sustain and make more resilient the livelihoods of the local communities. * To support the dissemination of knowledge on oases, strengthens political dialogue and facilitates the adoption of strategies and plans for sustainable oasis management at both national and regional levels. * To encourage the use of an integrated approach for the management of oasis agro-ecosystem, with a deeply involvement of all stakeholders in the decision process.
RE-ACTIVATE and PEDEL programs.	Are projects concerning oasis farming system and initiated by German Corporation for International Cooperation GmbH (GIZ). This organization works in Morocco since 1975 and supports projects about sustainable development, water harvesting, renewable energy, climate change, good governance and health.
DéLIO program	Oriental Integrated Local Development Program (DéLIO) is a 2008-2011 program that concerns, among other areas, the Figuig Oasis. It focuses on the valorization of 5 local products (dates and the most important handicraft), improving and integrating their production chain.
Program of Mohammed VI Foundation for the Protection of the Environment	For the benefit of the Marrakesh palm grove and has resulted in the planting of 561,000 palm trees and the maintenance of 81,000 adults in this palm grove.

(Mainguet et *al.*, 2011; Dahan et *al.*, 2013; Zucca et *al.*, 2016; Pegna et *al.*, 2017).

Cultural Practices

In order to mitigate the deleterious effects of drought and salinity on agriculture, due to limitation of erratic rainfall and increase of evaporation rates over and within years, traditional and modern cultural practices were applied in Morocco. Among the cultural practices applied for agriculture production in arid and semi-arid regions is the creation of tens to hundreds of kilometers of canals as well as thousands basin or reservoirs to store water (Manuel et *al.*, 2018). Those canals are subterranean tunnels and were implemented in order to carry and to drain a large amount of water from springs, dams or upland areas to the dry, arid, and semi-arid regions. Additionally, the use of groundwater wells and drip irrigation systems has been encouraged for the rational exploitation of water (Kinicki et *al.*, 2017).

The sustaining agriculture in oasis has succeeded due to the establishment and development of palm grove which has the ability of making barrier against the advancement of the desert (Karmaoui et *al.*, 2015). Palm trees leaves form umbrellas that could involve in the formation of a mild microclimate and protect other fruit threes (e.g., figs, almonds, olives, pomegranates) and crops (e.g., barley, wheat, sorghum, alfalfa and various vegetables) from excessive light (Figure 4) (Pegna et *al.*, 2017).

Figure 4. The palm tree plantation in Aoufouss Oasis (Tafilalet region, Morocco). Three levels of plantations: Palm trees, fruits threes, and crops.
(Mainguet et al., 2011)

Advanced Researches

The selection of tolerant plants as well as the production of biological and organic fertilizers have been proposed to alleviate the negative effects of drought and salinity, due among others to climate variation, in order to improve oasis agriculture.

Tolerant Plants

Many strategies were applied in Morocco, based on the advances in research, in order to exploit the land and to solve the problem of salinity as well as water stress. Choosing and encouraging either salt or drought plant tolerant genotypes are among the applied strategies. Those plants have the ability to make different responses to handle abiotic stress through morphological and physiological changes and the accumulation of anti-stressful component "osmolytes" that involved in detoxification of reactive oxygen species and stabilization of membranes (Serraj and Sinclair, 2002; Chaves and Oliveira, 2004).

As shown in Table 2, there are many stress tolerant genotypes of oasian crops to provide the subsistence for local populations. For example, there is a lot of interest to cultivate saffron (*Crocus sativus* L.) (the most expensive spice in the world), barley (*Hordeum vulgare* L.) and wheat (*Triticum durum* Desf) (the important cereals for human and animal feed), Alfalfa (*Medicago sativa* L.) (the forage crop), faba bean (*Vicia faba* L.) (the grain legumes) and sugarcane (*Saccharum* sp.) (the main source of sugar). Those crops were cultivated for centuries in desert places due to their ability to adjust the morphology and the physiology of plant under stressful conditions by increasing number and area of leaves, increasing root system, and maintaining the relative water content. Additionally, the greater accumulation of proline, soluble sugars and phenolic compounds has the pivotal role in reducing the oxidative damage caused by stress (Kumar et *al.*, 2017) and in osmotic adjustment (Lamaoui et *al.*, 2018; Koźmińska et *al.*, 2019).

Biological and Organic Fertilizers

Recently, scientists are focusing on the use of microorganisms (AMF and PGPR) and organic amendments (manure and compost) to improve soil fertility and to promote growth development under harmful conditions due to climate change.

Arbuscular Mycorrhizal Fungi

Arbuscular mycorrhizal fungi (AMF) are known as a soil-borne fungi that have the ability to improve plant mineral nutrient uptake, especially phosphorus, to increase

Table 2. Salt and drought tolerant cultivars used in oasis area

Crop	Cultivars	Stress	Effects	Anti-stressful Components	Reference
Saffron	Not mentioned	Salt stress	1 g.l^{-1} of NaCl increased number and area of leaves, height of plants, aerial dry biomass, number and weight of corms and stigmata yield, and kept relative water content.	Accumulation of proline, soluble sugars, phenolic content and malondialdehyde (MDA) compound.	Mzabri et *al.* (2017).
Barley	Tamelalt	Salt stress	5 g/l had no impact on germination of seeds, roots and shoots lengths, fresh and dry biomass and relative water content.	Increase of the level of hexoses phosphates, TCA cycle intermediates, and metabolites involved in cellular protection.	El Goumi et *al.* (2014). Patterson et *al.* (2009)
Alfalfa	Tata	Salt stress	100 mM of NaCl had no effect on fresh weight and seedling height	Accumulation of proline, soluble sugars and inorganic ions	Farissi et *al.* (2011).
			200 mM of NaCl increased shoot and root dry biomass and the relative water content	Increase the activity of acid phosphatase	Farissi et *al.* (2014).
Faba bean	Aguadulce	Water stress	Soil moisture at 40% of substrate field capacity increased leaf mass area	Accumulation of proline, glycine betaine, MDA, hydrogen peroxide (H_2O_2) in leaves and roots. Increase of the activity of ascorbate peroxidase, guaiacol peroxidase, catalase, and polyphenol peroxidase.	Kabbadj et *al.* (2017).
Wheat	Altay2000, 14IWWYTIR-19 UZ-11CWA-8	Salt stress	100 mM of NaCl had a positive effect on germination, seedling and mature field plant and maintained stable osmotic potential and chlorophyll intensity.	-	Oyiga et *al.* (2016)
Durum wheat	Marzak	Water stress	Increase of the mesophyll thickness and of the intercellular space.	-	Assem et *al.* (2017).
Sugarca-ne	NCo310	Salt stress	Maintain the relative growth rate and increase K^+ and Cl^- concentration at 34 mM.	Increase proline levels.	Gandonou et *al.* (2005) Gandonou et *al.* (2012)

water assimilation, and to help plant to adapt in biotic and abiotic stress through the establishment of symbiosis network (Augé, 2001; Smith and Read, 2008; Porcel et *al.*, 2011; Augé et *al.*, 2015; Balestrini et *al.*, 2015). Many researchers have reported the successful role of AMF in mitigating water and salt stress. Some research showed the growth promotion of the date palm and olive tree under the application of either Aoufous complex, *Glomus monosporus,* native *Rhizophagus manihotis* or non-native *Funneliformis mosseae* (see Table 3). Moreover, Ait El Mokhtar et *al.* (2019c) reported that the application of Aoufous mycorrhizal consortium has significantly improved the shoot and dry biomass (Figure 5 A and B), the photosynthetic efficiency (Fv/Fm) (Figure 5 C) and the leaf relative water content (Figure 5 D) of palm date under salt stress. Moreover, Meddich et *al.* (2015) showed that the use of the same consortium was able to promote growth of date palm under water stress conditions by improving the uptake of many ions such as P, Ca, Mg, Na, K, Cu and Mn and increasing the level of phenolic compounds and antioxidation enzymes, like

Table 3. Bio-fertilizers and organic amendment applied to alleviate salt and water stresses on oasis crops

Biological and Organic Fertilizers		**Tested Plants**	**Stress**	**Effects**	**Reference**
Bio-fertilizers	Aoufous complex and *Glomus monosporus*	Date palm	Water stress	Increased number of produced leaves, height of areal part, relative water content, ionic (P, Ca, Mg, Na, K, Cu, Mn) and phenols content, and polyphenoloxydase.	Meddich et *al.* (2015)
Arbuscular mycorrhizal fungi (AMF)	Native *Rhizophagus manihotis* and non-native *Funneliformis mosseae*	Olive	Water stress	Enhanced the protection against oxidative stress induced by water deficit. Increased shoot height, root length, fresh weight of shoot and root, dry weight of shoot and root. Decreased the level of H_2O_2, MDA and electrolyte leakage (EL). Enhanced the activity of superoxide dismutase, catalase, guaiacol peroxidase, and ascorbate peroxidase.	Fouad et *al.* (2014)
Symbiotic nitrogen fixing bacteria (PGPR)	*Sinorhizobium meliloti* and *S. medicae*	Alfalfa	Water and salt stress	Tolerant to NaCl (>513 mM), water stress (-1.5 MPa), high temperature (40°C), low pH (3.5), tolerance to heavy metals (Cd, Mn and Zn).	Elboutahiri et *al.* (2010)
	MC0415 strain	Chickpea	Water stress	Improved dry weight of shoot, root and nodules Increased nitrate reductase activity and K^+ accumulation under water deficit.	Khadraji et *al.* (2017)
Organic amendment / Compost (5 and 10 t ha^{-1})	From olive mill waste	Quinoa and Pea	Water stress	Increased total dry biomass, dry seed yield, and crop water productivity	Hirich et *al.* (2014a) Hirich et *al.* (2014b)

Figure 5. Changes in (A) shoot and (B) root dry weight, (C) Fv/Fm and (D) leaf relative water content (LRWC) of date palm (Phoenix dactylifera L.) inoculated and non-inoculated with AMF and grown under non-saline (0 mM NaCl) or saline conditions (240 mM NaCl)
(Ait-El-Mokhtar et al., 2019c)

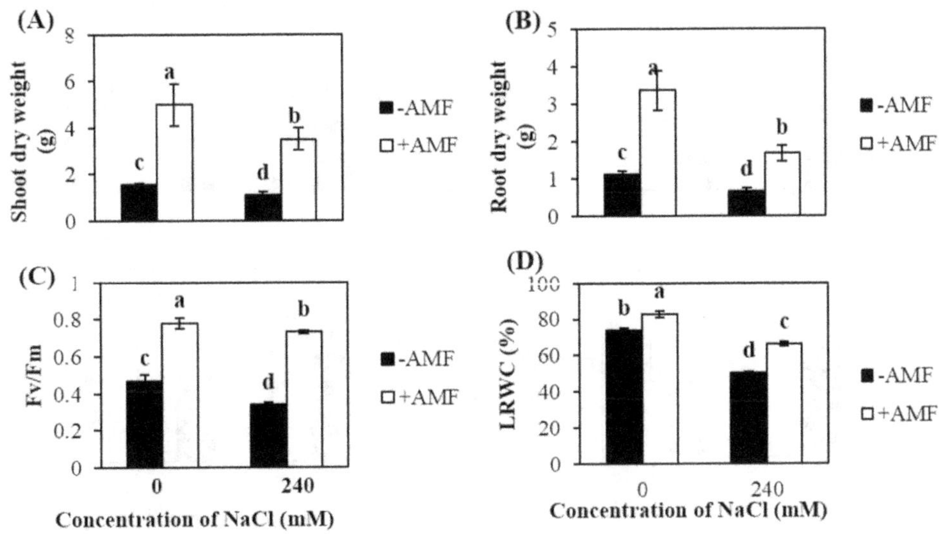

polyphenol oxydase. Thus, it would be interesting to apply AMF in oasis in order to reinforce the development of palm grove and other crops.

Plant Growth-Promoting Rhizobacteria

Besides AMF, plant growth-promoting rhizobacteria (PGPR) are symbiotic bacteria that interact with plant and have a positive influence on its growth (Beneduzi et *al.*, 2012) (Table 3). PGPR play a crucial role in providing the essential nutrients to plants, by increasing the mobilization of minerals and organic nutrients and the fixation of atmospheric di-nitrogen (Di Benedetto et *al.*, 2017). Moreover, PGPR can enhance salt and drought tolerance through several mechanisms (Vurukonda et *al.*, 2016; Ilangumaran and Smith, 2017). Ben-Laouane et *al.* (2019) reported that the rhizobia strain (RhLO1) could increase shoot and root biomass as well as Fv/Fm of alfalfa salt stressed plants (Figure 6 A and C). Those bacteria could overcome osmotic stress by inducing accumulation of osmolyte (e.g., sugar, see Figure 6 C), enhancing nutrient uptake and biosynthesis of phytohormones (e.g., indole acetic acid (IAA)) (Di Benedetto et *al.*, 2017). On the other hand, it was reported that the combination of AMF with PGPR could be superior and have a greater effect on

Figure 6. Changes in (A) shoot and (B) root dry weight, (C) Fv/Fm and (D) sugar content in root of alfalfa inoculated and non-inoculated with PGPR and grown under non-saline (0 mM NaCl) or saline conditions (120 mM NaCl)
(Ben-Laouane et al., 2019).

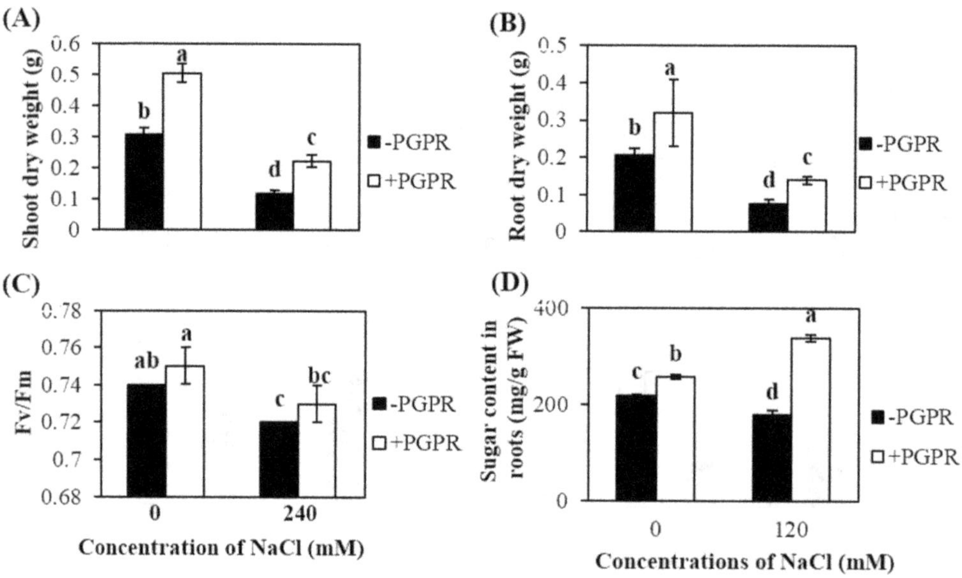

plants survive in harsh desert conditions than the application of a single symbiotic (Ben-Laouane et *al.*, 2019).

Organic Amendments

The use of organic amendments to increase soil fertility is taking much interest in Morocco. The application of animal manure and human sewage to soil was promoted the growth of many crops (Mohamed et *al.*, 2018). Furthermore, the second approach applied in Morocco is the biosynthesis of compost, a high-quality soil amendment, from organic horticultural wastes that is estimated up to 55.2 million tons per year (Belmakki et *al.*, 2015). Composting of organic waste represents one of the most practical, economical, and recycling management options (Pérez-Piqueres et *al.*, 2006). Azim et *al.* (2017) demonstrated that only 37.5% of compost is made by treating tomato wastes, whilst the rest is synthesized via treating manure and olive mill wastes. Compost is most frequently used in agriculture to improve the growth of crops grown in normal and stressful conditions (see Table 3). The beneficial effect of compost in agriculture is due to its ability to increase soil organic carbon, to provide essential micro- and macro-nutrients, to increase water retention capability and to improve microbial activities (Scotti et *al.*, 2015).

Use of Remote Sensing and Geographic Information System Technologies

Remote sensing (RS) and Geographic Information System (GIS) figure among technologies that may improve agriculture under climate change. RS is one of the assets used in collecting and monitoring environmental information that can be used in smart agriculture (Mulla, 2013). This technology can be used for collecting crops growth parameters like size, height and stem diameter, crop damage from pest infestation, and all environment parameters. In the other hand, GIS is a technology that allows gathering and sharing information to provide a very precise insight of the farming conditions, and measuring and monitoring the effects of land management practices (ESRI, 2017). GIS enable the farmers to map and project current and future fluctuations in precipitation, temperature, and crop output to provide a better perspective of their operations. When combined with RS technology, GIS can be efficient in determining and controlling inputs, reducing expense and allowing the best land use. Thus, farm producers use GIS to visualize their land, crops and management practices (Brion and Balahadia, 2017).

Nowdays, GIS and RS could be considered as two promising tools to use in mitigating the impacts of climate change on oasis agriculture. These strategies are currently being utilized in evaluating several land issues such as soil erosion, land exploitation and soil deterioration (Brion and Balahadia, 2017). The application of these technologies may provide suitable environmental information to the government to initiate new projects to support the farmers in vulnerable oasis area.

GIS and RS can also be used in assessing the environmental sensitivity to desertification in arid zones (Hadeel et al., 2010). Many information are required to estimate the desertification sensitivity index such as thematic layers of soil, vegetation, climate, and extent of sand movement (Brion and Balahadia, 2017). This index could be very useful to prevent land desertification and therefore to improve land exploitation and agriculture production.

CONCLUSION

Climate change is a major concern for all countries and thus only a global vision can deal with this problem. Mediterranean countries such as Morocco are the most affected by climate variation effects and must put in place a response strategy and means of adaptation. They must provide urgent scientific, technical and financial support in the context of international cooperation within the framework of sustainable development of all sectors affected by changing climate. Agriculture is the sector probably most at risk to climate change. Thereby, most adaptation actions should be initiated within the agricultural sector and its related components, especially in areas characterized by water scarcity and poor soil quality, such as the oases area. Rehabilitation programs and traditional cultural practices are not sufficient to face this issue. Other means based on scientific progress should be adopted like agriculture biotechnology and Remote sensing (RS) and Geographic Information System (GIS) could be more efficient and may help us to make better decisions in land management, environmental protection and restoration solution and thus to mitigate the deleterious impacts of climate change on oasian agriculture.

REFERENCES

Agovino, M., Casaccia, M., Ciommi, M., Ferrara, M., & Marchesano, K. (2019). Agriculture, climate change and sustainability: The case of EU-28. *Ecological Indicators*, *105*, 525–543. doi:10.1016/j.ecolind.2018.04.064

Ait El Mokhtar, M., Anli, M., Ben Laouane, R., Boutasknit, A., Boutaj, H., Draoui, A., Zarik, L., & Fakhech, A. (2019a). Food Security and Climate Change. In *Handbook of Research on Global Environmental Changes and Human Health* (pp. 53–73). IGI Global. doi:10.4018/978-1-5225-7775-1.ch004

Ait El Mokhtar, M., Ben Laouane, R., Anli, M., Boutasknit, A., Fakhech, A., Wahbi, S., & Meddich, A. (2019b). Climate Change and Its Impacts on Oases Ecosystem in Morocco. In *Climate Change and Its Impact on Ecosystem Services and Biodiversity in Arid and Semi-Arid Zones* (pp. 217–245). IGI Global. doi:10.4018/978-1-5225-7387-6.ch012

Ait El Mokhtar, M., Ben-Laouane, R., Anli, M., Boutasknit, A., Wahbi, S., & Meddich, A. (2019c). Use of mycorrhizal fungi in improving tolerance of the date palm (*Phoenix dactylifera* L.) seedlings to salt stress. *Scientia Horticulturae*, *253*, 429–438. doi:10.1016/j.scienta.2019.04.066

Ait Hmida, A. (2003). Systèmes de production et stratégies des agriculteurs dans les oasis de la région d'Errachidia au Maroc. *New Medit*, *2*, 37–43.

Ait Houssa, A., Drissi, S., Asehraou, A., Asfers, A., Oubaki, L., & Chraibi, H. (2017). Changements climatiques au Maroc: Quels systèmes de culture et quelles biotechnologies pour s'y adapter? *Revue Marocaine des Sciences Agronomiques et Vétérinaires*, *5*(3).

Akesbi, N. (2012). Une nouvelle stratégie pour l'agriculture marocaine: Le Plan Maroc Vert. *New Medit*, *11*, 12–23.

Alcalá, F. J., Martínez-Valderrama, J., Robles-Marín, P., Guerrera, F., Martín-Martín, M., Raffaelli, G., de León, J. T., & Asebriy, L. (2015). A hydrological–economic model for sustainable groundwater use in sparse-data drylands: Application to the Amtoudi Oasis in southern Morocco, northern Sahara. *The Science of the Total Environment*, *537*, 309–322. doi:10.1016/j.scitotenv.2015.07.062 PMID:26282765

Ali, S., Rizwan, M., Qayyum, M. F., Ok, Y. S., Ibrahim, M., Riaz, M., Arif, M. S., Hafeez, F., Al-Wabel, M. I., & Shahzad, A. N. (2017). Biochar soil amendment on alleviation of drought and salt stress in plants: A critical review. *Environmental Science and Pollution Research International*, *24*(14), 12700–12712. doi:10.100711356-017-8904-x PMID:28374202

Almeida-Garcıa, F., & Chahine, S. (2015). Morocco, tourism. In Encyclopedia of Tourism. Academic Press.

Arrus, R., & Rousset, N. (2007). *L'agriculture du Maghreb au défi du changement climatique: Quelles stratégies d'adaptation face à la raréfaction des ressources hydriques?* Academic Press.

Assem, N., El Hafid, L., & Lamchouri, F. (2017). Impact of water stress on the leaf tissues of two Moroccan varieties of durum wheat (Triticum durum Desf). *Journal of Materials*, *8*(10), 3583–3487.

Augé, R. M. (2001). Water relations, drought and vesicular-arbuscular mycorrhizal symbiosis. *Mycorrhiza*, *11*(1), 3–42. doi:10.1007005720100097

Augé, R. M., Toler, H. D., & Saxton, A. M. (2015). Arbuscular mycorrhizal symbiosis alters stomatal conductance of host plants more under drought than under amply watered conditions: A meta-analysis. *Mycorrhiza*, *25*(1), 13–24. doi:10.100700572-014-0585-4 PMID:24831020

Azim, K., Komenane, S., & Soudi, B. (2017). Agro-environmental assessment of composting plants in Southwestern of Morocco (Souss-Massa Region). *International Journal of Recycling of Organic Waste in Agriculture*, *6*(2), 107–115. doi:10.100740093-017-0157-7

Badraoui, M. (2014). The Green Morocco Plan: an innovative strategy of agricultural development. *Arab environment: Food security. Annual report of the Arab forum for environment and development*, 34-35.

Balestrini, R., Lumini, E., Borriello, R., & Bianciotto, V. (2015). Plant-soil biota interactions. In Soil Microbiology, Ecology and Biochemistry. London: Academic Press. doi:10.1016/B978-0-12-415955-6.00011-6

Belmakki, M., Bartali, E. H., Xiaoru, H., & Bjerre, A. B. (2015). Identification and characterization of organic waste in Morocco, an important step towards the valorization of waste. *Revue Marocaine des Sciences Agronomiques et Vétérinaires*, *3*(3), 37–45.

Ben Laouane, R., Meddich, A., Bechtaoui, N., Oufdou, K., & Wahbi, S. (2019). Effects of Arbuscular Mycorrhizal Fungi and Rhizobia Symbiosis on the Tolerance of Medicago Sativa to Salt Stress. *Gesunde Pflanzen*, *71*(2), 135–146. doi:10.100710343-019-00461-x

Ben Salem, A., Messouli, M., & Yacoubi-Khebiza, M. (2011). *Developing an oasis-based water management tool: Ecohydrologic approach and weap software for a large arid catchment in Morocco*. Academic Press.

Beneduzi, A., Ambrosini, A., & Passaglia, L. M. (2012). Plant growth-promoting rhizobacteria (PGPR): Their potential as antagonists and biocontrol agents. *Genetics and Molecular Biology*, *35*(4), 1044–1051. doi:10.1590/S1415-47572012000600020 PMID:23411488

Benmohammadi, A., Benmohammadi, L., Ballais, J. L., & Riser, J. (2000). Analysis of anthropic and natural interrelations: Their impact on the recrudescence of the phenomena of sand deposition and desertification in south east Morocco (Drâa valley and Ziz valley). *Sécheresse*, *11*(4), 297–308.

Bor, M., Özdemir, F., & Türkan, I. (2003). The effect of salt stress on lipid peroxidation and antioxidants in leaves of sugar beet Beta vulgaris L. and wild beet Beta maritima L. *Plant Science*, *164*(1), 77–84. doi:10.1016/S0168-9452(02)00338-2

Born, K., Fink, A. H., & Paeth, H. (2008). Dry and wet periods in the northwestern Maghreb for present day and future climate conditions. *Meteorologische Zeitschrift (Berlin)*, *17*(5), 533–551. doi:10.1127/0941-2948/2008/0313

Brion, J. D., & Balahadia, F. F. (2017). Application of remote sensing and GIS for climate change and agriculture in Philippines. *IEEE 15th Student Conference on Research and Development (SCOReD)*, 229-233. 10.1109/SCORED.2017.8305388

Brooks, T. M., Mittermeier, R. A., da Fonseca, G. A., Gerlach, J., Hoffmann, M., Lamoreux, J. F., ... & Rodrigues, A. S. (2006). Global biodiversity conservation priorities. *Science, 313*(5783), 58-61.

Chaves, M. M., & Oliveira, M. M. (2004). Mechanisms underlying plant resilience to water deficits: Prospects for water-saving agriculture. *Journal of Experimental Botany*, *55*(407), 2365–2384. doi:10.1093/jxb/erh269 PMID:15475377

Chelleri, L., Minucci, G., Ruiz, A., & Karmaoui, A. (2014). Responses to Drought and Desertification in the Moroccan Drâa Valley Region: Resilience at the Expense of Sustainability? *International Journal of Climate Change: Impacts & Responses*, *5*(2), 17–33. doi:10.18848/1835-7156/CGP/v05i02/37199

Dahan, R., Boughala, M., Mrabet, R., Laamari, A., Balaghi, R., & Laiouad, L. (2012). *A review of available knowledge on land degradation in Morocco OASIS—combating dryland degradation. International Center for Agricultural Research in the Dry Areas*. ICARDA.

De Haas, H. (2001). *Migration and agricultural transformations in the oases of Morocco and Tunisia*. KNAG.

De Jong, C., Makroum, K., & Leavesley, G. (2006). *Developing an oasis-based irrigation management tool for a large semi-arid mountainous catchment in Morocco*. Academic Press.

Di Benedetto, N. A., Corbo, M. R., Campaniello, D., Cataldi, M. P., Bevilacqua, A., Sinigaglia, M., & Flagella, Z. (2017). *The role of plant growth promoting bacteria in improving nitrogen use efficiency for sustainable crop production: a focus on wheat*. Academic Press.

Diekkrüger, B., Busche, H., Klase, A., Klase, S., Rademacher, C., & Schulz, O. (2010). Impact of global change on hydrology and soil degradation–scenario analysis for the semi-arid Drâa catchment (South Morocco). *The global dimension of change in river basins: Threat, linkages and adaptation*, 5-13.

Driouech, F. (2010). *Distribution des précipitations hivernales sur le Maroc dans le cadre d'un changement climatique: descente d'échelle et incertitudes* (Thèse de doctorat). de l'Univers, de l'Environnement et de l'Espace, Université de Toulouse.

Driouech, F., Deque, M., & Mokssit, A. (2009). Numerical simulation of the probability distribution function of precipitation over Morocco. *Climate Dynamics*, *32*(7-8), 1055–1063. doi:10.100700382-008-0430-6

El Goumi, Y., Fakiri, M., Lamsaouri, O., & Benchekroun, M. (2014). Salt stress effect on seed germination and some physiological traits in three Moroccan barley (Hordeum vulgare L.) cultivars. *Journal of Materials and Environmental Science*, *5*(2), 625–632.

El Khoumsi, W., Hammani, A., Kuper, M., & Bouaziz, A. (2014). Deterioration of groundwater in arid environments: what impact in oasis dynamics? Case study of Tafilalet, Morocco. World Academy of Science, Engineering and Technology, International Journal of Environmental, Chemical, Ecological, Geological and Geophysical Engineering, 8(11), 764-770.

El Khoumsi, W., Hammani, A., Kuper, M., & Bouaziz, A. (2017). La durabilité du système oasien face à la détérioration des ressources en eaux souterraines: Cas de la palmeraie de Tafilalet. *Revue Marocaine des Sciences Agronomiques et Vétérinaires, 5*(1), 41–51.

Elboutahiri, N., Thami-Alami, I., & Udupa, S. M. (2010). Phenotypic and genetic diversity in Sinorhizobium meliloti and S. medicae from drought and salt affected regions of Morocco. *BMC Microbiology, 10*(1), 15. doi:10.1186/1471-2180-10-15 PMID:20089174

ESRI. (2017). *GIS Solution for Agriculture*. Retrieved from https://www.esri.com/library/brochures/pdfs/gis-sols-for-agriculture.pdf

Farissi, M., Bouizgaren, A., Faghire, M., Bargaz, A., & Ghoulam, C. (2011). Agro-physiological responses of Moroccan alfalfa (Medicago sativa L.) populations to salt stress during germination and early seedling stages. *Seed Science and Technology, 39*(2), 389–401. doi:10.15258st.2011.39.2.11

Farissi, M., Faghire, M., Bargaz, A., Bouizgaren, A., Makoudi, B., Sentenac, H., & Ghoulam, C. (2014). *Growth, nutrients concentrations, and enzymes involved in plants nutrition of alfalfa populations under saline conditions*. Academic Press.

Fertah, M., Belfkira, A., Taourirte, M., & Brouillette, F. (2017). Extraction and characterization of sodium alginate from Moroccan Laminaria digitata brown seaweed. *Arabian Journal of Chemistry, 10*, S3707–S3714. doi:10.1016/j.arabjc.2014.05.003

Fouad, M. O., Essahibi, A., Benhiba, L., & Qaddoury, A. (2014). Effectiveness of arbuscular mycorrhizal fungi in the protection of olive plants against oxidative stress induced by drought. *Spanish Journal of Agricultural Research, 12*(3), 763–771. doi:10.5424jar/2014123-4815

Fox, H. R., Wilby, R. L., & Moore, H. M. (2001). The impact of river regulation and climate change on the barred estuary of the Oued Massa, southern Morocco. *Regulated Rivers: Research & Management: An International Journal Devoted to River Research and Management, 17*(3), 235–250. doi:10.1002/rrr.613

Gandonou, C., Abrini, J., Idaomar, M., & Skali-Senhaji, N. (2005). Effects of NaCl on growth and ion and proline accumulation in sugarcane (Saccharum sp.) callus culture. *Belgian Journal of Botany*, 173–180.

Gandonou, C. B., Gnancadja, L. S., Abrini, J., & Skali-Senhaji, N. (2012). Salinity tolerance of some sugarcane (Saccharum sp.) cultivars in hydroponic medium. *International Sugar Journal*, *114*(1359), 190.

Geng, S. M., Yan, D. H., Zhang, T. X., Weng, B. S., Zhang, Z. B., & Qin, T. L. (2015). Effects of drought stress on agriculture soil. *Natural Hazards*, *75*(2), 1997–2011. doi:10.100711069-014-1409-8

Gibelin, A. L., & Déqué, M. (2003). Anthropogenic climate change over the Mediterranean region simulated by a global variable resolution model. *Climate Dynamics*, *20*(4), 327–339. doi:10.100700382-002-0277-1

GIEC. (2007). Résumé à l'intention des décideurs. In Bilan 2007 des changements climatiques: Impacts, adaptation et vulnérabilité. Contribution du Groupe de travail II au quatrième Rapport d'évaluation. Cambridge University Press.

Gong, L., Ran, Q., He, G., & Tiyip, T. (2015). A soil quality assessment under different land use types in Keriya river basin, Southern Xinjiang, China. *Soil & Tillage Research*, *146*, 223–229. doi:10.1016/j.still.2014.11.001

Haddouch, M. (1996). Situation actuelle et perspectives de développement du palmier dattier au Maroc. *Options Méditerranéennes*, *28*, 63–79.

Hadeel, A., Jabbar, M., & Chen, K. (2010). Application of remote sensing and GIS in the study of environmental sensitivity to desertification: A case study in Basrah Province, southern part of Iraq. Published at Springerlink.com. *Appl Geomat*, *2*(3), 101–112. doi:10.100712518-010-0024-y

Hannah, L., Midgley, G. F., & Millar, D. (2002). Climate change-integrated conservation strategies. *Global Ecology and Biogeography*, *11*(6), 485–495. doi:10.1046/j.1466-822X.2002.00306.x

Heidecke, C., & Heckelei, T. (2010). Impacts of changing water inflow distributions on irrigation and farm income along the Drâa River in Morocco. *Agricultural Economics*, *41*(2), 135–149. doi:10.1111/j.1574-0862.2009.00431.x

Hirich, A., Choukr-Allah, R., & Jacobsen, S. E. (2014). Deficit irrigation and organic compost improve growth and yield of quinoa and pea. *Journal Agronomy & Crop Science*, *200*(5), 390–398. doi:10.1111/jac.12073

Hirich, A., Choukr-Allah, R., & Jacobsen, S. E. (2014). The combined effect of deficit irrigation by treated wastewater and organic amendment on quinoa (Chenopodium quinoa Willd.) productivity. *Desalination and Water Treatment*, *52*(10-12), 2208–2213. doi:10.1080/19443994.2013.777944

Ilangumaran, G., & Smith, D. L. (2017). Plant growth promoting rhizobacteria in amelioration of salinity stress: A systems biology perspective. *Frontiers in Plant Science*, *8*, 1768. doi:10.3389/fpls.2017.01768 PMID:29109733

Jilali, A., & Zarhloule, Y. (2015). Structural control of the hydrogeology in Figuig Oasis in the eastern High Atlas of Morocco. *Proceedings of the Geologists' Association, 126*(2), 232-243. 10.1016/j.pgeola.2014.12.003

Kabbadj, A., Makoudi, B., Mouradi, M., Pauly, N., Frendo, P., & Ghoulam, C. (2017). Physiological and biochemical responses involved in water deficit tolerance of nitrogen-fixing Vicia faba. *PLoS One*, *12*(12), e0190284. doi:10.1371/journal.pone.0190284 PMID:29281721

Kabiri, L. (2004, June). Impact des changements climatiques et anthropiques sur les ressources en Eau dans les Oasis de Sud Marocain: Cas de Ferkla (Tinjdad, Errachidia, Maroc). In *Colloque International, "Quaternaire et Changements Globaux: Bilan et Perspectives" en Hommage au Professeu Hugues FAURE, Orléans, France* (*Vol. 3*). Academic Press.

Kabiri, L. (2005). Impact des changements climatiques et anthropiques sur les ressources en eau dans l'Oasis de Ferkla (Tinjdad, Goulmima, Errachidia, Maroc). Academic Press.

Karmaoui, A. (2019). Drought and desertification in Moroccan Pre-Sahara, Draa valleys: Exploring from the perspective of young people. *Geoenvironmental Disasters*, *6*(1), 2. doi:10.118640677-019-0118-8

Karmaoui, A., Ifaadassan, I., Messouli, M., & Khebiza, M. Y. (2015). Characterization of common environmental indicators of the Moroccan Oasean Biome, Pilot study in the reserve biosphere of Oases in Southern Morocco. *Advances in Research*, *5*(5), 1–15. doi:10.9734/AIR/2015/19369

Karmaoui, A., Ifaadassan, I., Messouli, M., & Khebiza, M. Y. (2015). Sustainability of the Moroccan oasean system (case study: Middle Draa Valley). *Global Journal of Technology and Optimization*, *6*(01), 170. doi:10.4172/2229-8711.1000170

Karmaoui, A., Messouli, M., Khebiza, Y. M., & Ifaadassan, I. (2014). Environmental vulnerability to climate change and anthropogenic impacts in dryland,(pilot study: Middle Draa Valley, South Morocco). *Journal of Earth Science & Climatic Change*, (S11), 1.

Karmaoui, A., & Moumane, A. (2016). Changes in the environmental vulnerability of Oasean system (desert oasis), Pilot study in Middle Draa Valley, Morocco. *Expert Opinion on Environmental Biology*, *5*(3). Advance online publication. doi:10.4172/2325-9655.1000135

Khadraji, A., Mouradi, M., & Ghoulam, C. (2017). Growth and Mineral Nutrition of the Chickpea (Cicer arietinum L.)-Rhizobia Symbiosis under Water Deficit. *Brazilian Archives of Biology and Technology*, 60.

Kinicki, R., Eddy, B., Shockey, I., Boecker, S., Ghahremani, S., Kuck, M., & O'Dell, S. (2017). Organic Farming in Morocco: An Investigation of Agricultural Extension Centers. Academic Press.

Knippertz, P., Christoph, M., & Speth, P. (2003). Long-term precipitation variability in Morocco and the link to the large-scale circulation in recent and future climates. *Meteorology and Atmospheric Physics*, *83*(1-2), 67–88. doi:10.100700703-002-0561-y

Koźmińska, A., Wiszniewska, A., Hanus-Fajerska, E., Boscaiu, M., Al Hassan, M., Halecki, W., & Vicente, O. (2019). Identification of Salt and Drought Biochemical Stress Markers in Several Silene vulgaris Populations. *Sustainability*, *11*(3), 800. doi:10.3390u11030800

Kuhn, A., Heidecke, C., Roth, A., Goldbach, H., Burkhardt, J., Linstädter, A., Kemmerling, B. & Gaiser, T. (2010). *II-5.1 Importance of resource management for livelihood security under Climate Change in Southern Morocco*. Academic Press.

Kumar, D., Al Hassan, M., Naranjo, M. A., Agrawal, V., Boscaiu, M., & Vicente, O. (2017). Effects of salinity and drought on growth, ionic relations, compatible solutes and activation of antioxidant systems in oleander (Nerium oleander L.). *PLoS One*, *12*(9), e0185017. doi:10.1371/journal.pone.0185017 PMID:28922384

Lamaoui, M., Jemo, M., Datla, R., & Bekkaoui, F. (2018). Heat and drought stresses in crops and approaches for their mitigation. *Frontiers in Chemistry*, *6*, 26. doi:10.3389/fchem.2018.00026 PMID:29520357

Lamqadem, A., Saber, H., & Pradhan, B. (2018). Quantitative assessment of desertification in an arid Oasis using remote sensing data and spectral index techniques. *Remote Sensing*, *10*(12), 1862. doi:10.3390/rs10121862

Lamqadem, A. A., Saber, H., & Rahimi, A. (2017). Spatiotemporal changes of vegetation in the Middle Draa Valley Oasis: A study case of M'hamid El Ghizlane Oasis (Morocco). *European Scientific Journal, 13*(24), 115–132. doi:10.19044/esj.2017.v13n24p115

Mainguet, M., Dumay, F., Kabiri, L., & Rémini, B. (2011). Prodromes of desertification in the Oasis of Tafilalet (Morocco) and specific local solutions. In *Coping with Global Environmental Change, Disasters and Security* (pp. 901–919). Springer. doi:10.1007/978-3-642-17776-7_53

Manuel, M., Lightfoot, D., & Fattahi, M. (2018). The sustainability of ancient water control techniques in Iran: An overview. *Water History, 10*(1), 13–30. doi:10.100712685-017-0200-7

Mariotte, P., Robroek, B. J., Jassey, V. E., & Buttler, A. (2015). Subordinate plants mitigate drought effects on soil ecosystem processes by stimulating fungi. *Functional Ecology, 29*(12), 1578–1586. doi:10.1111/1365-2435.12467

Meddich, A., Jaiti, F., Bourzik, W., El Asli, A., & Hafidi, M. (2015). Use of mycorrhizal fungi as a strategy for improving the drought tolerance in date palm (Phoenix dactylifera). *Scientia Horticulturae, 192*, 468–474. doi:10.1016/j.scienta.2015.06.024

Mohamed, B., Mounia, K., Aziz, A., Ahmed, H., Rachid, B., & Lotfi, A. (2018). Sewage sludge used as organic manure in Moroccan sunflower culture: Effects on certain soil properties, growth and yield components. *The Science of the Total Environment, 627*, 681–688. doi:10.1016/j.scitotenv.2018.01.258 PMID:29426193

Mulla, D. J. (2013). Twenty five years of remote sensing in precision agriculture: Key advances and remaining knowledge gaps. *Biosystems Engineering, 114*(4), 358–371. doi:10.1016/j.biosystemseng.2012.08.009

Mzabri, I., Legsayer, M., Kouddane, N., Boukroute, A., & Berrichi, A. (2017). *Salt stress effects on some morphological, physiological and biochemical parameters of saffron plant (Crocus sativus L.) in Eastern Morocco*. Academic Press.

Neset, T. S., Wiréhn, L., Opach, T., Glaas, E., & Linnér, B. O. (2019). Evaluation of indicators for agricultural vulnerability to climate change: The case of Swedish agriculture. *Ecological Indicators, 105*, 571–580. doi:10.1016/j.ecolind.2018.05.042

ONEE. (2015). *Environmental and social impact study of solar photovoltaic project of Tafilalt, Morocco: site Zagora*. Author.

Ouraich, I., & Tyner, W. E. (2014). Moroccan agriculture, climate change, and the Moroccan Green Plan: A CGE analysis. *African Journal of Agricultural and Resource Economics, 13*(311-2019-681), 307-330.

Oyiga, B. C., Sharma, R. C., Shen, J., Baum, M., Ogbonnaya, F. C., Léon, J., & Ballvora, A. (2016). Identification and characterization of salt tolerance of wheat germplasm using a multivariable screening approach. *Journal Agronomy & Crop Science*, *202*(6), 472–485. doi:10.1111/jac.12178

Patterson, J. H., Newbigin, E. D., Tester, M., Bacic, A., & Roessner, U. (2009). Metabolic responses to salt stress of barley (Hordeum vulgare L.) cultivars, Sahara and Clipper, which differ in salinity tolerance. *Journal of Experimental Botany*, *60*(14), 4089–4103. doi:10.1093/jxb/erp243 PMID:19666960

Pegna, F. G., Bartolini, P., El Rhaffari, L., Fahim, S., Bonaiuti, E., Le, Q. B., & Zucca, C. (2019). Sustaining Moroccan oasis agricultural system through small mechanization inputs. *JUNCO| Journal of UNiversities and international development COoperation, 1*(1).

Pérez-Piqueres, A., Edel-Hermann, V., Alabouvette, C., & Steinberg, C. (2006). Response of soil microbial communities to compost amendments. *Soil Biology & Biochemistry*, *38*(3), 460–470. doi:10.1016/j.soilbio.2005.05.025

PNUD: Programme des Nations Unies pour le Développement. (2011a). *Adaptation au changement climatique pour les oasis résilientes. Etude « évaluation du changement climatique futur au niveau des zones oasienne marocaine»*. Author.

PNUD: Programme des Nations Unies pour le Développement. (2011b). *Evaluation prospective des vulnérabilités et risques climatiques aux horizons 2030 et 2050. Projet d'évaluation de la vulnérabilité et des impacts du changement climatique dans les oasis du Maroc et structuration de stratégies territoriales d'adaptation*. Author.

Porcel, R., Aroca, R., & Ruiz-Lozano, J. M. (2012). Salinity stress alleviation using arbuscular mycorrhizal fungi. A review. *Agronomy for Sustainable Development*, *32*(1), 181–200. doi:10.100713593-011-0029-x

Rath, K. M., & Rousk, J. (2015). Salt effects on the soil microbial decomposer community and their role in organic carbon cycling: A review. *Soil Biology & Biochemistry*, *81*, 108–123. doi:10.1016/j.soilbio.2014.11.001

Rochdane, S., Bounoua, L., Zhang, P., Imhoff, M., Messouli, M., & Yacoubi-Khebiza, M. (2014). Combining satellite data and models to assess vulnerability to climate change and its impact on food security in Morocco. *Sustainability*, *6*(4), 1729–1746. doi:10.3390u6041729

Rowell, D. P. (2005). A scenario of European climate change for the late twenty-first century: Seasonal means and interannual variability. *Climate Dynamics*, *25*(7-8), 837–849. doi:10.100700382-005-0068-6

Royaume du Maroc. (2010). *Seconde communication nationale à la Convention-Cadre des Nations Unies sur les Changements Climatiques*. Ministère de l'Environnement.

Sati, V. P. (2015). Climate change and socio-ecological transformation in high mountains: An empirical study of Garhwal Himalaya. *Change and Adaptation in Socio-Ecological Systems*, *2*(1). Advance online publication. doi:10.1515/cass-2015-0005

Schilling, J., Freier, K. P., Hertig, E., & Scheffran, J. (2012). Climate change, vulnerability and adaptation in North Africa with focus on Morocco. *Agriculture, Ecosystems & Environment*, *156*, 12–26. doi:10.1016/j.agee.2012.04.021

Scotti, R., Bonanomi, G., Scelza, R., Zoina, A., & Rao, M. A. (2015). Organic amendments as sustainable tool to recovery fertility in intensive agricultural systems. *Journal of Soil Science and Plant Nutrition*, *15*(2), 333–352. doi:10.4067/S0718-95162015005000031

Serraj, R., & Sinclair, T. R. (2002). Osmolyte accumulation: Can it really help increase crop yield under drought conditions? *Plant, Cell & Environment*, *25*(2), 333–341. doi:10.1046/j.1365-3040.2002.00754.x PMID:11841674

Simonneaux, V., Cheggour, A., Deschamps, C., Mouillot, F., Cerdan, O., & Le Bissonnais, Y. (2015). Land use and climate change effects on soil erosion in a semi-arid mountainous watershed (High Atlas, Morocco). *Journal of Arid Environments*, *122*, 64–75. doi:10.1016/j.jaridenv.2015.06.002

Smith, S. E., & Read, D. J. (2008). *Mycorrhizal symbiosis* (3rd ed.). Academic.

Tripathi, S., Kumari, S., Chakraborty, A., Gupta, A., Chakrabarti, K., & Bandyapadhyay, B. K. (2006). Microbial biomass and its activities in salt-affected coastal soils. *Biology and Fertility of Soils*, *42*(3), 273–277. doi:10.100700374-005-0037-6

Vurukonda, S. S. K. P., Vardharajula, S., Shrivastava, M., & SkZ, A. (2016). Enhancement of drought stress tolerance in crops by plant growth promoting rhizobacteria. *Microbiological Research*, *184*, 13–24. doi:10.1016/j.micres.2015.12.003 PMID:26856449

Waldrop, M. P., & Firestone, M. K. (2006). Response of microbial community composition and function to soil climate change. *Microbial Ecology*, *52*(4), 716–724. doi:10.100700248-006-9103-3 PMID:17061172

Zucca, C. (2016). *Agropastoral System Diagnosis: Description of Agricultural Production Systems and Restoration Interventions with a Stress on their Weaknesses and Constraints*. Academic Press.

KEY TERMS AND DEFINITIONS

Agriculture: Is the process of producing food, forage, fiber, and various other products by the cultivation the soil and the rearing of domesticated animals.

AMF: Type of fungi which form symbiosis with roots of vascular plants.

Climate Change: Refers to all variations in climatic parameters of the global climate of the Earth or its various regional climates over time.

Compost: Refers to an organic matter that has been decomposed in a process called composting.

Geographic Information Systems: Is a technology designed to capture, store, manipulate, analyze, manage, and present spatial or geographic data.

Oases: An ecosystem consisting of a spot of isolated vegetation in a desert found near a water source.

PGPR: Are symbiotic bacteria that form symbiosis with plant and have a positive influence on its growth and development.

Remote Sensing: Is the technology of obtaining information about objects or phenomenon from a distance.

Sustainability: The ability to be maintained at a certain rate or level over a period of time.

Vulnerability: Refers to the inability to resist the effects of a hostile environment.

APPENDIX: LIST OF ABBREVIATIONS

ADPS: Agency for the Economic and Social Promotion and Development of the Moroccan Southern Provinces
AMF: Arbuscular mycorrhizal fungi
ANDZOA: National Agency for the Development of the Oasian and Argan Zones
COP22: 22nd COP United Nations Climate Change Conference
DéLIO: Oriental Integrated Local Development Program
DOP: Draa Oasis Program
DPE: Provincial Directorate of Equipment
EL: Electrolyte leakage
FAO: Food and Agriculture Organization of the United Nations
Fv/Fm: Photosynthetic efficiency
GEF: Global Environment Facility
GDP: Annual gross domestic product
GIZ: German Corporation for International Cooperation GmbH
GMP: Green Moroccan Plan
IAA: Indole acetic acid
INRA: National Institute of Agronomic Research
LRWC: Leaf relative water content
MDA: Malondialdehyde
PGPR: Plant growth-promoting rhizobacteria
SOP: Southern Oasis Program
TOP: Tafilalet Oasis Program

Section 2

Productive Ecosystems: Challenges and Solutions

Chapter 5
Impact of Thermal Stress on the Moroccan Argan Agroecosystem

Issam Ifaadassan
Cadi Ayyad University, Morocco

Ahmed Karmaoui
https://orcid.org/0000-0003-3881-4029
Southern Center for Culture and Sciences, Zagora, Morocco

Mohammed Messouli
Cadi Ayyad University, Morocco

Houssam Ayt Ougougdal
https://orcid.org/0000-0002-5681-0681
Cadi Ayyad University, Morocco

Mohammed Khebiza Yacoubi
Cadi Ayyad University, Morocco

Abdelaziz Babqiqi
Regional Observatory of the Environment and Sustainable Development, Morocco

ABSTRACT

The argan tree is exclusively endemic in the drylands of Southwest Morocco, an agroecosystem of great ecological, cultural, and economic importance. The argan agroecosystem is already damaged. It is particularly vulnerable to climate change as well as the harsh natural conditions aggravated by the current population growth

DOI: 10.4018/978-1-7998-3343-7.ch005

and the exploitation in excess of the production capacities. Unfortunately, during the 20th century, its area has been reduced by half. Current projections indicate an increase in temperature under climate scenarios. Anticipated climate change could accelerate this trend resulting in the argan tree degradation. To assess the climate change impact, the authors used the SDSM model at the argan agroecosystem scale and the thermal stress model to assess its vulnerability and estimate its tolerance response in relation to temperature stress for a projected climate in the near term (2010-2025 years). In this chapter, the authors explored the impact of climate change on the argan tree regeneration.

INTRODUCTION

The arganeraie (argan agroecosystem) includes an important rate of biodiversity offering many ecosystem services (Karmaoui, 2016). Various parts of the argan tree are globally used from traditional medicine for common human and animal ailments (Aabd et al., 2014). The traditional health care was scientifically validated (Charrouf & Guillaume, 2010).

In the last 30 years, this arganeraie is among the most vulnerable ecosystems of the arid and semi-arid lands. In regard to argan area, it lost almost 9900 hectares, an average of 550 ha/year between the years of 1969-1986, the Argan plain lost (EL Yousfi, 1988). In Haha region (western High Atlas) for example, de-densification was estimated to 600 ha per year (Bouzemouri, 2007). The decline in the argan area is attributed to both the lack of regrowth and loss of trees (De Waroux & Lambin, 2012). Currently, this ecosystem continues to be destroyed with all its components of biodiversity (El Mrabet et al., 2014). It has been exploited as firewood, timber, as forage for goats and sheep (Alados, 2008). Increasing aridity reported at a regional-scale by Esper et al., (2007). In fact, increasing aridity may be attributed to climate change (Nouaim, 2005). Climate change causes more droughts and more heat in the argan region, making the argan tree less adapted (Lybbert et al., 2011). During drought years the amount of surface water may drop 20 times (SIWM, 2001). This decrease may influence argan primary production. Consequently, climatic stress can be a key risk factor that accelerates degradation produced by the local population.

This chapter aims:

- To assess the climate change impact, the Statistical Downscaling Model (SDSM) was used at argan agroecosystem scale.
- To explore the thermal stress for a projected climate in near term (2010-2025) using a thermal stress model.

Because of climate change, large areas of argan groves could then be created on other continents in regions most appropriate to the argan tree (Lybbert et al., 2011). The study designed to look at the relationship between climate change and the argan regeneration. The authors look to confirm or refute this affirmation basing on modeling methods.

MATERIALS AND METHODS

Study Area

This study was carried out on the Argan Biosphere Reserve in south east of Morocco, with a geographical area of 950 000 ha, the most part was located in the Souss-Massa

Figure 1. Overview of the argan biome in south east of Morocco: Land cover
Source: the authors

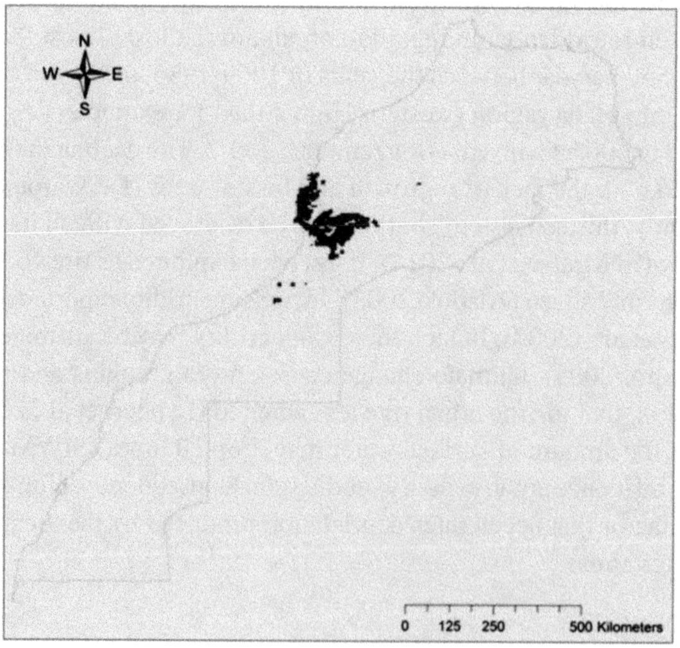

region before 2015 (Figure 1). This biome was classified as UNESCO Biosphere Reserve in 1998 (El Benni & Reviron, 2009).

Methodology

Climatic Trends in Argan Agroecosystem Area Under IPCC-A2 and B2

In this chapter, downscaling is achieved using the Statistical Down-Scaling Model (SDSM), developed by the SEI (Stockholm environment institute). The SDSM model is used to show good results for maximum and minimum temperature for historical climate simulation. SDSM develops an empirical relationship between a few selected large-scale predictor variables (i.e. mean sea level pressure, wind velocity) and local scale predictands (i.e. precipitation and temperature) (Koukids & Berg, 2009). The SDSM is a popular method because of the reduced computational effort and its ease of use. Temperature data, minimum (T min) and maximum (T max) were used for the period of 1961-2000 under IPCC-A2 and B2 scenarios. Agadir and Essaouira stations are the two stations used conduct the assessment in the Biosphere Reserve of Argan.

Thermal Function Stress

There are two thermal stress types as function of temperature minimum or maximum (Sellers et al., 1996):

1. Maximum thermal stress $f\left(Tc\right)_{\max}$:

$$f\left(Tc\right)_{\max} = 1 \Big/ 1 + e^{\alpha * \left(T\max - Tc\max\right)} \quad (1)$$

Where:

- T: High temperature measured in k of each time period of 2010, 2015, 2020, and 2025;
- T c max: Critical high temperature characterized each biome or species in K;
- α: High temperature factor photosynthesis in 1/K.

2. Minimum thermal stress ($\pi_{T\min}$):

$$f\left(Tc\right)_{\min} = \frac{1}{1+e^{\alpha*\left(T\min - Tc\min\right)}} \tag{2}$$

Where:

- T (k): low temperature measured of each time period of 2010, 2015, 2020 and 2025,
- T cmin (k): critical low temperature characterized each biome or species,
- α (1/k): low temperature factor photosynthesis.

According to Sellers et al., (1996), K represents the photosynthesis inhibition sloping, the half of this parameter is fixed to 0, 5 for the calculation of maximum stress and 0,2 for the minimum stress.

Figure 2. a: Maximum and minimum temperatures projections in a, Agadir station under IPCC-A2 scenario; b, in Agadir under IPCC-B2 scenario; c, in Essaouira station under IPCC-A2 scenario; and d, in Essaouira station under IPCC-B2 scenario
Source: The authors

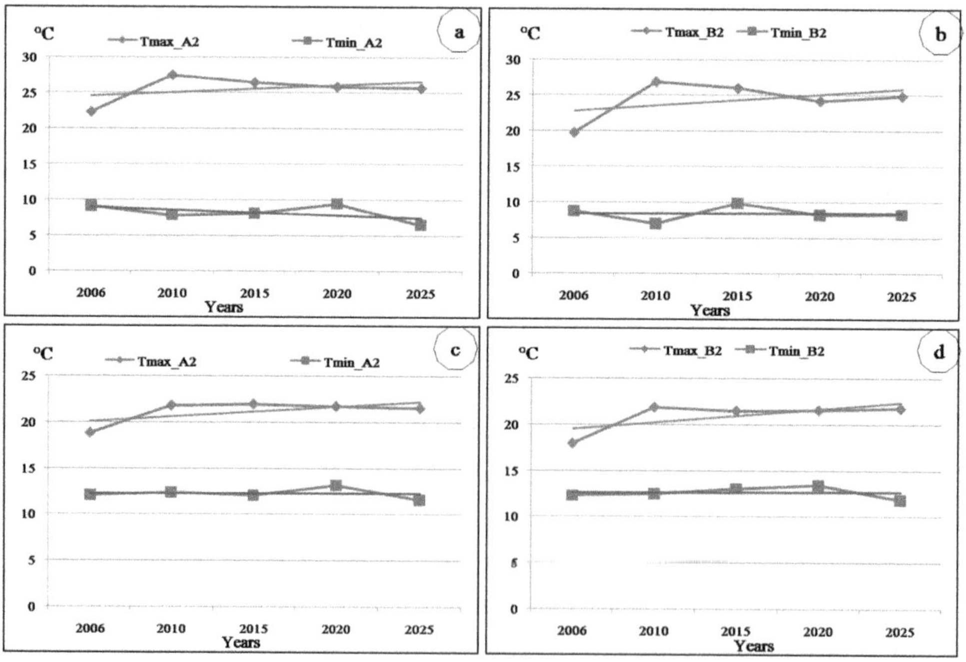

RESULTS

Impact of Climate Change on Argan Tree Productivity

Climate Change in Argan Agroecosystem

The use of data allows representing the trend of minimum and maximum temperatures under climate change scenarios in the two stations of the Biosphere Reserve of Argan, Agadir and Essaouira stations (Figure 2).

According to figure 2 (a, b, c and d), maximum and minimum temperature in the two stations of the biosphere reserve have rising trends for the two IPCC scenarios (A2 and B2). The values of the maximum and minimum temperature in both scenarios were used to assess the current and future thermal stress of the argan tree.

Thermal Stress (Maximum and Minimum Temperatures) of Argan Under A2 and B2 IPCC Scenarios

For the maximum thermal stress, there is no effect on the productivity of argan trees as shown in Figure 3 (b and d) under the two IPCC scenarios (A2 and B2). In fact, we can note that the values of maximum thermal stress fluctuate closely around zero, as mentioned in figure (6b and 6d)). This means that climate change is going to restore the range of optimal growth for the argan tree but its limitation is due to other factors, may be linked to the substrate, anthropogenic stresses (anthropogenic stress) or to varietal selection. For the other argan groves like Taroudant and Safi, we found the same behavior toward the maximum and minimum temperatures for the two scenarios (IPCC-A2 and B2).

DISCUSSION AND CONCLUSION

Several regeneration programs of argan trees were initiated since 2000, unfortunately, they have not yielded the expected results because the techniques are not adapted to soil and climatic conditions (El Mrabet et al., 2014). To explore the impact of the climatic conditions, two synoptic stations were used around the study area namely *Essaouira and Agadir stations*. These are selected based on the availability of data as well as the length and completeness of the record. The data of these two stations were used as input to the SDSM model to trace the future climate change trend. The projected change in minimum and maximum temperature may affect negatively the agroecosystem primary production and mainly the water resources (decrease in water reserves). The results of the current study show that there is no

Figure 3. a, Minimum thermal stress under IPCC-A2 scenario; b, Maximum thermal stress under IPCC-A2 scenario; c, Minimum thermal stress under B2 scenario; and d, Maximum thermal stress under IPCC-B2 scenario
Source: the authors

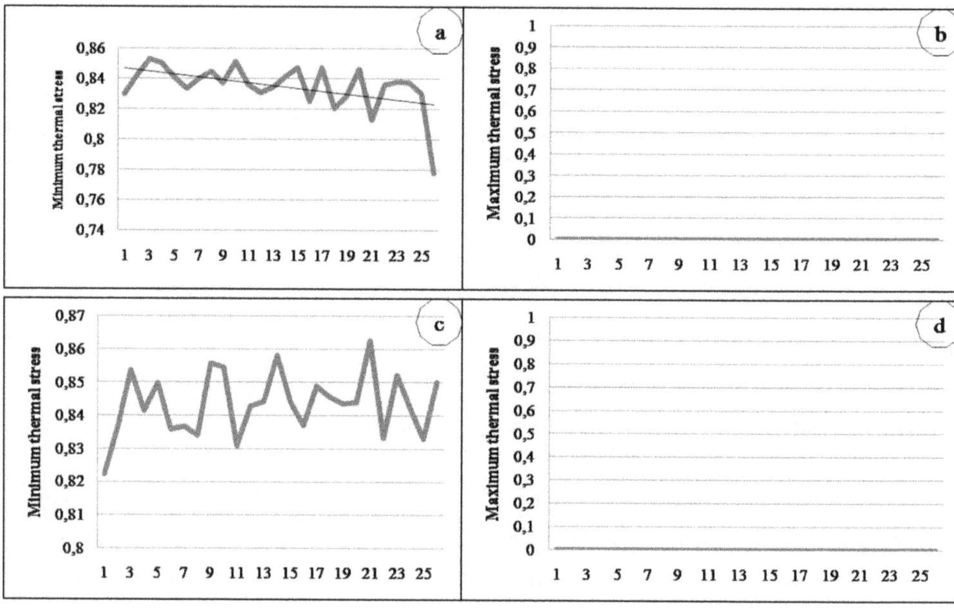

direct correlation between climate change and tree generation. However, Bouzoubaà, (2002), found that the seeds of the Argan tree can germinate even at temperatures from 35 to 40°C, but the germination of the different progenies is limited by the lowest temperatures (0,5 and 1°C) and highest temperatures (40°C). Germination is clearly affected below 15°C. Indeed, the Argan tree in the mountains hardly withstands the night frosts and temperatures around 0°C. In this context, Esper et al., (2007) reported that increasing aridity reported at a regional-scale can also have an impact on germination and seedling survival (Nouaim, 2005; see also Stour & Agoumi, 2009). Otherwise, aridity has shown to increase developmental instability which amounts to decreasing resilience (Alados & El Aich, 2008). Yann, 2012 advanced that in addition to the climatic impacts, the argan agroecosystem degradation is accelerated by social factors. It appears that there are indirect effects of climate change on argan agroecosystem are strongly dependent on interactions with the human disturbances, especially overgrazing, deforestation, and clearing. M'hirit et al., (1998) reported that droughts may exacerbate grazing pressures on existing resources. In fact, experiments on argan confirmed that regrowth is successful when grazing is excluded (Culmsee, 2005; Taleb et al., 2007). Clearing of trees and understory also causes erosion that along with soil degradation (in the form of salinization and loss of organic matter, mostly in irrigated areas) may render some soils unfavorable for germination (Nouaim, 2005). Elwahidi et al., 2014 reported also that clearings on the edge of the forest stands for expansion of agricultural land caused a loss surface of about 19.4%.

To conclude the argan ecosystem (Southwest Morocco) is of great ecological, socio-economical importance, due to the medicinal benefits and high nutritional value for the local wellbeing. But in the thirty last years, the argan agroecosystem is in continuous degradation and but regeneration efforts have not given the expected results. In this chapter, the impact of climate change on the Argan tree regeneration was explored. The SDSM model at argan agroecosystem scale and a thermal stress model were used in order to assess the vulnerability of the argan agroecosystem, and estimate the tolerance response in relation to temperature stress for a projected climate in near term (2010-2025 years). The results of this study show that heat stress has no effect on argan productivity in the two IPCC scenarios (A2 and B2). This means that climate change is going to restore the range of optimal growth for the argan tree but its limitation is due to other factors.

REFERENCES

Aabd, N. A., Msanda, F., & El Mousadik, A. (2014). Evaluation of variability in argan oil content through different environments and preselection of elite genotypes. *Euphytica*, *195*(2), 157–167. doi:10.100710681-013-0969-0

Alados, C. L., & Aich, A. (2008). Stress assessment of argan (Argania spinosa (L.) Skeels) in response to land uses across an aridity gradient: Translational asymmetry and branch fractal dimension. *Journal of Arid Environments*, *72*(4), 338–349. doi:10.1016/j.jaridenv.2007.06.015

Bouzemouri, B. (2007). Problems of conservation and development of the argan tree. In The argan tree: lever for human development of rural Morocco symposium internationalorganisé by the association Ibn Al Baytar at the Faculty of sciences Rabat.

Charrouf, Z., & Guillaume, D. (2010). Should the amazigh diet (regular and moderate argan-oil consumption) have a beneficial impact on human health? *Critical Reviews in Food Science and Nutrition*, *50*(5), 473–477. doi:10.1080/10408390802544520 PMID:20373191

Culmsee, H. (2005). Vegetation and pastoral use in the Western High Atlas Mountains (Morocco) An assessment of sustainability from the geobotanical perspective. In M. Aït Hamza, & H. Popp (Eds.), Pour une nouvelle perception des fonctions des montagnes du Maroc (pp. 67-80). Rabat: Actes du 7e colloque maroco-allemand.

De Waroux, Y. L. P., & Lambin, E. F. (2012). Monitoring degradation in arid and semi-arid forests and woodlands: The case of the argan woodlands (Morocco). *Applied Geography (Sevenoaks, England)*, *32*(2), 777–786. doi:10.1016/j.apgeog.2011.08.005

El Benni, N., & Reviron, S. (2009). Geographical Indications: review of seven case-studies world wide. Swiss National Centre of Competence in Research.

El Mrabet, S., Ouahmane, L., El Mousadik, A., Msanda, F., & Abbas, Y. (2014). The Effectiveness of Arbuscular Mycorrhizal Inoculation and Bio-Compost Addition for Enhancing Reforestation with Argania spinosa in Morocco. *Open Journal of Forestry*, *4*(01), 14–23. doi:10.4236/ojf.2014.41003

Elwahidi, Bellefontaine, Ponette, & Defourny. (2014). Dynamics of change between the arganeraie on-purpose and social change: a socio-ecological balance opportunity? *Journal of Agriculture and Environment for International Development, 108*(2), 109-133 DOI: 10.12895/jaeid.20142.210

Esper, J., Frank, D., Büntgen, U., Verstege, A., Luterbacher, J., & Xoplaki, E. (2007). Long-term drought severity variations in Morocco. *Geophysical Research Letters*, *34*(17), L17702. doi:10.1029/2007GL030844

Lybbert, T. J., Aboudrare, A., Chaloud, D., Magnan, N., & Nash, M. (2011). Booming markets for Moroccan argan oil appear to benefit some rural households while threatening the endemic argan forest. *Proceedings of the National Academy of Sciences of the United States of America*, *108*(34), 13963–13968. doi:10.1073/pnas.1106382108 PMID:21873185

M'Hirit, Benchekroum, El Yousfi, & Bendaanoun. (1998). The argan tree: a forest species with multiple uses. Mardaga.

Nouaim, R. (2005). *The argan tree in Morocco: between myths and re'alite's.* L'Harmattan.

Sellers, P. J., Randall, D. A., Collatz, G. J., Berry, J. A., Field, C. B., Dazlich, D. A., Zhang, C., Collelo, G. D., & Bounoua, L. A. (1996). Revised Land Surface Parameterization (SiB2) for Atmospheric GCMS. Part I: Model Formulation. *American Meteorological Society*. Advance online publication. doi:10.1175/1520-0442(1996)009<0676:ARLSPF>2.0.CO;2

SIWM. (2001). *Integrated management of water resources project in the Souss Massa report Achievement of Pilot Projects*. USAID Government of Morocco.

Taleb, S., Bellaka, M., & Achour, A. (2007). Effect of the fence on biodiversity and the dynamics of argan tree. *Annals of Forest Research in Morocco, 38*, 171-185.

Yousfi, E. L. M. (1988). Degradation of forest cover in southern Morocco: Case of Admine arganeraie. Rabat: 3rd cycle Memory IAV Hassan II.

Chapter 6
Sustainable Ecosystem Management:
Challenges and Solutions

Joan Nyika
https://orcid.org/0000-0001-8300-6990
Technical University of Kenya, Kenya

ABSTRACT

Ecosystems are part of human wellbeing and their sustainable management is essential for the survival of the human race and biodiversity. This chapter explores the concept of sustainable ecosystem management (SEM), its principles, elements, faces, and implementation. SEM is defined as environmentally sensitive, ecosystem-based, and eco-regional based. Its successful implementation is therefore complex due to the different priorities of stakeholders, the scope of ecosystems, some of which are transboundary, and the ever-changing nature of these areas amidst environmental uncertainties. These aspects are vulnerable to political changes and reconciling them is difficult. This chapter proposes a five-step implementation plan on SEM that is pegged on adaptive management and holistic consideration of ecological resources. Using documented case studies, SEM is a proposed solution to ecosystem challenges of modern-day amidst hindrances of rising resource demand, population increase, and climate variability.

DOI: 10.4018/978-1-7998-3343-7.ch006

INTRODUCTION

An ecosystem describes an environment consisting of living things and their associations with abiotic factors including minerals, water and air as they occupy space (McLeod & Leslie, 2009). Tsujimoto, Kajikawa, Tomita and Matsumoto (2017) describe an ecosystem as a biological system consisting of organisms (dead or alive) that interact with each other and with the physical environment that they are sited. For ultimate survival, the wide range of native ecosystems must be maintained and their habitats protected by creating buffer zones in their particular reserves and preparing them for resilience in events of disturbances such as is the case with effects of climate variability and change (Leech, Wiensczyk & Turner, 2009). These preconditions have led to ideologies skewed towards creating adequate space for such habitats even to the extent of traversing administrative boundaries. Implementation of these ideologies require coordination and cooperation of involved stakeholders (Tallis et al., 2009). Additionally, the awareness of ecosystem degradation and biodiversity loss due to resource over-extraction is growing and consequently, necessitating ecosystem conservation while considering the social-economic needs of humans. Ravy - Jonsen (2009) highlighted that adopting ecosystem approaches in resource conservation aims at balancing various societal goals by taking into account the human, biotic and abiotic uncertainties, interactions and knowledge in an ecologically sound and integrated manner. It is from the need to balance the societal goals that the concept of ecosystem management is derived.

Ecosystem management (EM) refers to the use of ecological science to manage natural and manmade resources with the motive of enhancing ecosystem sustainability while concurrently, delivering goods and services for human welfare improvement (Rosenberg & McLeod, 2005). It moves beyond compartmentalised focus on some species and regions to holistic management of natural resources. The aim of the concept is to protect targeted and non-targeted species, restore and preserve high integrity of their habitats and ecosystem services (Tsujimoto et al., 2017). Such precepts require sound socioeconomic, biological and governance outlooks. These outlooks are significant considering that land and water resources use by humans is capable of altering ecosystems by disturbing their habitats. Modifications of such ecosystems through anthropogenic activities reduces their resilience, productivity and health. As such, ecosystems must be managed to ensure the availability of their specific services, which is the heart of sustainable ecosystem management (SEM). SEM emphasizes on balancing long-term sustainability of ecosystems with human needs. It agrees with the typical concept of "meeting the needs of the present and local population without compromising the ability of future generations or populations in other locations to meet their needs" (Leech et al., 2009, 2). This chapter focuses on SEM considering that the concept in modern day has been challenged by the rise of human activities such as over-exploitation of natural resources such as land and water, deforestation, desertification, climate change and pollution. Consequently, the habitats of ecosystems have been destroyed and their resilience to stresses such as climate change has depleted. Ultimately, water, food and energy insecurity resulting to poverty is on the rise despite the ever-rising population and demand for these resources.

CLIMATE CHANGE AND ITS IMPACTS ON ECOSYSTEMS

Climate change is a change in the condition of climate, which is identifiable through changes in the variability and/or mean of its characteristics and remains unchanged for long periods beyond decades (United Nations Framework Convention on Climate Change (UNFCCC), 2011). These changes occur due to natural forces or human-based activities that modify land uses and atmospheric composition. Climate change triggers extreme weather (droughts and floods) and can result to damaging events including cyclones, mudslides, tornadoes, wildfires and waterspouts among others (Zhao et al., 2018). This phenomenon along with its anthropogenic stressors is a driver to ecosystem changes as summarised in Figure 1.

Figure 1. The effects of climate change on ecosystem functions and services

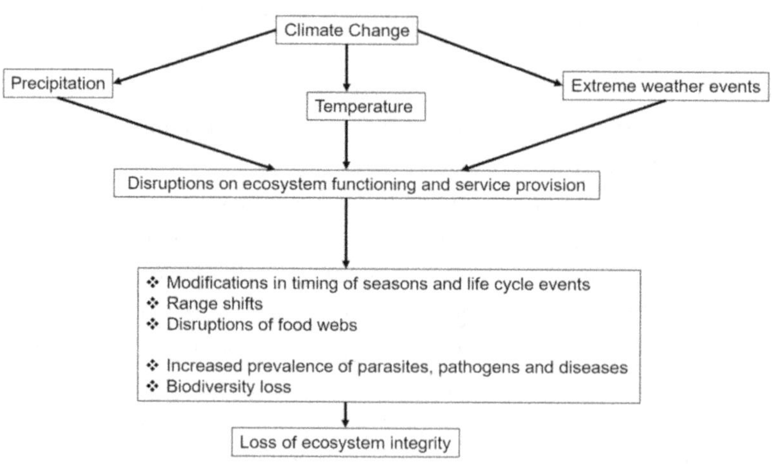

As a result of climate change, life cycle processes of some species such as reproduction, blooming and migration have been disrupted as a result of milder and shorter winters. For instance, migratory birds nest earlier in the east coast of

USA, while the migration of butterflies and the northeastern birds occur earlier due to disrupted seasons of the year (Settele et al., 2014). These seasonal changes have induced asynchronous interaction of affected species with their habitats leading to low food availability, pest avoidance and breeding that have negative effects on ecosystem productivity.

As temperatures rise, wild animals are moving to high altitude areas in search of pasture as a result to land degradation owing to extreme weather events. Although to some animals this is advantageous due to expanded rangelands, to others this phenomenon has increased competition for food and facilitated movement to hostile habitats resulting to extinction of both animal and plant species as Horton et al. (2014) noted. A case example is the projected 47% loss of habitat for the cold-water fish by 2080 with the rise of global warming and the loss of tundra ecosystems due to boreal forest invasion (Horton et al., 2014). Climate change also disrupts food webs and chains by compromising the provision of food, water and habitats for a wide range of species.

Climate change effects are supporting the spreads of diseases and their causatives, which affects aquatic, human and agricultural productivity. The situation is aggravated by land and water resources pollution, degradation and loss, which results to habitat destruction. According to Settele et al. (2014), 20-30% of animal and plant species are at risk of extinction as a result of climate change-induced heat stress. Animals residing in mountainous environments and those of polar climate are the most vulnerable.

Solutions and Mitigations to Climate Change Effects

A number of solutions have been suggested to mitigate climate change effects on ecosystems and their associated integrity. Some of these are summarised in Table 1. These solutions collectively constitute to SEM since they involve the management of natural and manmade resources to enhance and maintain the integral functioning of ecosystems. Additionally, the actions towards climate change mitigations have some principles of SEM such as the involvement of many actors, the advocacy on efficient resource use and the management of sensitive ecosystems such as forests. Therefore implementing SEM is a viable approach to mitigating climate change effects.

Conceptualising Ecosystem Management

Definitions of SEM and Its Role in Modern Day

SEM management is an all-inclusive view of social-economic and natural systems unlike the traditional view (which is still widely used) where these systems were

Table 1. Solutions to climate change effects on the ecosystem

Suggested Solutions	Reference
-Trees based systems to promote agroforestry and sequester carbon. -Adoption of integrated animal farm waste management strategies for renewable energy and proper animal husbandry. -Reduced greenhouse gas emissions to prevent global warming induced biodiversity loss.	Smith et al., 2007
-Integrated management of natural (land and water) resources. -Efficient use of natural resources and their fair allocation. -Soil and water conservation practices. -Integrated pathogen and pest management. -Climate forecasting to reduce its associated risks and for disaster preparedness. -Introduction of precision agriculture using improved and drought tolerant animal breeds and plant varieties. -Forest conservation and agroforestry practices.	Howden et al., 2007

distinct in their objectives and modes of operation. Viewing systems as distinct leads to exploitative management of available resources as each sub-system seeks to optimize its activities without assessing their effects on the others (Leech et al., 2009). Many authors have tried to define the SEM concept as outlined in Table 2.

From these definitions, it is evident that SEM has some overarching elements and principles that define the methodology of its implementation, which will be discussed in the following sub-sections of this chapter.

The Faces of SEM

Although SEM is widely researched and in many cases applied on ad hoc basis, the conceptualisation of an ecosystem-based management approach remains less understood. There is lack of unanimity and certainty on the elements and principles of SEM to apply based on site-specificities of an ecosystem. To alleviate these problems, Yaffee (1999) proposed three concepts to understand and explain this concept better particularly its application in resource management in a multi-disciplinary and multi-sectoral environment. These are the three faces of SEM and will be discussed in the following sub-sections.

Face 1: SEM is environmental sensitive and involves multiple uses

This face recognises that the primary aim of SEM is to provide human needs through enhanced production of outputs that human beings value. However, the limitations and complexities of ecosystems in meeting these needs should be equally considered (Yaffee, 1999). In the face, spatial composition with greater recognition of

Table 2. Some definitions of SEM

Source	Definition
Rowe (1992)	The management, conservation and restoration of local and regional ecosystem landscape in an integrated approach.
SAF[1] (1993)	A coordinated strategy to maintain full forest functions and values at landscape level.
FEMAT [2](1993)	A plan to manage all organisms as interrelated rather than as distinct species.
ESFHAT[3] (1993)	Optimal consideration of technological, economic and societal expectations and values in reference to ecological potentials.
AFPAFRB[4] (1993)	A resource management plan whose aim is to enhance and maintain ecosystem productivity and health and at the same time produce essentials that meet human desires and needs with limited and acceptable economic, biological and social risks.
SAF (1993)	An ecological approach that focuses on maintaining complex interdependencies, pathways and processes of forest ecosystems for their long-term soundness and resilience to environmental shocks.
Thomas (1996)	Integrated and holistic management of natural resources whose scope goes beyond compartmentalised management of forests to include physical, biological and human dimensions.
Brussard, Reed, & Tracy (1998)	Managing regions at scales that accommodate preservation of biological resources, ecosystem services and sustains human livelihoods and uses of those resources.
Pirot, Meynell & Elder (2000)	The process of regulating ecosystem use to reap its optimal benefits and at the same time, preserve its basic functions despite the expected stresses.
Quinn (2002)	Adoptive methods that guide human activities in a collaborative, inter- and multi-disciplinary approach to sustain the ecology and support sustainable and healthy communities.
CIT[5] (2004)	A strategy to adopt human activities that enhance coexistence of fully functional and healthy human communities and ecosystems.
Leech et al. (2009)	The science of public acceptance and awareness of environmental values, enhanced insight on ecological systems, increased awareness of ecological issues related to biodiversity and aimed at regulating resource use, improving productivity and realizing sustainability.

the biota structure and distribution is precedence. As such, SEM should be executed along land-ownership and administrative boundaries though specific issues should consider the spatial scope especially for ecosystems, which are transboundary. This calls for an overlaid management of habitats on administrative units particularly when protecting endangered and old species. In this face, resources have multiple uses and their sustained production (which may not be necessarily optimal) with minimal environmental effects (spatially and temporally) and with protection of spatially integrity is primary. The face also assesses the economic implications of implementing alternative SEM management regimes through participatory decision-making so that only optimal management action plans are taken up. According to Berkes (2016), this face summarizes SEM as the inclusion of social-ecological systems to biophysical ones with attention to management of shared resources, institutional cooperation, resilience theory of resource use and good governance.

Face 2: Ecosystem-Based Resource Management

Although the aforementioned principles and elements of SEM are used in land and water resources management, rarely does the scope of implementation extend to an ecosystem. In this case, the ecosystem refers to the inherent complexities and interactions of a geographical region. Yaffee (1999) in explaining this face sought to shift SEM from maximizing resource use to optimising the integrity and health of ecosystems subject to existent environmental constraints. With this definition, trade-offs are essential in balancing ecosystems to human wants/needs though the former is priority. In the face, maintenance and restoration of ecological processes such as hydrological flow, nutrient cycling and regime disturbance are crucial factors that influence diversity and composition of species. It is therefore necessary for ecological managers to match and merge ecosystems with similar characteristics. For instance, water resources should be managed at basin scales, which in most cases are transboundary. Jenkins (2018) supports the views of Yaffe (1999) to prioritize ecosystems in SEM and merge those of similar characteristics by highlighting their three complexities. First, ecosystem operate at diverse hierarchy levels of different spatial expansions that defy the boundaries notion. Second, ecosystems are multi-components, which react to one another predictably or unpredictably. Third, ecosystems have an adaptive nature in that they reorganise themselves in response to environmental disruptions or stresses. With these complexities of ecosystems, their sustainable management must predict imminent uncertainties accurately to enhance the resilience of ecosystems and their biodiversity. Biggs et al. (2012) share a similar view claiming that SEM will promote diversity maintenance, manage ecosystem connectivity, foster adaptive management of complex socio-economic-ecological systems and broaden participation and learning.

Face 3: Eco-Regional-Based Management

This face is closely related to the ecosystem-based approach but pays attention to the landscape-scope of SEM. In this context, landscapes are geographical units described by their floristic distribution, abiotic community and species components (Leech et al., 2009). An eco-regional approach to SEM incorporates decentralisation and organised management of resources. Additionally, conservation legislations and their applicability differ from one eco-region to another. According to Mason

(2011), these preconditions make SEM implementation difficult due to economic and political incompatibilities of eco-regions and their technical and financial constraints. Yermolaev, Lisetskii, Marinina and Buryak (2015) differ with these views claiming that eco-regional resource management promotes natural resources resilience and productivity. To deal with these different views, Yaffee (1999) advocated for site-specific management, which considers the uniqueness of ecosystems at regional level.

Methodology for SEM

With the background knowledge that SEM though complex is a necessity owing to the rise in the demand for human needs and uncertainties exposed to ecosystems, this research proposes a five-step methodology towards its implementation. These steps are summarised in Figure 2. The first step involves identification and analysis of stakeholders in a given ecosystem. According to Hein, Koppen, Groot and Ierland, a "stakeholder is any group or individual who can affect or is affected by the ecosystem's services" (2006, p. 213). In SEM, four stakeholder groups exist: 1) those who influence an ecosystem indirectly, 2) direct beneficiaries of an ecosystem, 3) those who burden and affect an ecosystem negatively and 4) those who directly influence ecosystem services (Demeyer & Turkelboom, 2014). Examples of category 1-4 include civil society organisations advocating for efficient natural resources use, water/land users, tree loggers/ polluters and landowners in respective order. Identification, analysis and management of stakeholders in an ecosystem is critical in SEM in that it promotes transparency, knowledge sharing, good governance and collaborative management (Hauck, Stein, Schiffer & Vandewalle, 2013). Stakeholders are defined within a given ecosystem area. According to the CBC (2004), an appropriate scale and size of an ecosystem during stakeholder definition must meet a scientific criterion, have an appropriate management experience, knowledge and capacity, understand that management is transitional and consider cultural, legal and administrative boundaries. After definition of stakeholders, their interests and the areal scope of an ecosystem, links between them are made to generate a mosaic relationship in which a variety of stakeholders manage an ecosystem at varied intensities. The interrelationships between ecosystems and stakeholders promote adaptive management of these areas in events of fluctuation and disturbance of ecosystem services as Demeyer and Turkelboom (2014) highlighted.

Ecosystem functions and structures also need to be understood in relation to the services and goods they provide. This step is core in SEM and identifies ecosystems under threat due to over-exploitation of resources. Lu et al. (2015) terms the process of understanding ecosystems as an integrity analysis that gauges the ability of ecosystems to maintain their sound functioning through multi-sectoral involvement. The analysis can be done through complementary knowledge sharing

Figure 2. The roadmap to sustainable ecosystem management implementation

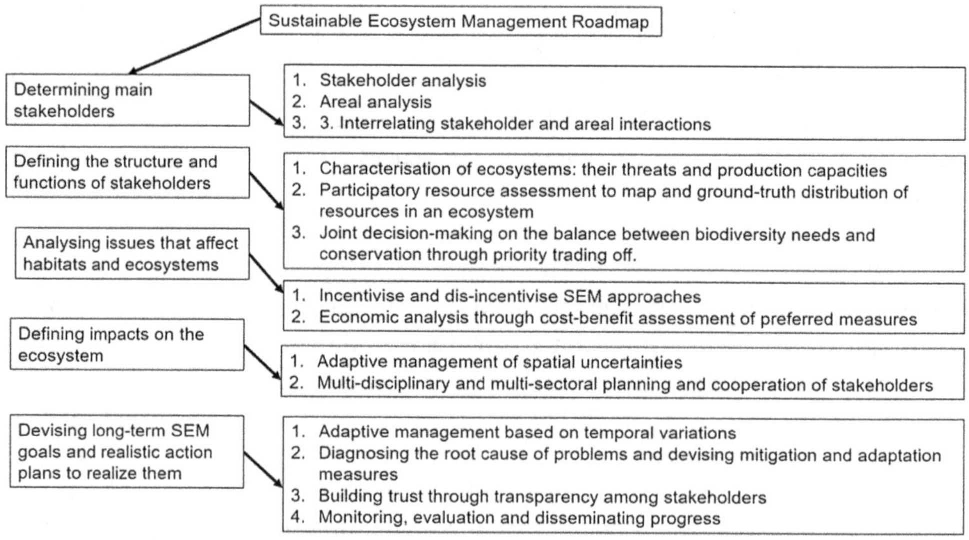

of locals and scientists in a participatory resource management and decision-making process that embraces realism. Lu et al. (2015) emphasized that ecosystem integrity analysis should be realistic considering that these areas are sensitive to stress and that management approaches should be practical to apply, have a strong scientific foundation, be quantifiable and focused to meet a certain goal. The realism of SEM becomes more relevant through trading off preferences to ensure management is applied at the lowest appropriate levels (CBC, 2004).

Following the assessment of ecosystem health, issues affecting inhabitants and SEM need to be identified. This step is crucial in reducing stresses and distortions that negatively affect biodiversity while formulating and strengthening incentives on effective biodiversity use. According to Muchapondwa et al. (2012), economic incentives are promising boosters of SEM that change behaviour of stakeholders with respect to the choices they make and their influence to biodiversity. These incentives bind stakeholders and support practices, norms and standards that promote persistence of ecosystems. Some of the applied incentives include rewarding ecosystem conservation practices, sharing the proceeds from ecosystem components with conservation actors, providing grants, tax concessions and subsidies to locals as rewards of ecosystem conservation, creating conservation easements and biodiversity offsets (Muchapondwa et al., 2012). These incentives change as economic realities and markets evolve. Therefore, cost benefit analysis is crucial in these incentives and disincentives since it enables managers to access their realism and long-term merits or demerits to ecosystems.

The recognition that ecosystems do not provide predictable returns especially after disturbances such as climate variability necessitate the concept of adaptive management. The adaptive management of one ecosystem is said to influence the positive management of adjacent ones. Adaptive management of ecosystems assesses cross-scale trade-offs of ecosystem services production and minimises negative effects resulting from growing human demands (Birge, Allen, Garmestani & Pope, 2016). In addition, adaptive SEM takes into account externalities such as reduced funding for ecosystem conservation projects. Adaptive management of ecosystems occur in two phases: - the deliberative and iterative phases (Williams & Brown, 2014). In the former, stakeholders are involved to redefine objectives and alternative measures to improve ecosystems and in the iterative phase trade-offs are made to make decisions, monitor and assess their success (Williams, 2010). In the deliberative step, multiple stakeholders are involved in coordinated problem analysis and collective action planning (multi-planning). In the iterative step, stakeholders are mobilised in analysing the costs affiliated with suggested action plans to select the most appropriate, based on its viability and available finances. At this step, stakeholder cooperation and mutual trust is required as compromises have to be made.

The last step in implementing SEM is devising realistic long-term action plans towards its realisation. This step requires strategic planning, implementation of action plans, their monitoring and assessment with the transparent involvement of stakeholders. Long-term plans are important in understanding ecosystems and their associated processes, risks and limitations. Planning is part of adaptive management of ecosystems and it must diagnose challenges and mitigation measures based on the highlighted principles of SEM. A valuable component in all these steps apart from their monitoring and assessment is providing feedback on progress made and bottlenecks experienced considering that SEM is a learning processes, which can be replicated between ecosystems (Williams & Brown, 2014; Birge et al., 2016).

Challenges and Solutions in Implementing SEM

The complex nature of ecosystem processes amidst a change-prone environment are hindrances to its sustainable management. Leech et al. (2009) highlighted that ecological disturbances, stresses and the lack of adequate funding and human capacity are the key impediments to successful SEM. The discrepancy in natural boundaries from social and political ones is a challenge in adopting this concept as ecosystems in natural landscapes are currently managed autonomously by various agencies whose management ideologies, policies and agendas are divergent. Rosenberg and McLeod (2005) claimed that managing ecosystems as distinct entities limited to a particular administrative boundary cannot realize SEM since some ecosystems are transboundary and stakeholders involved in management have different interests, which sometimes are irreconcilable. Additionally, no single region or sector is fully equipped with information on the broad focus to manage ecosystems while taking in consideration all stakeholders of interest and dealing with uncertainties of such habitats.

Therefore, multi-sectoral collaboration and stakeholder recognition is valuable in SEM so that the right individuals are involved in the process from the onset. Considering ecosystems as place-based, local communities and the public should be involved to realize common understanding and goal formulation. According to Quinn (2002), this is the link between collaborative and adaptive ecosystem management where planning for SEM recognizes that biodiversity is interconnected, extends beyond administrative boundaries and its success is based on multi-stakeholder participation. To overcome boundary hindrances in SEM, Leech et al. (2009) proposes collaboration of stakeholders and institutions to build trust and push such initiatives to their full implementation. Additionally, considering the biases and preferences of organizations, individuals and management is key in unlocking administrative and political boundaries that hinder holistic ecosystem management (Jenkins, 2018).

Another challenge to overcome in SEM is the conditions of ecosystems that operate over extended temporal and spatial scales. Spatially, ecosystems could cover large areal extensions that include many components with diverse needs and preferences. Temporally, ecosystems change and with these modifications, site-specific management approaches are required. With these challenges, no single time or space scaled management framework is fit for SEM. To bridge these gaps, time and space-considerate adaptive management is essential according to Williams and Brown (2014). This concept of management rely on availability of baseline data regarding ecosystem mapping, biodiversity distribution and consumption patterns. The data can be used to make accurate predictions on future ecosystem scenarios (disturbances and stresses) and develop best management practices. Although data inadequacy is not a justification for inaction towards SEM, its availability is a crucial component of adaptive management particularly the setting of realistic goals and targets and in the monitoring of success and sharing feedback on milestones made (Leech et al., 2009).

Stakeholder collaboration during SEM implementation is a great challenge since their benefits from an ecosystem and the perceived value of such regions differ from stakeholder-to-stakeholder and even for similar stakeholders at different time and spatial scales (Hein et al., 2006). Under these circumstances, stakeholders in a given ecosystem seek to optimize the benefits of an ecosystem without considering others and effects to other ecosystems particularly those that are transboundary. Some stakeholders often resist SEM since it has negative effects on the benefits they reap from such areas as Leech et al. (2009) noted. For example, tree loggers only prioritise the economic benefits they reap from deforestation without considering the benefits of trees in carbon sequestration, provision of wildlife habitats and attracting rainfall towards climate change adaptation. On the other hand, industry owners in most cases prioritise on their manufacturing activities and neglect the resultant effects in the form of wastewater production or the release of greenhouse gases that result to environmental pollution. Inadequate knowledge sharing, mistrust among stakeholders and lack of public awareness on the benefits of SEM results to resistance. To overcome this problem, managers of ecosystem-based projects must involve all stakeholders, use their skills and knowledge at all stages of SEM to work with them effectively. Additionally, they should invest in training them on the relevance of such initiatives to promote acceptability and participation in ecosystem-based projects. Hein et al. (2006) uphold these ideas claiming that stakeholder involvement in SEM projects supports sound action-plans and enhance holistic ecosystem goods and services valuation.

The ever-increasing demand for ecosystem good and services from an ever-rising population is a key challenge in SEM. Population rise is a driver to ecosystem change despite the uneven distribution of resources and the fact that vulnerable populations are the ones most affected. According to Rashid, Scholes and Ash (2005), the situation is best described as, 'living beyond our means' where everyone depends on ecosystem services for life. Although these ecosystems have improved the lives of many, their capacity to provide goods and services in the future has been weakened to make them unsustainable. Degradation of freshwater, aquatic food and the rise of climate variability, hunger and disease prevalence are some of the consequences of weakened ecosystems. The precautionary principle therefore must be applied in SEM to ensure available ecosystems are used sparingly to prevent depletion through adaptive management as Birge et al. (2016) advised. Lu et al. (2015) also recommended the need to develop healthy ecosystems through human population control claiming that it will regulate consumption patterns and offset the existent heterogeneity in distribution of ecosystem goods and services globally.

The dynamisms in funding, resource planning, on-the-ground practices and policy formulation is another challenge in SEM. Usually, these aspects are influenced by political regimes, which are transient though SEM is long-term and its scope in terms of time and space goes beyond the political regimes (Leech et al., 2009). To address this problem, policies on SEM should be flexible enough to recognize its long-term nature and enable funding even with changes in governance. Evenly, SEM goals should be reconciled with political timeframes that enhance their feasibility through funding. According to Curticle, Dunn, Roberts, Carr and Halpin (2012), most SEM initiatives are implemented by researchers who have funding limitations. The authors noted that the implementation of the initiatives without adequate financing would lead to their ultimate failure, as tools to optimise them will be underused. Ways that can enhance SEM funding include grants from nongovernmental organizations, internal government funds, fee-for-services, endowments and through funding skunkworks projects (Curticle et al., 2012).

Case Studies

Using the aforementioned principles and elements, SEM has been implemented successfully in some parts of the world. In the following sub-sections, three case studies showing the applicability of this concept are discussed.

The Case of Indonesia's Raja Ampat

The Raja Ampat Islands are found in Indonesia and neighbour New Guinea to the west. The islands are known for their biological diversity. Although the area is sparsely populated, its marine ecosystem is under pressure due to overfishing, corallivore outbreaks, pollution and natural resources over-exploitation (Bailey & Sumaila, 2009). Commercial fisheries and artisanal activities are the main economic contributors though logging, tourism, agriculture and pearl farming contribute to the economy (Tallis et al., 2010). Ecosystem management of the area is under regent authority since the 1990s though area residents prefer a devolved system considering the diffuse population and the highlighted ecosystem issues. A combined effort between the Raja Ampat regency government and several nongovernmental organisations has developed an initiative to protect marine areas and safeguard the livelihoods of fishers for holistic economic growth in the area. The acceptance of the initiative by local stakeholders has enabled them to collaborate with the NGOs and government to implement a SEM plan towards overcoming the challenges of inadequate institutional commitment and poor governance as Grieve and Short (2007) noted.

In planning, three NGOs namely World Wildlife Fund (WWF), Conservation International and The Nature Conservancy were involved in defining the objectives of the initiative and formal project scoping. Although data on ecosystem health in the area was scanty, increased financing of ecosystem-based projects of the region by the government and NGOs saw a successful ecosystem characterisation in a participatory stakeholder engagement (Donnelly, Neville & Mous, 2002). Using available data to establish ecosystem conditions and the fish biomass threshold, Raja Ampat ecosystem was reconstructed. Indicators such as fish and turtles were shown to decline due to a rising population density and overharvesting of these ecosystem services (Palomares & Heymans, 2006). Historical information regarding the ecosystem was collected through interviews and the data was used to conduct a formal risk assessment, which concluded the need for action towards SEM (Ainsworth, Pitcher & Rotinsulu, 2008).

Actions toward SEM started with primary data collection to assess the status of the ecosystem. This was done using a combination of techniques including oceanographic monitoring, aerial coastal surveys, community interviews, dive transects and sampling of fish stomachs. An ecological path to SEM was designed using Ecosim model to devise sustainable fish harvesting approaches (Ainsworth et al., 2008). Additionally, multi-stakeholders used primary data alongside historical data to assess Raja Ampat ecosystem effects on fisheries (Palomares & Heymans, 2006). Although the highlighted actions resulted to better fishing habits and partial restoration the marine ecosystems towards SEM, a lack of capacity to monitor indicators such as the fish population, coral bleaching, invertebrate grazing and coral cover hinders accurate feedback processing towards sustainability. Incidents of illegal fishing, exploitation of forests and artisanal fishing are reported and without an effective monitoring system, gains made might be reversed (Tallis et al., 2010). The area also grapples with poor institutionalisation of ecosystem conservation efforts hence the lack of appropriate data management. Therefore, there is a need to incorporate all economic sectors of the area's ecosystem in future including mining and forestry sectors to ensure that SEM is holistic and multi-disciplinary (Tallis et al., 2010).

The Case of Venezuela's Guacharo National Park

The Guacharo national park located in Venezuela and 12 km from Monagas town is an essential ecosystem to residents. The park faced several ecosystem challenges including declined fish survival, overfishing, excessive logging and destruction of the pristine environment particularly along the Turimirique river basin (Smith & Maltby 2003). A combination of reserve protection, restoration and management measures were developed towards SEM. In response to these challenges, the government approved the enlargement of the park from its original 15, 000 ha to 62, 000 ha after an ecosystem characterisation in aspects of its value from sink holes, caves and pristine environments of the park (Smith & Maltby 2003). The national oil company in collaboration with area stakeholders initiated the Cerro Negro project to reforest 2000 ha of the park that was previously destroyed by wildfire. Through a collaboration of the University of Oriente, Audubon de Venezuela that is a NGO and the national oil company, a SEM project in Middle basin was started to research on the park ecosystem; protect the forest and some of its endangered birds and vertebrate species (Smith & Maltby 2003). The awareness of ecosystem conservation through this project was extended to local schools. A south basin programme was initiated to characterise the ecosystem and promote alternative economic activities such as agriculture (coffee growing) to regulate overfishing practices.

The above projects in Guacharo national park highlight some aspects of SEM in their characteristics. The efforts of the projects was to use resources of the area sustainably through fair sharing as addressed in the south basin programme. Participation of multi-stakeholders (such as nongovernmental organisations, research institutions, locals and the government), portray participatory and collaborative management of resources through shared knowledge and cooperative decision-making (Smith & Maltby 2003). The fact that ecosystem management has an extended scope from marine habitats to forests and agricultural biodiversity depicts efforts towards holistic management that addresses cross cutting ecological issues towards SEM. The success of the projects in this region were pegged on multi-stakeholder involvement throughout the SEM lifecycle (Smith & Maltby 2003). Lack of clearly defined ecosystem services and goods due to inadequate data, which hindered adaptive resource management, was a common challenge in the four projects.

The Case of Kenya's Maasai Land

The Maasai are a pastoral community who occupy parts of northern Tanzania and southern Kenya commonly known as the Maasailand in East Africa (Homewood, Kristjanson & Trench, 2009). The community lives in arid areas prone to rainfall variability and climate change has made the situation worse by increasing the severity of extreme weather (floods and drought). The nomadic livelihood of the community has made it vulnerable to unpredictable drought and destruction of biodiversity through the death of the domestic animals. With climate variability, ecosystems of the Maasailand have been destroyed and the challenges that pastoralists face amplified by loss of herding spaces, privatization of land, death of animals, loss of culture by their sedentarisation and periodic shock as a result of political and economic instability (Osano et al., 2013). Additionally, Maasailand fertility has been compromised because of extended drought leading to desertification and adverse effects of flooding (Kimaro et al., 2018).

Apart from pastoralism, the Maasai benefits from other dryland ecosystem services including ecotourism and wildlife support. With this recognition and in light of the challenges they face, the Maasai have devised several adaptive management strategies to preserve the integrity of their ecosystem. Examples are partnerships between commercial tourism enterprises and Maasailand owners to develop wildlife conservancies as Osano et al. (2013) noted. Some examples are in Kitengela, Mara and Ol Kiramatian areas of Kenya. These conservancies though privately owned have received government support particularly in Kenya to expand biodiversity and extend protected areas. Initiatives focusing on payment of ecosystem services (PES) have been initiated in the area to enhance the income benefits from direct access of ecosystems and reduce poverty. Some of the PES programs adopted in Maasailand include contracts between pastoralists, NGOs or government conservation agencies and incentivising conservation practices to promote cultural integrity and eco-tourism. The Maasai Mara wildlife conservation project, which is funded by the government in collaboration with nongovernmental conservation agencies, promotes restoration of the Maasailand ecosystem through eco-tourism and has been instrumental in reducing poverty among the residents by serving as an incentive (Osano, 2011).

From the perspective of SEM, adoption of PES and the rise of conservancies in Maasailand is crucial in adaptive management in relation to climate change stresses. These initiatives enhance community resilience to drought by diversifying their income sources. For instance, wildlife conservation as an alternative income source buffers the nomads from fluctuating livestock values during extreme drought, which helps to overcome liquidity constraints from such ecosystem stresses (Osano et al., 2013). Additionally, these initiatives are synergies that promote the trade-offs for nomadic livelihoods by restricting their land over-exploitation practices. Through SEM, these benefits can be optimised.

CONCLUSION

SEM is a growing concept in ecological conservation amidst rising population and environmental uncertainties particularly from climate change. This chapter explains the principles of the concept including the need for holistic and adaptive management whose scope is transboundary. Elements of sustainability and trading off between different human and ecological needs is explored. A 5-step process to implement SEM is proposed. The chapter describes the challenges of SEM including divergent political ideologies, stakeholder resistance and limited funding. To overcome these challenges, the chapter proposes the investment in ecological characterisation and conservation through accurate baseline data collection, enhanced funding for SEM projects and better collaboration of multi-disciplinary stakeholders including local communities. Using success stories of three case studies, implementation of SEM is shown to be feasible and realistic even in developing countries though some bottlenecks must be dealt with.

REFERENCES

Ainsworth, C., Pitcher, T., & Rotinsulu, C. (2008). Evidence of fishery depletions and shifting cognitive baselines in Eastern Indonesia. *Biological Conservation, 141*(3), 848–859. doi:10.1016/j.biocon.2008.01.006

American Forest and Paper Association Resources Board (AFPARB). (1993). *Sustainable Forestry Principles and Implementation Guidelines. Ecosystem Management: A new Approach to Federal Forest Management and Planning*. American Forest and Paper Association Resources Board.

Bailey, M., & Sumaila, U. (2008). Towards ecosystem-based management in the birds head functional seascape of Papua, Indonesia: the economic sub-project. In M. Bailey & T. J. Pitcher (Eds.), *Ecological and economic analyses of marine ecosystems in the Bird's Head seascape, Papua, Indonesia: II. Fisheries Centre Research Reports 16(1)*. Fisheries Centre, University of British Columbia.

Berkes, F. (2016). Sustainability policy considerations for ecosystem management in Central and Eastern Europe. *Ecosystem Health and Sustainability*, *2*(8), e01234. doi:10.1002/ehs2.1234

Biggs, R., Schluter, M., Biggs, D., Bohensky, E., BurnSilver, S., Cundill, G., Dakos, V., Daw, T. M., Evans, L. S., Kotschy, K., Leitch, A. M., Meek, C., Quinlan, A., Raudsepp-Hearne, C., Robards, M. D., Schoon, M. L., Schultz, L., & West, P. C. (2012). Toward principles for enhancing the resilience of ecosystem services. *Annual Review of Environment and Resources, 37*(1), 421–448. doi:10.1146/annurev-environ-051211-123836

Birge, H., Allen, C., Garmestani, A., & Pope, K. (2016). Adaptive management for ecosystem services. *Journal of Environmental Management, 183*, 343–352. doi:10.1016/j.jenvman.2016.07.054 PMID:27460215

Brussard, P., Reed, J., & Tracy, R. (1998). Ecosystem management: What is it really? *Landscape and Urban Planning*, *40*(1-3), 9–20. doi:10.1016/S0169-2046(97)00094-7

Coast Information Team. (2004). *Ecosystem based management planning handbook*. Report to Coast Information Team. www.citbc.org/

Convention of Biological Diversity (CBC). (2004). *The ecosystem approach*. CBC.

Curticle, C., Dunn, D., Roberts, J., Carr, S., & Halpin, P. (2012). Why ecosystem-based management may fail without changes to tool development and financing. *Bioscience*, *62*(5), 508–515. doi:10.1525/bio.2012.62.5.13

Demeyer, R., & Turkelboom, F. (2014). The ecosystem services stakeholder matrix. In *OpenNESS manual: Stakeholder analysis for environmental decision-making at local level. EC FP7 Grant Agreement no. 308428*. INBO.

Donnelly, R., Neville, D., & Mous, P. (2003). *Report on a rapid ecological assessment of the Raja Ampat Islands, Papua, Eastern Indonesia*. The Nature Conservancy Southeast Asia Centre for Marine Protected Areas.

Eastside Forest Health Assessment Team (ESFHAT). (1993). Eastside Forest Health Assessment, Vol. 1, Executive Summary. U.S. Department of Agriculture, National Forest Service.

Forest Ecosystem Management Assessment Team (FEMAT). (1993). Forest Ecosystem Management: An Ecological, Economic, and Social Assessment. Report of the Ecosystem Management Assessment Team (A Federal Agency consortium), Portland, OR.

Grieve, C. & Short, K. (2007). *Implementation of ecosystem-based management in marine capture fisheries. Case studies from World Wildlife Fund's marine ecoregions*. Academic Press.

Hauck, J., Stein, C., Schiffer, E., & Vandewalle, M. (2015). Seeing the forest and the trees: Facilitating participatory network planning in environmental governance. *Global Environmental Change*, *35*, 400–410. doi:10.1016/j.gloenvcha.2015.09.022

Hein, L., Koppen, K., Groot, R., & Ierland, E. (2006). Spatial scales, stakeholders and the valuation of ecosystem services. *Ecological Economics*, *57*(2), 209–228. doi:10.1016/j.ecolecon.2005.04.005

Homewood, K., Kristjanson, P., & Trench, P. C. (Eds.). (2009). Staying Maasai? Livelihoods, Conservation and Development in East Africa's Rangeland. Springer. doi:10.1007/978-0-387-87492-0

Horton, R., Yohe, G., Easterling, W., Kates, R., Ruth, M., Sussman, E., Whelchel, A., Wolfe, D., & Lipschultz, F. (2014). Ch. 16: Northeast. Climate Change Impacts in the United States: The Third National Climate Assessment. U.S. Global Change Research Program.

Howden, M., Soussana, J. F., Tubiello, F. N., Chhetri, N., Dunlop, M., & Meinke, H. (2007). Adaptation strategies for climate change. *Proceedings of the National Academy of Sciences of the United States of America*, *104*(50), 19691–19698. doi:10.1073/pnas.0701890104 PMID:18077402

Jenkins, V. (2018). Sustainable management of natural resources: Lessons from Wales. *Journal of Environmental Law, 30*(3), 399–423. doi:10.1093/jel/eqy012

Kimaro, E., Mor, S., Jenny-Ann, L., & Toribio, M. (2018). Climate change perception and impacts on cattle production in pastoral communities of northern Tanzania. *Pastoralism Research. Policy & Practice, 8*, 19.

Leech, S., Wiensczyk, A., & Turner, J. (2009). Ecosystem management: A practitioner's guide. *BC Journal of Ecosystems and Management, 10*(2), 1–12.

Lu, Y., Wang, R., Zhang, Y., Su, H., Wang, P., Jenkins, A., Ferrier, R. C., Bailey, M., & Squire, G. (2015). Ecosystem health towards sustainability. *Ecosystem Health and Sustainability, 1*(1), 2. doi:10.1890/EHS14-0013.1

Mason, R. (2011). Ecoregional planning. *Journal of Planning Literature, 26*(4), 405–419. doi:10.1177/0885412211411091

McLeod, K., & Leslie, H. (2009). Why ecosystem-based management? In K. L. McLeod & H. M. Leslie (Eds.), *Ecosystem-Based Management for the Oceans* (pp. 3–12). Island Press.

Muchapondwa, E., Biggs, H., Driver, A., Matose, F., Mungatana, E., & Scheepers, K. (2012). Providing economic incentives for biodiversity conservation in an emrging bioregional context. *Journal of Sustainable Development, 5*(11), 118–129. doi:10.5539/jsd.v5n11p118

Osano, P. (2011). *Life at the Crossroads: How Climate Change Threatens the Existence of the Maasai*. Centre for International Governance Innovation. Available at www.africaportal.org/articles/2011/09/06/life-crossroads-how-climate-change-threatensexistence-maasai

Osano, P., Said, M., Moiko, S., Ole Kaelo, D., Schomers, S., Birner, R., & Ogutu, J. (2013). Pastoralism and ecosystem-based adaptation in Kenyan Maasailand. *International Journal of Climate Change Strategies and Management, 5*(2), 198–214. doi:10.1108/17568691311327596

Palomares, M., & Heymans, J. (2006). Historical Ecology of the Raja Ampat Archipelago, Papua Province, Indonesia. Vancouver, Canada: Fisheries Centre Research Reports 14(7), Fisheries Centre, University of British Columbia.

Pirot, J. Y., Meynell, P., & Elder, D. (2000). *Ecosystem management: Lessons from around the world – A guide for development and conservation practitioners*. IUCN. data.iucn.org/dbtw-wpd/edocs/2000-051.pdf

Quinn, M. S. (2002). Ecosystem-based management. In D. Thompson (Ed.), *Tools for environmental management* (pp. 370–382). New Society Publishers.

Rashid, H., Scholes, R., & Ash, N. (2005). Ecosystems and human well-being: Current State and Trends: Vol. 1. *Findings of the Condition and Trends Working Group of the Millennium Ecosystem Assessment Washington*. Island Press.

Ravy-Jonsen, L. (2009). Ecosystem management a management view. Department of Environmental and Business Economics. IME Working Paper 86/09.

Rosenberg, A., & McLeod, K. (2005). Implementing ecosystem-based approaches to management for the conservation of ecosystem services. *Marine Ecology Progress Series*, *300*, 241–296. doi:10.3354/meps300270

Rowe, J. S. (1992). The ecosystem approach to forestland management. *Forestry Chronicle*, *68*(2), 222–224. doi:10.5558/tfc68222-2

Settele, J., Scholes, R., Betts, R., Bunn, S., Leadley, P., Nepstad, D., Overpeck, J. T., & Taboada, M. A. (2014). Terrestrial and Inland Water Systems. In Climate Change (2014) Impacts, Adaptation and Vulnerability. Part A: Global and Sectoral Aspects. Contribution of Working Group II to the Fourth Assessment Report of the Intergovernmental Panel on Climate Change. Cambridge University Press.

Smith, P., Martino, D., Cai, Z., Gwary, D., Janzen, H., Kumar, P., McCarl, B., Ogle, S., O'Mara, F., Rice, C., Scholes, B., & Sirotenko, O. (2007). Agriculture. In *Climate Change 2007: Mitigation. Contribution of Working Group III to the Fourth Assessment Report of the Intergovernmental Panel on Climate Change*. Cambridge University Press.

Smith, R., & Maltby, E. (2003). *Using the ecosystem approach to implement the convention on biological diversity: Key issues and case studies*. International Union for Conservation of Nature and Natural Resources. doi:10.2305/IUCN.CH.2003.CEM.2.en

Society of American Foresters. (1993). *Sustaining Long Term Forest Health and Productivity*. Society of American Foresters.

Tallis, H., Levin, P., Ruckelshaus, M., Lester, S., McLeod, K., Fluharty, D., & Halpern, B. (2009). The many faces of ecosystem-based management: Making the process work today in real places. *Marine Policy*, *34*(2), 340–348. doi:10.1016/j.marpol.2009.08.003

Thomas, J. W. (1996). Forest Service perspective on ecosystem management. *Ecological Applications*, *6*(3), 703–705. doi:10.2307/2269465

Tsujimoto, M., Kajikawa, Y., Tomita, J., & Matsumoto, Y. (2017). A review of the ecosystem concept—Towards coherent ecosystem design. *Technological Forecasting and Social Change*, *136*, 49–58. doi:10.1016/j.techfore.2017.06.032

United Nations Framework Convention on Climate Change (UNFCCC). (2011). *Fact sheet: climate change science-the status of climate change science today*. https://unfccc.int/files/press/backgrounders/application/pdf/press_factsh_science.pdf

Williams, B. (2010). Adaptive management of natural resources-framework and issues. *Journal of Environmental Management*, *92*(5), 1346–1353. doi:10.1016/j.jenvman.2010.10.041 PMID:21075505

Williams, B., & Brown, E. (2014). Adaptive management: From the talk to real action. *Environmental Management*, *53*(2), 465–479. doi:10.100700267-013-0205-7 PMID:24271618

Yaffee, S. (1999). Three faces of ecosystem management. *Conservation Biology*, *13*(4), 713–725. doi:10.1046/j.1523-1739.1999.98127.x

Yermolaev, O., Lisetskii, F., Marinina, O., & Buryak, A. (2015). Basin and eco-regional approach to optimise the use of water and land resources. *Biosciences Biotechnology Research Asia*, *12*(2), 145–158. doi:10.13005/bbra/2185

Zhao, C., Chen, J., Du, P., & Yuan, H. (2018). Characteristics of climate change and extreme weather from 1951 to 2011 in China. *International Journal of Environmental Research and Public Health*, *15*(11), 2540. doi:10.3390/ijerph15112540 PMID:30428540

ENDNOTES

1. Society of American Foresters- It is an educational and scientific nongovernmental organization, which promotes best forestry management practices through education and science.
2. An association of the federal government of Oregon State charged with the management of forest ecosystems.
3. A division of the national forest service of the USA charged with management of forest habitats and resources.
4. A national trade group of paper and wood products industry that advocates for sustainable use of forest goods and services.

5 A multidisciplinary and autonomous group supported by British Columbia's provincial government, communities, forest industry and national government to implement coastal land resources plans.

Chapter 7
Climate Change and Its Impact on Terrestrial Ecosystems

Banwari Dandotiya
Jiwaji University, Gwalior, India

Harendra K. Sharma
Jiwaji University, Gwalior, India

ABSTRACT

This chapter provides a general overview of the effects of climate change on the terrestrial ecosystem and is meant to set the stage for the specific papers. The discussion in this chapter focuses basically on the effects of climatic disturbances on terrestrial flora and fauna, including increasing global temperature and changing climatic patterns of terrestrial areas of the globe. Basically, climate disturbances derived increasing temperature and greenhouse gases have the ability to induce this phenomenon. Greenhouse gases are emitted by a number of sources in the atmosphere such as urbanization, industrialization, transportation, and population growth, so these contributing factors and its effects on climatic events like temperature rise, change precipitation pattern, extreme weather events, survival and shifting of biodiversity, seasonal disturbances, and effects on glaciers are relatively described in this chapter.

INTRODUCTION

This chapter provides a general overview of the effects of climate change on terrestrial ecosystem and is meant to set the stage for the specific papers. The discussion in this chapter focuses initially on individual factors as affected by climate change or, for

DOI: 10.4018/978-1-7998-3343-7.ch007

instance, choice of adaptation strategy, followed by some discussion of how these factors might interact with each other. The northernmost terrestrial regions of Norway and Sweden are experiencing an increasingly warmer and wetter climate and the rate of warming is projected to increase in the future (IPCC 2013). Climate change in concert with land-use change and other anthropogenic drivers like air pollution, of change are already rapidly affecting northern ecosystems with consequences for living beings, infrastructure and economic condition (Arctic Council 2013). The 1992 Framework Convention on Climate Change (FCCC) includes climate changes connected directly or indirectly to increasing anthropogenic activities on the globe. The impact that climate change has on natural terrestrial and other ecosystems, on human society and economies could be disastrous. The potential effects range from sea level rise and melting ice at higher latitudes and altitudes, to changing regional weather patterns. The increase in the atmospheric abundance of greenhouse gases alters the energy balance of the climate system of terrestrial ecosystems, and causes a variety of natural disturbances such as increased desertification, extreme droughts, floods, tropical cyclones, more frequent wildfires, rising sea levels and melting glaciers (IPCC, 2007).

While there is a range of factors contributing to long- and short-term climate variability, it is now clear that changes during recent decades are at least partly related to the accumulation of anthropogenic greenhouse gases (GHGs) accumulating in the atmosphere (Crowley 2000; Karoly et al. 2003; Kerr 2001; Mann and Jones 2003; Stott et al. 2000), and while terrestrial and marine systems may act as natural sinks, helping to buffer the total GHG accumulation (Field and Fung 1999; Fung 2000), there will be a net increase in atmospheric GHGs over time (Falkowski et al. 2000). The Intergovernmental Panel on Climate Change (IPCC) has recently projected a global temperature increase of 1.4–5.8 °C by 2100 (IPCC 2001a). The global average surface temperature increased during the twentieth century, with the 1990s the warmest on record. Snow and ice cover have decreased, global average sea level has risen, and the heat content of the oceans has increased (IPCC 1997). Other aspects of climate phenomenon have during the recent century, such as: changes in precipitation and cloud cover, low and high temperature periods, more frequent, persistent and intense episodes of the El Nino and related adverse effects on weather in many areas and an increase in drought and severe wet periods (IPCC 1997). Some climate and weather related phenomenon, including tornadoes or tropical storms, do not shown any change in their patterns (IPCC 1997).

Climate change has been largely induced through land-use changes, deforestation and reductions in soil organic matter levels related to soil tillage. Over the past several millennia, large portions of Asia and Europe were deforested, generally to convert forest areas into agricultural fields. During the 18th and 19th centuries, the clearing of northern hemisphere forests, largely North American, for agriculture was the largest agent of change in the carbon cycle (Houghton et al. 1999). Existing global temperature rise is already affecting the development and behavior of many organisms of terrestrial ecosystem (Jensen 2003). As long as the increased temperatures are not extreme, they can lead to increased crop yields (Lobell and Asner 2003). However, in small grain cereals crops, increasing temperatures can lead to accelerated development and lower crop yields (Batts et al. 1997).Climate change has already had an observed impact on biodiversity and natural ecosystems. Climate change induced loss of biodiversity has directly or indirectly changed the pattern and dynamics of flow of energy and material circulation in a number of ecosystems including terrestrial ecosystem (Zhong and Wang 2017). The impact of change in climatic phenomenon upon avian biodiversity of terrestrial and wetland complexes is more evident than upon waterfowl (IPCC 2002).

International agencies have identified a range of potentially adverse impacts for wildlife likely to occur in a climate changing world. As the (IPCC) Intergovernmental Panel on Climate Change has observed, animals and plants can reproduce, survive and grow only within specific ranges of environmental and climatic conditions (Rosenzweig et al. 2007). The climatic changes alter beyond the tolerances level of animal and plant species, then they may respond accordingly for survival. Changes in global terrestrial ecosystems under climate change may themselves exert further effects on regional climate. Terrestrial ecosystem induced changes in concentrations of GHGs, such as CO_2 and CH_4 will exert global scale impacts.

CONTRIBUTING FACTORS

Waste Burning

Forest fires and the intentional burning of biomass in the terrestrial ecosystem are an integral part of Earth's system since they occur. In the terrestrial ecosystem, fires influence the composition of flora and fauna species and hence indirectly, the type of biome (Brimblecombe 1996), change surface energy fluxes, and affect the atmospheric water cycle with removal of vegetation and changes to vegetation types (Ekirch 1993). While they can recycle nutrients to ecosystems located downwind they also damage ecosystems, property, infrastructure, and human life. Trace gases and aerosol fire emissions influence atmospheric chemistry, radiative processes, and cloud formation (Boyle 1992), and can contribute to local, regional, and global level. Predicting the occurrence and magnitude of fires is the subject of significant research activity, with advanced tools and methods continually being improved to refine the understanding to predict fire occurrence in a global changing climate.

Transportation

While transportation is essential for the economic well-being of any society, the losses associated with transportation, including the erosion of non-renewable resources and environmental pollution, are considerable. In the recent several years, the amount of automobiles increases rapidly with the increasing needs of traffic. The unburned gaseous pollutants discharged by automobiles are one of the most important factors that cause the air pollution around the main roads in urban areas. The pollution sources can be divided into two kinds, stationary and mobile pollution sources. Automobiles in the urban areas of the developing and developed cities considered as a major air pollution and GHG source. With the rapidly increasing number of automobiles in the world in public and private sector responsible for rapidly increasing concentration of GHGs (Dandotiya et al. 2020).

Industrialization

Industrialization period of many countries also known as industrial revolution in the World, was the first stage of beginning of air pollution in the atmosphere of the earth. The impact of industrialization on terrestrial ecosystem is well known. The terrestrial communities and other living beings that had so far depended solely on agriculture and forests as means of livelihood are now facing the challenge of machine technology. The industrialization has resulted net only in the change in their socio-religious life but also in pattern of settlement and health status. Industrialization has brought in large number of alien diseases. On the whole industrialization is one of the major factors in the destabilization and therefore, terrestrial population needs special attention. Industrialization, urbanization, deforestation and immigration have resulted in Deteriorations of ecology. Social and health status of terrestrial communities faced a number of changes in surrounding ecosystems after the above mentioned activities. Continuous increase in industrial air pollutants emission will lead the ratio of population suffering chronicle diseases, among which respiratory diseases occupy a significant proportion, to rise. The harmful effects of air pollution on public health not to be comparable with the positive effect of economic benefits (Dandotiya et al. 2019).

Urbanization

Urbanization leads to develop a new kind of ecosystem on the earth called as urban ecosystem. Urban ecological risk caused by rapid urbanization means potential threat to terrestrial ecosystem structure, pattern and services. The scales of ecological risk assessment have been expanded from individual organisms to watersheds and terrestrial regions. Urban areas consist of natural and manmade elements, corresponding to the urban green infrastructure and the built infrastructure, respectively. The Green Infrastructure includes natural and semi-natural areas, designed and managed to protect biodiversity and deliver ecosystem services, whereas the Build Infrastructure consists of artificial elements such as buildings, roads, pavements, bridges and brown fields (Bai and Schandl 2010; European Commission 2013). The urbanization trend in combination with scarcity of land creates pressure for an expansion driven by price rather than environmental considerations (European Commission, 2011a) and this can lead to changes in the structure and function of urban ecosystems and the services provided (Bolund and Hunhammar 1999; Maes et al. 2016).

Energy Consumption

In the 1970s the primary focus was on the relationship between energy and economics. At that time, the linkage between energy and the environment did not receive as much attention (Xueliang et al. 2015). Several environmental problems associated deeply with energy use span a growing spectrum of ambient air pollutants, hazards, and accidents and degradation of quality of environmental factors and natural ecosystems. The rapid growth of developing economy has massive implications to the energy demand and associated environmental impacts such as climate change, global warming and ecosystem disturbances. Modernization of human society has established a rich material civilization through mass production made possible by consuming a great quantity of resources (Omer 2008). However, due to the huge consumption of energy, many environmental problems have occurred, which exerts a serious influence on the existence of mankind and destroys the regeneration circulation function of nature and exceeds the purification ability of natural terrestrial ecosystem.

The increase of energy consumption associated with mass production and mass consumption increases the air pollution problem is caused mainly by the consumption of coal and petroleum.

Air Pollution

Every year, more than 3.5 million people die on the globe due to air pollution, a figure which increased by 4 percent from 2005 to 2010 (UNEP 2014). Both short-term and long-term exposures to ambient gaseous and particulate air pollutants have been shown to increase the risk for respiratory diseases (Sharma et al. 2017, Dandotiya et al. 2019). In the last few decades, air pollution caused by transportation, urban development and growing industrialization has caused air pollution to become a major emission source for atmospheric air quality degradation (Banwari Dandotiya, 2019). The main cause concerned with climate change and other related disturbed

Figure 1. World Energy Consumption 2015
(Source; Workbook, SRWE, London 2016)

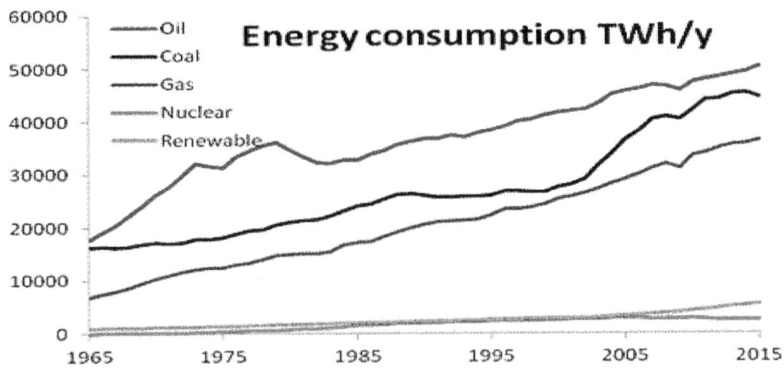

atmospheric events was increasing concentration of air pollutants in the atmosphere (Dandotiya et al. 2018). Many air pollutant sources also emit dominant anthropogenic greenhouse gases like carbon dioxide. The two main contributors to non-attainment, Ozone and particulate matter, interact with solar radiation, forcing climate change, contributes in warming by absorbing sunlight (e.g., carbon particles) or cools by scattering (e.g., sulfates) and interacts with atmospheric clouds; these radiative and microphysical interactions can induce changes in precipitation quality, pattern and regional circulation system (Arlene M. Fiore et al, 2015). Processes disturbed by the increased concentration of green house gases disturb atmospheric processes, structure, distributions and normal atmospheric episodes.

Deforestation

Forest cutting and other land use changes have profoundly altered the natural distribution and composition of global forest types, which are strongly correlated with environmental degradation and green house gas emission (Graumlich and Davis 1993, Kimmerer 1989). Vegetation can reduce the concentration of gaseous and particulate pollution by absorption and other structural factors, thus control overall concentration of green house gases from different sources (Dandotiya et al, 2019). In particular, terrestrial vegetation growth absorbs carbon from the atmosphere while deforestation responsible for increase in carbon concentration to the atmosphere. Human activities for their greed also alter ecosystems richness through land clearing for agriculture, farm abandonment, and introduction of invasive species (Bonan 2008). It is also observed in studies in recent decade that vegetation of terrestrial ecosystems provide significant feedback to climate and that natural and anthropogenic changes (Bonan 2008).

Figure 2. Changes with global warming
Source: Hunt JCR et al. 2018)

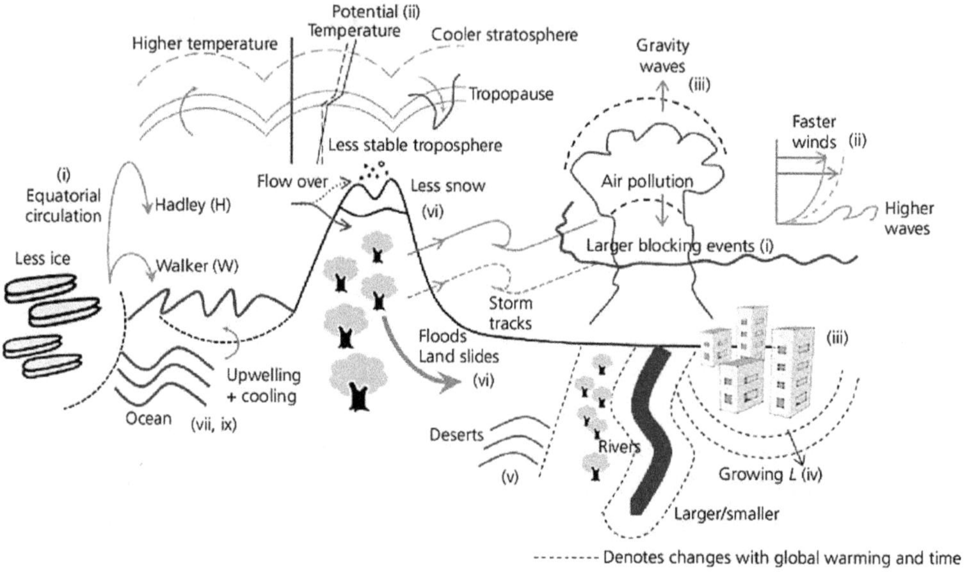

IMPACTS OF CLIMATE CHANGE

Temperature

There is a range of model predictions available for the future, but most indicated that global average surface temperatures will rise by 1.5–4.5 °C over the next 100 years (IPCC 2013). The increases will be smallest at the equator and greatest at the poles (Vaughan et al. 2001).

Seasons

At higher latitudes, where the length of the growing season is set by the time of last spring frost and first fall frost, the potential growing seasons will be longer. As night temperatures are increased more than the day temperatures, and killing frosts occur at night, season lengths will increase faster than average daily temperatures. However, the higher night temperatures will increase plant respiratory consumption of photosynthate disproportionately (Amthor 1997). In addition, winter temperatures are increasing more rapidly than summer temperatures (IPCC 2001a), and temperature increases are more rapid at higher latitudes. All in all, climate-change effects will be most dramatic for higher latitude nations.

Figure 3. Projected greenhouse gas levels and temperature increases on possible future emissions scenarios.
(Source: Climate Interactive December 9, 2011. www.ClimateScoreboard.org)

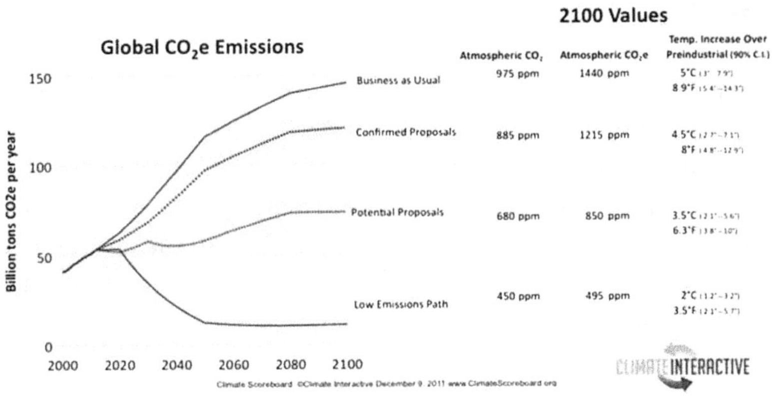

Glaciers and Rivers

Glaciers around the world retreated an average of about 30% during the 20th century (Anonymous 1998). The glaciers of the Himalayan mountains are retreating at a rate on the order of 30 m per year (McDowell 2002); many of these glaciers feed the large rivers of Asia, which provide water to produce the crops that feed a very large number of people. Populations have developed along many of these rivers, based on this extra water (from glacier melting) and the resulting greater food production, and reductions in river flow, when the glacier water additions decline, will lead to hardship. Similar events are occurring in terrestrial areas. A large amount of irrigated agriculture in Alberta is based on the many small rivers that run out of the Rocky Mountains and onto the prairies. These rivers are also glacier fed, and many of these glaciers are shrinking, for instance the Peyto glacier in Alberta has lost 70% of its mass during the last few decades.

Extreme Weather

Climate change will lead to more frequent episodes of drought and high temperature (Easterling et al. 2000). Increased frequencies of drought will hamper crop production in many areas of the world. Some semiarid areas, like the Paliser triangle discussed above, will become unable to support crop production. We will have more extreme El Niños and more often (Fedorov and Philander 2000; Kerr 1999; Timmermann et al. 1999). Tropical storms will be more frequent, stronger, and more destructive (Goldenberg et al. 2001). Collectively, these changes will make crop production more unpredictable and more difficult, and crops and cropping systems will have to be adapted to these new circumstances (Shugart 1998; Walker et al. 1999).

Sea Level Rise

Most models predict a sea level rise of about 50 cm by 2100 (Gregory and Oerlemans 1998) because of a mixture of glacier and ice-sheet melting, and thermal expansion of sea water. This will lead to the loss of agricultural land because of flooding by sea water and salinization in areas that are newly coastal. River deltas are some of the most productive agricultural lands, and these will be the areas most extensively flooded, with Bangladesh (Clarke 2003; Huq 2001) and some of the small island nations being most affected.

People

There is good evidence of past climate shifts causing social disruptions (Hodell et al. 2001), including migrations (deMenocal 2001). In North America, there will be a northward migration of crop production. For instance, the Paliser triangle may become too dry for annual crop production, while more northerly areas of the Canadian prairies will become warm enough for crop production. There is an area, largely in Northern Saskatchewan and Alberta that could be brought into production; this area is roughly as large as the Paliser triangle, but the soils are younger and less fertile. A northward movement of agricultural activities will require the development of rail infrastructure in the Canadian North. At the same time, as conditions warm, it may become more feasible to ship larger amounts of grain from the port of Churchill, perhaps even across the Northwest Passage to Asia. There will be a need for new crop production and shipping infrastructure in northern North America. Similarly, in Finland, models predict poleward migration of cereal production at 100– 130 km per 1 °C increase in average yearly temperature (Carter and Saarikko 1996). Presumably, this northward migration of agriculture will be accompanied by a parallel migration of agriculturalists.

Biodiversity

Anthropogenic sources of emissions of greenhouse gases including carbon dioxide and methane are considered the main cause of an observed 0.8 °C increase in average surface temperature of the globe since pre-industrial times (IPCC 2013). These increasing changes in greenhouse gas concentrations have a number of implications not only for temperature, but also for precipitation patterns, ice-sheet dynamics, sea levels, ocean acidification and extreme weather events such as floods and droughts etc. (IPCC 2013). Such changes are already starting to have significant impacts on terrestrial biodiversity and many ecosystems types, including varied species' distributions, inter-specific relationships and life history events, and are predicted to intensify into the future (Warren et al. 2013). With continued alarming greenhouse gas emissions (Jackson et al. 2016)

Flora

Long-term changes in temperature, precipitation pattern, atmospheric CO_2 concentration, and the chemistry of precipitation alter the adaptive capacity for vegetation growth (Livia 2014). The carbon budget of terrestrial forests has important role in the global carbon cycle (Prentice et al. 2001). Inter-annual variability in the terrestrial carbon cycle has a large impact on the global carbon cycle responding to climate variations (Bousquet et al. 2000). These inter-annual variations are induced by changes in photosynthesis and respiration due to climate variations, understanding of the responses of terrestrial ecosystems to environmental changes is not fully understood (Fearnside 2004).

Fauna

Climatic change are considered to have alarmingly influenced distribution of terrestrial and amphibian species, particularly for cold blooded animal species, due to their reliance on external conditions for efficient physiological processes (Vitt and Caldwell 2014). There is limited information regarding distribution patterns and ecological requirements restricted species. Climate change highly increases the vulnerability of these species, due to their likely low dispersal capabilities and narrow physiological tolerance ranges on terrestrial ecosystems (Botts et al. 2015). There are often limited species-specific data available for such restricted species.

Biological Evidence

Natural biological events are some of the most potent indicators of climate change. Phenological studies are now viewed as sources of primary data of great value in tracking the impact, scale and rate of climatic alteration like disturbance in synchrony between temperature and photoperiod. Pathogens and insects show individual patterns of response to temperature, CO_2 and photoperiod, the result is a loss of evolved phasing. This changes the relationships between plants and the environment. The biological evidence demonstrated in a study for woody perennials such as *Rhododendron* spp. and cultivars by Dixon & Biggs (1996) for acclimation to temperature change following the cessation of dormancy. Some more studies presents data from Salisbury, Wiltshire containing 64000 records dating from 1950 identified that the growing season for many fungi has increased as ambient temperatures have risen (Gange et al., 2007). Over this 50-year period of temperature increase, fungi doubled the length of their reproductive period from 33 days to 74 days, Some are now reproducing twice per year instead of once (Gange et al., 2007). It is suggested that changes seen in the growing season of fungi are the most substantial for any group of organisms on Earth for biological evidence (Gange et al., 2007).

CONCLUSION

Climate Change is extremely important issue of our time and we are at a determining moment. From changing weather patterns that is a menace food production, to increasing sea levels that increase the risk of calamitous flooding, the impacts of climate change are global in scope and unprecedented in scale. Without far-reaching action today, adapting to these impacts in the future will be more difficult. Our country is the fourth largest economy and fifth largest greenhouse gas (GHG) emitter, globally, accounting for about 5% of global emissions. Rapidly growing population and their needs, increase the green house gas emission in the atmosphere in a number of ways. Climate change is a reality of life. Urgent actions are needed if we are to mitigate an irreversible build-up of GHGs which induced global warming at a potentially huge cost to the economy and human society globally. The emission of greenhouse gases in the earth's atmosphere, much of it driven by anthropogenic activity, is already affecting the global climate. Under current projections, concentrations of green house gases will continue to increase into the indefinite future, entailing a process of continued global warming. The probable effects of global warming and climate change can be faced by human society include more frequent wildfires, longer periods of drought in some regions and an increase in the number, duration and intensity of tropical storms. Changes in management practices may be helpful in achieving mitigation in climate change such as making equipment more energy efficient, adopting greener technologies and renewable energies. Climate change mitigation generally involves reductions in anthropogenic emissions of greenhouse gases. Fossil fuels account for about 70% of GHG emissions in developed and developing countries (Kurokawa et al. 2013). Mitigation measures should be taken for climate changes are as follows, enhancing the energy efficiency of houses and a reduction of green house gases, make housing emission free. Mitigation can be implemented by the energy tax system, incentives for electric vehicles through tax reduction, more efficient and sustainable transport. Subsidy schemes also helpful for mitigation such as renewable energy use, funding for innovation aimed at hydrogen and other sustainable fuels, Phasing out coal-fired electricity generation and incentives for climate-friendly food consumption, etc.

REFERENCES

Amthor, J. S. (1997). Plant respiratory responses to elevated carbon dioxide partial pressure. Advances in Carbon Dioxide Effects Research. In *Proceedings of a Symposium*. American Society of Agronomy. 10.2134/asaspecpub61.c2

Anonymous. (1998). Himalayan glacier backing off 1998. *Science*, *281*(5381), 1277. doi:10.1126cience.281.5381.1277a

Arctic Council. (2013). *Arctic Resilience Interim Report 2013*. Stockholm Environmental Institute and Stockhol Resilience Centre.

Bai, X., & Schandl, H. (2010). *Urban Ecology and Industrial Ecology: The Routledge Handbook of Urban Ecology*. Routledge Publications.

Batts, G. R., Morison, J. I. L., Ellis, R. H., Hadley, P., & Wheeler, T. R. (1997). Effects of CO_2 and temperature on growth and yield of crops of winter wheat over several seasons. *European Journal of Agronomy*, *7*(1-3), 43–52. doi:10.1016/S1161-0301(97)00022-1

Bolund, P., & Hunhammar, S. (1999). Ecosystem Services in Urban Areas. *Ecological Economics*, *29*(2), 293–301. doi:10.1016/S0921-8009(99)00013-0

Bonan, G. (2008). *Ecological climatology: Concepts and application*. Cambridge University Press. doi:10.1017/CBO9780511805530

Botts, E. A., Erasmus, B. F. N., & Alexander, G. J. (2015). Observed range dynamics of South African amphibians under conditions of global change. *Austral Ecology*, *40*(3), 309–317. doi:10.1111/aec.12215

Bousquet, P. (2000). Regional changes in carbon dioxide fluxes of land and oceans since 1980. *Science*, *290*(5495), 1342–1346. doi:10.1126cience.290.5495.1342 PMID:11082059

Boyle, R. (1992). A General History of the Air. Academic Press.

Brimblecombe, P. (1996). Attitudes and responses towards air pollution in medieval England. *Journal of the Air Pollution Control Association*, *26*(10), 941–945. doi:10.1080/00022470.1976.10470341 PMID:789426

Carter, T. R., & Saarikko, R. A. (1996). Estimating regional crop potential in Finland under a changing climate. *Agricultural and Forest Meteorology*, *79*(4), 301–313. doi:10.1016/0168-1923(95)02287-2

Clarke, T. (2003). Delta blues. *Nature*, *422*(6929), 254–256. doi:10.1038/422254a PMID:12646891

Crowley, T. J. (2000). Causes of climate change over the past 1000 years. *Science*, *289*(5477), 270–277. doi:10.1126cience.289.5477.270 PMID:10894770

Dandotiya, Jadon, & Sharma. (2018). Effects of meteorological parameters on gaseous air pollutant concentrations in urban area of Gwalior City, India. *Environmental Claims Journal*, *1507508*. Advance online publication. doi:10.1080/10406026.2018

Dandotiya, B. (2019). *Health Effects of Air Pollution in Urban Environment*. Climate Change and Its Impact on Ecosystem Services and Biodiversity in Arid and Semi-Arid Zones., doi:10.4018/978-1-5225-7387-6.ch006

Dandotiya, B., Sharma, H. K., & Jadon, N. (2019, January). Role of Urban Vegetation in Particulate Pollution Control in Urban Areas of Gwalior City with Special Reference to SPM. *Adv. Biores*, *10*(1), 97–103.

Dandotiya, B., Sharma, H. K., & Jadon, N. (2020). Ambient Air Quality and Meteorological Monitoring of Gaseous Pollutants in Urban Areas of Gwalior City India. *Environmental Claims Journal*, 1–16. Advance online publication. doi:10.1080/10406026.2020.1744854

deMenocal, P. B. (2001). Cultural responses to climate change during the late Holocene. *Science*, *292*(5517), 667–673. doi:10.1126cience.1059827 PMID:11303088

Dixon, G. R., & Biggs, M. P. (1996). The impact of environmental change on acclimation and dormancy in *Rhododendron*. *Aspects of Applied Biology*, *45*, 93–100.

Easterling, D. R., Meehl, G. A., Parmesan, C., Changnon, S. A., Karl, T. R., & Mearns, L. O. (2000). Climate extremes: Observations, modeling, and impacts. *Science*, *289*(5487), 2068–2074. doi:10.1126cience.289.5487.2068 PMID:11000103

Ekirch, A. A. Jr. (1993). *Man and Nature in America*. University of Nebraska Press.

European Commission. (2013). *Green Infrastructure (GI) Enhancing Europe's Natural Capital*. COM249. https://ec.europa.eu/environment/nature/ecosystems/docs/green_infrastructures/1_EN_ACT_part1_v5

Fearnside, P. M. (2004). Are climate change impacts already affecting tropical forest biomass? *Global Environmental Change*, *14*(4), 299–302. doi:10.1016/j.gloenvcha.2004.02.001

Fedorov, A. V., & Philander, S. G. (2000). Is El Niño changing? *Science*, *288*(5473), 1997–2002. doi:10.1126cience.288.5473.1997 PMID:10856205

Field, C. B., & Fung, I. Y. (1999). Enhanced: The not-so-big U.S. carbon sink. *Science*, *285*(5427), 544–545. doi:10.1126cience.285.5427.544

Fiore, A. M., Naik, V., & Leibensperger, E. M. (2015). Air Quality and Climate Connections. *Journal of the Air & Waste Management Association*, *65*(6), 645–685. doi:10.1080/10962247.2015.1040526 PMID:25976481

Fung, I. (2000). Variable carbon sinks. *Science*, *290*(5495), 1313. doi:10.1126cience.290.5495.1313 PMID:17787235

Gange, A. C., Ganage, E. G., Sparks, T. H., & Boddy, L. (2007). Rapid and recent changes in fungal fruiting patterns. *Science*, *316*(5821), 71.

Garrett, D. S. (1957). *Biology of Root Invading Fungi*. Cambridge University Press.

Goldenberg, S. B., Landsea, C. W., Mestas-Nunez, A. M., & Gray, W. M. (2001). The recent increase in Atlantic hurricane activity: Causes and implications. *Science*, *293*(5529), 474–479. doi:10.1126cience.1060040 PMID:11463911

Graumlich, L. J., & Davis, M. B. (1993). Holocene variation in spatial scales of vegetation pattern in the upper Great Lakes. *Ecology*, *74*(3), 826–839. doi:10.2307/1940809

Gregory, J. M., & Oerlemans, J. (1998). Simulated future sea-level rise due to glacier melt based on regionally and seasonally resolved temperature changes. *Nature*, *391*(6666), 474–476. doi:10.1038/35119

Hodell, D. A., Brenner, M., Curtis, J. H., & Guilderson, T. (2001). Solar forcing of drought frequency in the Maya lowlands. *Science*, *292*(5520), 1367–1370. doi:10.1126cience.1057759 PMID:11359010

Houghton, R. A., Hackler, J. L., & Lawrence, K. T. (1999). The U.S. carbon budget: Contributions from land-use change. *Science*, *285*(5427), 574–578. doi:10.1126cience.285.5427.574 PMID:10417385

Hunt, J. C. R., Aktas, Y. D., Mahalov, A., Moustaoui, M., Salamanca, F., & Georgescu, M. (2018). Climate change and growing megacities: Hazards and vulnerability. *Proceedings of the Institution of Civil Engineers. Engineering Sustainability, 171*(6), 314–326. doi:10.1680/jensu.16.00068

Huq, S. (2001). Climate change and Bangladesh. *Science, 294*(5547), 1617. doi:10.1126cience.294.5547.1617 PMID:11721020

IPCC. (1997). *The Regional Impacts of Climate Change*. IPCC.

IPCC. (2001a). *Climate change 2001: the scientific basis*. Contribution of Working Group I to the Third Assessment Report of the IPCC. Cambridge University Press. Available from http://www.ipcc.ch

IPCC. (2002). *Climate change and biodiversity*. IPCC Technical Paper V. https://www.ipcc.ch/pub/tpbiodiv.pdf

IPCC. (2007). *Climate Change 2007: The Physical Science Basis: Contribution of Working Group I to the Fourth Assessment Report of the Intergovernmental Panel on Climate Change*. Cambridge University Press.

IPCC. (2013). Summary for policy makers. In *Climate Change 2013: The Physical Science Basis. Contribution of Working Group I to the Fifth Assessment Report of the Intergovernmental Panel on Climate Change* (pp. 1–32). Cambridge University Press.

IPCC. (2013). *Working Group I Contribution to the IPCC Fifth Assessment Report Climate Change 2013:The Physical Science Basis Summary for Policymakers*. Cambridge University Press.

Jackson, R. B., Canadell, J. G., Le Quéré, C., Andrew, R. M., Korsbakken, J. I., Peters, G. P., & Nakicenovic, N. (2016). Reaching peak emissions. *Nature Climate Change, 6*(1), 7–9. doi:10.1038/nclimate2892

Jensen, M. N. (2003). Consensus on ecological impacts remains elusive. *Science, 299*(5603), 38. doi:10.1126cience.299.5603.38 PMID:12511629

Karoly, D. J., Braganza, K., Stott, P. A., Arblaster, J. M., Meehl, G. A., Broccoli, A. J., & Dixon, K. W. (2003). Detection of a human influence on North American climate. *Science, 302*(5648), 1200–1203. doi:10.1126cience.1089159 PMID:14615536

Kerr, R. A. (1999). Big El Niños ride the back of slower climate change. *Science, 283*(5405), 1108–1109. doi:10.1126cience.283.5405.1108

Kerr, R. A. (2001). It's official: Humans are behind most of global warming. *Science, 291*(5504), 566. doi:10.1126cience.291.5504.566A PMID:11229388

Kimmerer, R. W. (1989). Environmental determinants of spatial pattern in the vegetation of Pb-Zn mines in southwestern Wisconsin. *American Midlaiid Nattrralist, 121*(1), 134–143. doi:10.2307/2425664

Kurokawa, J., Ohara, T., Morikawa, T., Hanayama, S., Greet, J. M., Fukui, T., Kawashima, K., & Akimoto, H. (2013). Emissions of air pollutants and greenhouse gases over Asian regions during 2000–2008: Regional Emission inventory in Asia (REAS) version 2. *Atmospheric Chemistry and Physics, 13*(21), 11019–11058. doi:10.5194/acp-13-11019-2013

Lobell, D. B., & Asner, G. P. (2003). Climate and management contributions to recent trends in U.S. agricultural yields. *Science, 299*(5609), 1032. doi:10.1126cience.1077838 PMID:12586935

Maes, J., Zulian, G., Thijssen, M., Castell, C., Baro, F., Ferreira, A., & Melo, J. (2016). Mapping and Assessment of Ecosystems and Their Services, Urban Ecosystems: 4th Report- Final. Luxembourg: Office for Official Publications of the European Union.

Mann, M. E., & Jones, P. D. (2003). Global surface temperatures over the past two millennia. *Geophysical Research Letters, 30*(15), 5–4. doi:10.1029/2003GL017814

McDowell, N. (2002). Developing countries to gain from carbon trading fund. [London]. *Nature, 420*(6911), 4. doi:10.1038/420004a PMID:12422171

Omer, A. M. (2008). Energy, environment and sustainable development. *Renewable & Sustainable Energy Reviews, 12*(9), 2265–2300. doi:10.1016/j.rser.2007.05.001

Prentice, I.C., Farquhar, G.D., Fasham, M.J.R., Goulden, M.L., Heimann, M., Jaramillo, V.J., Kheshgi, H.S., Le Que're´, C., Scholes, R.J., & Wallace, D.W.R. (2001). The carbon cycle and atmospheric carbon dioxide. In Climate Change 2001: The Scientific Basis. Contribution of Working Group I to the Third Assessment Report of the Intergovernmental Panel on Climate Change. Cambridge University Press.

Rasche, L. (2014). The importance of climate variability changes for future levels of tree-based forest ecosystem services. *The International Journal of Biodiversity Science, Ecosystem Services & Management, 10*(3), 187–197. doi:10.1080/21513732.2014.939719

Rosenzweig, C (2007). Assessment of Observed Changes and Responses in Natural and Managed Systems. In *Climate Change 2007: Impacts, Adaptation and Vulnerability. Contribution of Working Group II to the Fourth Assessment Report of the Intergovernmental Panel on Climate Change* (p. 98). Cambridge University Press.

Sharma, H. K., Dandotiya, B., & Jadon, N. (2017). Exposure of Air Pollution and Its Health Effects in Traffic Police Persons of Gwalior City, India. *Environmental Claims Journal, 29*(4), 305–315. doi:10.1080/10406026.2017.1390357

Shugart, H. H. (1998). *Terrestrial ecosystems in changing environments*. Cambridge University Press.

Stott, P. A., Tett, S. F. B., Jones, G. S., Allen, M. R., Mitchell, J. F. B., & Jenkins, G. J. (2000). External control of 20th century temperature by natural and anthropogenic forcings. *Science, 290*(5499), 2133–2137. doi:10.1126cience.290.5499.2133 PMID:11118145

Timmermann, A., Oberhuber, J., Bacher, A., Esch, E., Latif, M., & Roeckner, E. (1999). Increased El Niño frequency in a climate model forced by future greenhouse warming. [London]. *Nature, 398*(6729), 694–697. doi:10.1038/19505

Vaughan, D. G., Marshall, G. J., Connolley, W. M., King, J. C., & Mulvaney, R. (2001). Devil in the detail. *Science, 293*(5536), 1777–1779. doi:10.1126cience.1065116 PMID:11546858

Vitt, L. J., & Caldwell, J. P. (2014). *Herpetology: an introductory biology of amphibians and reptiles*. Academic Press.

Walker, B., Steffen, W., Canadell, J., & Ingram, J. (1999). *The terrestrial biosphere and global change: implications for natural and managed ecosystems. Synthesis volume*. Cambridge University Press.

Warren, R., VanDerWal, J., Price, J., Walbergen, J. A., Atkinson, I., Ramirez-Villegas, J., Osborn, T. J., Jarvis, A., Shoo, L. P., Williams, S. E., & Lowe, J. (2013). Quantifying the benefit of early climate change mitigation in avoiding biodiversity loss. *Nature Climate Change, 3*(7), 678–682. doi:10.1038/nclimate1887

Yuan, X., Mu, R., Zuo, J., & Wang, Q. (2015). Economic Development, Energy Consumption, and Air Pollution: A Critical Assessment in China. *Human and Ecological Risk Assessment, 21*(3), 781–798. doi:10.1080/10807039.2014.932204

KEY TERMS AND DEFINITIONS

Air Pollution: Air pollution is degradation of air quality by anthropogenic or natural phenomenon.

Climate: Climate is the long-term seasonal weather conditions of more than 30 years.

Climate Change: Climate change is the change in pre-existed long term weather conditions.

Extreme Weather: Extreme weather means unexpected or unseasonal weather due to global warming or climate change. For example, change in rainfall pattern, rise in temperature, etc.

Global Warming: Global warming is the rise of global temperature, due to increasing concentration of greenhouse gases in the atmosphere.

Greenhouse Gases: Greenhouse gases are those gases releases by incomplete combustion and have comparatively high heat absorbing abilities from solar radiation.

Terrestrial Ecosystem: A terrestrial ecosystem is collective term for all land-based ecosystems including, grassland, forest, tundra, taiga, etc.

Chapter 8
Understanding Water-Food-Energy Nexus in the Climate Change Era and the Roadmap to Implementation in South Africa

Joan Nyika
https://orcid.org/0000-0001-8300-6990
Technical University of Kenya, Kenya

ABSTRACT

WFE nexus is an important aspect in building sustainable economies. Water is used in food production while water supply and food processing require energy. Understanding the interrelationships of the nexus components is a growing interest for researchers and policymakers towards sustainable development. This chapter analyses the in-depth meaning of the WFE nexus, its importance, and its involved processes. The chapter also evaluates the effects of climate change on the nexus using case examples in South Africa. It also proposes a road map to facilitate better management of the nexus by recommending useful action plans. These action plans prioritize on baseline data collection, optimization of WFE nexus processes and cooperative management of resources, and climate change adaptation.

DOI: 10.4018/978-1-7998-3343-7.ch008

INTRODUCTION

There is an inextricable relationship between the production, use and security of water, food and energy (WFE) (Cai et al., 2018). In exploring this intertwined relationship, Nhamo et al. (2018) claimed that water and energy are prerequisites of food production and water management requires energy where, hydro-energy is the essential power provider. Similarly, Weitz et al. (2017) exemplified the WFE interrelationship suggesting that drought results from climate change, which results to energy and food insecurity problems that are propagated by the lack of water. WFE are vital components of sustainable development, poverty reduction and human wellbeing. As the demand for WFE resources rises, the supply of these interrelated resources is reducing and so is there security. Zhang et al. (2018), who highlighted this issue, claimed that the global demand for WFE resources would rise by 50% by 2050 compared to the scenario in 2015. The ever-rising population, urbanization and climate change trends will further drive the demand for WFE resources. Consequently, the resources will be under additional pressure apart from constraints by competing needs under limited supply worldwide and increasing food demand.

Under these circumstances, the holistic management of the WFE relationships commonly referred to us the WFE nexus towards realising sustainable development is gaining more attention to policy formulators and decision makers as Weitz et al. (2017) highlighted. The concept advocates for integrated use of water and energy resources and its origin dates to the Harvard Water Program of the 1960s that focused on water management research in an interdisciplinary and multifaceted relationship to the environment and human society (Cai et al, 2018). Its formal introduction however only occurred during the World Economic Forum of 2008, when challenges of economic development were globally summarised under the WFE nexus perspective, although no consensus on its precepts was agreed upon (Smajgl et al., 2016; Zhang et al., 2018). Shifting the focus to WFE nexus is valuable as downplaying their interrelations has adverse consequences. For instance, bioenergy initiatives that were meant to mitigate climate change by avoiding the use of fossil fuels have since caused biodiversity loss, change in land-use and food crises as crops compete with food for land and water resources (Zhang et al., 2018). Similarly, the shift to desalination and water-conservation irrigation is likely to escalate energy consumption because of increased treatment processes and pumping that require power. With these considerations, this chapter seeks to demystify the concept of WFE nexus and the roadmap towards its implementation in the South African context.

DEFINITION OF THE WFE NEXUS

The WFE nexus is closely associated to integrated water resources management (IWRM) and has particular focus on water resources. It aims at using integrated approaches to assess investments and designs of multi-sectors while balancing objectives, values and motivations of competing parties. This affiliation however, is the source of the concept's criticism with authors purporting that it has skewed focus on water rather than natural resources, which demeans its holistic nature (Smajgl et al., 2015; Bohensky et al., 2009). Authors such as Shah (2009) and Muller (2015) question the applicability of the concept particularly in developing countries of Africa and Asia. In their argument, the authors claim that attempts to implement the WFE nexus in developing countries impose prescriptions on resources use rather than promoting deliberate discovery of processes to manage the resources effectively and efficiently. Consequently, the success of these initiatives is slow and even sometimes regressed by poor acceptability by implementers and targeted users (Muller, 2015).

The conceptualisation of the WFE nexus is under constant critique and according to Benson et al. (2015) who reported that there is no agreeable praxis on the concept since its definition changes based on the empirical focus, geopolitical context and focuses of specified integration sections. Zhang et al. (2018) however disagreed with these suggestions by highlighting the two most acceptable definition of the WFE nexus. The first definition describes the nexus as interactions of a variety of sub-systems in the nexus system. For instance, water-energy nexus consists of interdependencies between water and energy. Similarly, WFE systems can be subdivided into interdependencies of water, food and energy. In this case, the production of energy and food requires water and food can be used to generate biofuels. Cai et al. (2018) supported this definition by claiming that the WFE interconnections are through physicochemical processes. They have input-output relations and dominate markets, institutions and infrastructural processes. By demystifying the interlinkages of the WFE nexus, complex systems can be processed in a balanced approach. This means that the security concerns on WFE can be addressed holistically on understanding their interrelationships.

The second definition describes the WFE nexus as an analysis process, which quantifies the nexus nodes and links them. This is most emerging description and has been accepted by the United Nations (UN) and the Food and Agriculture Organization (FAO) who described the WFE nexus as systematic analysis of human-nature systems whose aim is to promote integrated management of resources at all scales and sectors while managing resultant trade-offs and building synergies (FAO, 2014). A deeper understanding of the nexus as a process is in its aims to enhance resource recovery, promote efficient use, manage and understand the dynamisms of its nodes (Keskinen et al., 2016; Zhang et al., 2018). With this definition, three aspects of the WFE nexus emerge; its methodology, governance and the view that it is an emerging discipline. Irrespective of differences in the two definitions, there is consensus that the WFE nexus is not like conventional decision-making as it considers the three sectors (water, food and energy) as integrated rather than distinct.

AN EXPANDED VIEW OF THE WFE NEXUS

The WFE nexus is characterised by three aspects 1) chemical, biophysical and physical interactions, 2) input and output of resources and 3) infrastructural, market and institutional processes. Interactions of these aspects can be represented as shown in Figure 1 and are derivatives of the first definition of the WFE nexus. Processes that drive the nexus ultimately affect the performance and dynamics of each of its components and the system as a whole through energy and mass exchanges. Water, which is naturally variable, has superior importance compared to the other two sectors in that it influences their dynamics (Scott and Sugg, 2015). Energy and food sectors shape the spatial boundaries of humans. In this case, people make decisions on where to site power plants and farms for crop and livestock production. Water

Figure 1. Interactions of the components of the WFE nexus
(Cai et al., 2018)

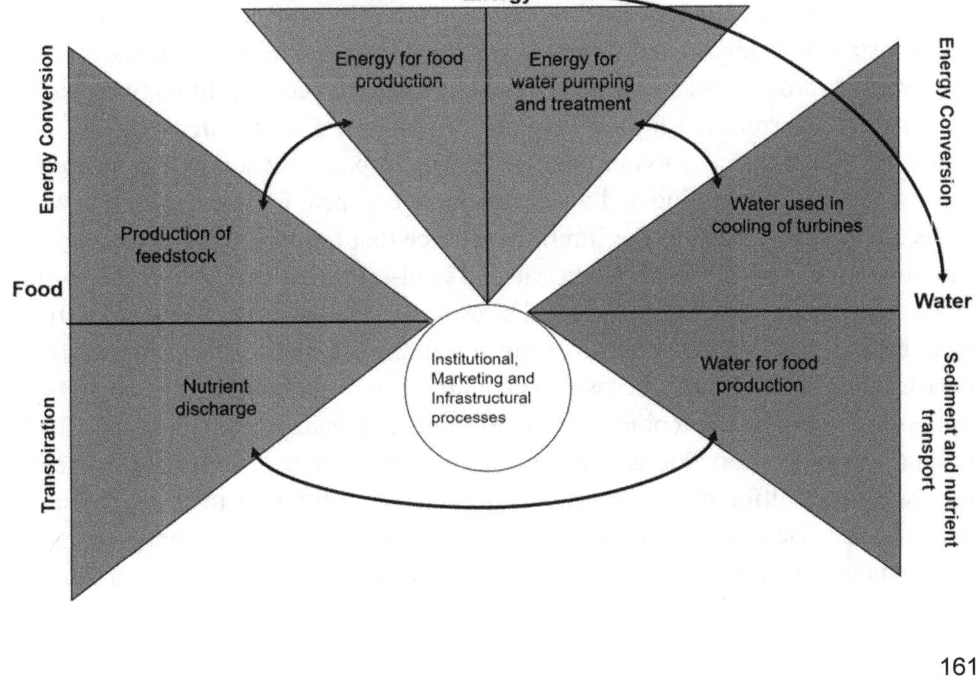

on the other hand, is a function of physical boundaries and can be divided into sub-basins, catchments or aquifers (Ray et al., 2015).

Figure 2. Interactions of the WFE nexus components and their interdependence on the same resources

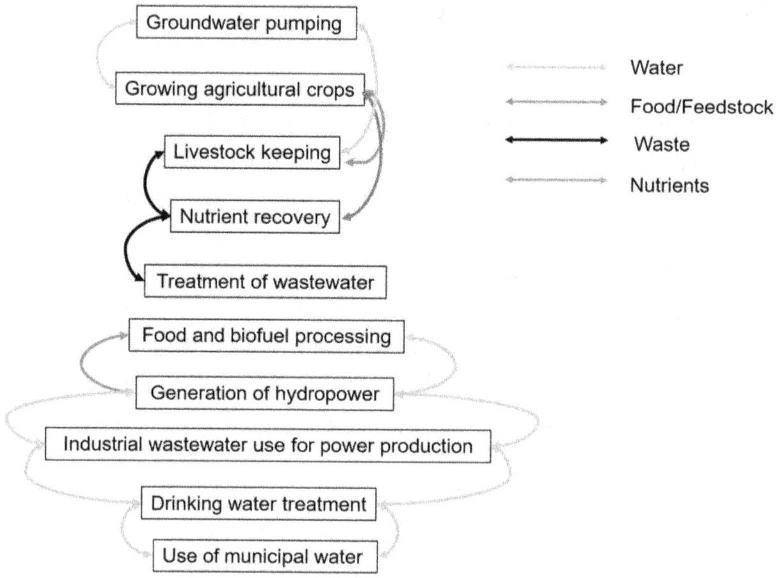

WFE are essential inputs used to generate other resources as shown in figure 2. For instance, water is used to grow crops that are used to feed livestock. Once fed, livestock produce waste, which is used in nutrient recovery. In addition to the input-output interrelationship, WFE sectors compete for resources locally and regionally. For instance, food production and hydropower generation compete for water while food processing and water treatment compete for energy. Of the WFE nexus components, water is the limiting resource that food and energy depend on. The situation is more predominant in arid and semi-arid areas that face chronic water shortage despite the high demand for the resource (Mekonnen & Hoekstra, 2016). WFE nexus is controlled by distinct but interlinked administrative, engineering, infrastructural and market processes. The multifaceted and intricate nature of these processes particularly their objectivity, operations and management make it difficult for their full integration. Successful WFE nexus depend on the coordination of these processes in the different nexus components and the willingness to compromise on their different views by implementers. The influence of socioeconomic changes, environmental modifications and technological innovations also play a role in influencing the WFE nexus by regulating the supply and demand of input-output resources. For instance, climate change has led to the depletion of water resources, which has resulted to reduced food production and increased energy costs.

From these interactions and input-output interlinkages of the WFE nexus, research issues in this concept can be sub-divided in three approaches as highlighted by Zhang et al. (2018). First, are the two-node sub-nexus such as water-land (Chen et al., 2018), food-energy (Walsh et al., 2018), energy-water (Li et al., 2012) and energy-irrigation (Sha et al., 2007). Second, are the three-node research issues including the WFE (Zhang et al., 2018), water-energy-climate (Mu & Khan, 2009), environment-water-climate (Groenfeldt, 2010) and land use-climate change-energy (Dale et al., 2011). Third, are the four-node nexus issues including the WFE-climate (Beck and Walker, 2013) and water-energy-land-climate (Hermann et al., 2012) interrelationships. These issues support the second definition of the WFE nexus, which emphasizes the analysis of inherent processes in the nexus nodes.

In these outlined two, three and four-node issues, climate and its associated changes is overarching and of great significance considering that a coordinated response measure to climate change in the WFE sectors is essential to socio-economic development and its effects are cross-sectoral (Pardoe et al., 2018). Addressing climate change in the nexus will enhance cross-sectoral synergies, trade-offs and enable feedback irrespective of institutional incompatibilities and governance challenges. Climate change adaptation will promote the convergence of inter-dependent WFE components and avoid significant stress from its effects. According to Pardoe et al. (2018), such a perspective will address the current resource management issues in developing countries. In Mekong basin of Asia, the WFE nexus was successfully implemented amidst the pressures on its resources, which have been exacerbated by climate change through adaptation (Smajgl et al., 2016). Similar attempts are underway in Tanzania (Pardoe et al., 2018).

The aforementioned nodes also emphasize the important role of water over the other components of the nexus. Food and energy production depend on water resources and with climate change, such resources have become insecure due to high demand to meet food-energy requirements. To emphasize the important role of water in the WFE nexus Simpson and Jewitt (2019) stated that unlike food and energy, it has a finite nature and has no alternatives and/or substitutes. As such, livelihoods of the energy and food sector depend on the availability of water. Bogardi et al. (2012) also stated that water moves a country's economy particularly food and energy sectors in the form of virtual water. In the light of climate change, water resources have been exploited because of prolonged drought and increased temperature compromising the capacity of its flow in the environment and deteriorating productivity of the food and energy sectors.

W-F-E NEXUS AND SUSTAINABLE DEVELOPMENT GOALS

Seventeen sustainable development goals (SDGs) were adopted in the United Nations General Assembly of 2015 (UN, 2015). These resolutions include goals specific to W-F-E as shown in Table 1. Implementing the WFE nexus would facilitate the successful implementation of these SDGs through planned development, a clearer alignment of these goals with planetary boundaries and a better framework for efficient use of resources (Salam et al., 2017). Gupta (2017) emphasized that all SDGs are directly or indirectly related to the W-F-E themes as both approaches seek to end poverty, empower the poor and marginalised and streamline production and consumption trends. From another perspective, energy and water accessibility improve food security and lead to development. Therefore, implementing the WFE nexus is a strategy to meeting these SDGs.

Table 1. SDGs directly associated with the WFE nexus.

Goal Number	Objective
2	Zero hunger through more food production, less wastage and enhanced support to farmers.
6	Clean water and sanitation through efficient and effective use of water
7	Cost effective and clean energy provision through innovative and renewable technologies.
13	Protect water, food and energy sources through climate action to mitigate climate change effects.
14	Protection of life below water to maintain their biodiversity and protect coastal livelihoods and food source.

INTERCONNECTEDNESS OF THE W-F-E NEXUS COMPONENTS

From the described expanded view of the WFE nexus (Figure 1-2), it is evident that understanding the connectedness and inherent processes of its components is essential. Although concepts such as IWRM have attempted to address some of the components, their interlinkages should be explored holistically. Some of the explored issues include water supply and management, wastewater treatment, irrigation and energy interconversions. The issues influence energy and food sectors significantly. Emerging fields such as sustainable ecosystem management, green energy production and eco-hydrology and are less explored though they enhance the understanding of the WFE nexus. However, issues to do with climate change and its relationship to hydrology, energy and agricultural sectors require much attention as literature documents gaps particularly in characterising their nexus interconnections (Cai et al., 2018). The following sections attempt to describe the processes interlinking the components of the WFE nexus.

Processes Connecting Water to Food Components

Water to Food

In food production, hydrodynamic processes such as subsurface flow, evapotranspiration and precipitation are integral. In particular, demystifying crop water requirement in the wake of climate change has not been fully exploited by eco-hydrologists (Nistor et al., 2017). To maximum crop production, drainage engineering and irrigation innovations to enable efficient water use have been developed and employed yet; their effectiveness is challenged by the extreme and erratic nature of weather resulting from climate change. Similarly, precipitation has been reported to vary temporally and spatially resulting to modified rainfall seasons, excessive dryness or wetness and erratic rainfall events (Kunkel et al., 2012). These phenomena influence land use, cover, and ultimately, affect food production decisions. For example, water that was previously stored during spring in the USA is now being used for irrigation during late summer due to extreme dryness (Zhang et al., 2018). The focus in providing water for food production is in making crops and livestock adapt to erratic moisture changes without affecting yields using engineering techniques.

Aquaculture is another source of food that uses water apart from irrigation for agricultural food. Although aquaculture is an important food source especially in coastal regions, its interconnections with hydrology are less known (Gephart et al., 2017). Several studies are documented to show that alterations of streamflow regimes affect aquatic lives and habitats, though more explicit linkages of terrestrial and aquatic systems through eco-hydrological indexing are needed (Endo et al., 2017). Schnier et al. (2016) also highlighted the need to enhance water storage and management operations that prioritize the interactions of aquatic habitats with hydrology considering their great impact on water and land use activities.

Food to Water

Food production consumes large amounts of water, fertilizer, land and chemical inputs and is a common environmental polluter worldwide. Excessive use of fertilizers and pesticides for higher food production is a known non-point source polluter of water that has negative effects to water supplies yet, it remains an unresolved environmental concern. The environmental protection agency reported that the growth of maize in the USA to make animal processed food and produce ethanol has resulted to deteriorated water quality downstream due to elevated phosphorous and nitrogen levels from fertilisers (EPA, 2015). Nutrient loading of water resources during agricultural food production has promoted vegetative growth in aquatic habitats and consequently, the costs of water treatment have risen. Pollution of downstream water resources through agricultural activities upstream have heighted inter-conflicts among involved stakeholders. In the relationships of food production to water, extensive research is needed to assess the flow of nutrients in soils, the effects of nutrient loading on water quality under human interferences and climate variability dynamics (Woo & Kumar, 2016). This is possible by devising best management practices to reduce chemical use in agriculture, recovery of nutrients during water treatment and their retention in soils as Jarvie et al. (2015) proposed. Food production activities influence natural flow regimes through flow returns during irrigation. The returns modify water balance accounting at catchment scale and influence its quality as Cai et al. (2018) noted. During development of water rights and determining water availability at basin scale, accurate determination of return flows is essential though, it is less understood (Grafton et al., 2012). Synchronising water flow and availability aspects will promote realistic food production plans with caution to use these depleting resources efficiently.

Processes Connecting Water to Energy Components

Water to Energy

Energy generation and its distribution is largely dependent on water resources. Water is used as a coolant in nuclear and thermoelectric power plants and is a raw material in hydropower generation. Although contemporary energy innovations are focusing on renewable energy production compared to conventional non-renewable technologies, some of these innovations are larger water users as Mekonnen et al. (2015) noted. For instance, the use of solar energy needs cooling that ends up using the same or more amounts of water used in nuclear and coal power plants (Bracken et al., 2015). Song et al. (2016) also noted that the production of biofuels through increased growth of biomass consumes large quantities of water. Fracking unconventional oil and gas uses utmost four times as much water during its cycling compared to traditional processing of such hydrocarbons (Clarke et al., 2013). Considering these examples, it is evident that both conventional and contemporary energy production methods use water and cost-benefit accounting before preferring a certain energy production approach is essential. These considerations should be prioritized in the prevalence of hydro-climatic changes that are interlinked to low energy outputs. Webber (2015) gave similar recommendations highlighting that power stations/plants ought to adopt operations that adapt extreme water conditions (drought and flooding) and temperature conditions (excess heat or cold). This is possible through accurate hydro-climatic forecasting on seasonal and spatial changes in the WFE nexus (Cai et al., 2018).

Energy to Water

The supply and distribution of water largely depends on energy. For instance, pumping groundwater, its treatment and wastewater processing consumes energy (Sanders & Webber, 2012). The withdrawal of water for consumptive uses and its discharge for energy production processes has compromised its availability and quality on the other extreme. Recent advances to develop bioenergy through hydrocarbon fracking have become threats to riparian health and water quality. For instance, the growth of corn for ethanol generation has put pressure on existing water and land resources used for growing corn for food and livestock feed (Simpson et al., 2008). EPA (2015) noted that these ethanol-processing plants are large water consumers and produce large quantities of wastewater, which increases the energy costs of pumping and treating the resource. Transportation spills, leaks through fractured rocks and well cases and discharge of drilling sites during natural gas exploration is a pollution threat to water resources. For instance, vicinities of stray gas exploration in the USA were found to have radium polluted surface waters as a result of its spillage (Vengosh et al., 2014). New energy production and supply pathways have compromised water resources through pollutant loading. In trying to explore alternative interlinkages, the use of bioenergy feedstock that is environment friendlier is on the rise (Smith et al., 2013). Hydrologists are also examining the water requirement of these second-generation feedstocks to enhance their efficient water use amidst expected shortages resulting from climate change (Housh et al., 2015).

Hydropower generation is a contentious IWRM issue with numerous advantages and disadvantages during hydro-climatic variability. Some of the benefits include enhanced water storage for use during dry seasons and improved flood regulation during extreme rainfall events. However, water resources for hydropower generation disrupt fish migration and reproduction, interfere with sediment transport and hinder recreation. Hydropower production also competes with other water demands including irrigation as noted by Zeng et al. (2017) after analysing the consumption patterns of water for agriculture and power production globally. Gunkel (2009) contested that hydropower is a green energy source as previously viewed attributing the decomposition in its production reservoirs as contributors to greenhouse gas emissions. The designing of water reservoirs for power production should consider the WFE nexus system holistically. This is crucial in balancing water supply for energy and aquatic habitats demand for the resource even with its flow variability and uncertainty due to climate variability as Suen et al. (2009) noted. This can be achieved by understanding the hydrodynamics of aquatic organisms and flow processes up-and-down-stream of power generation reservoirs (Zeng et al., 2017).

IMPACTS OF CLIMATE CHANGE ON THE WFE SECURITY IN SOUTH AFRICA

Climate change is becoming an inevitable process characterised by erratic rainfall events, shifted climate zones and rising temperatures (IPCC, 2013). These climate-affiliated shocks will put pressure on land and water resources and ultimately the WFE nexus will be negatively affected. In South Africa, the energy sector accounts for 83% of total greenhouse gas emissions, which are the main climate change drivers (Carter & Gulati, 2014). More than 86% of the country's electricity is derived from coal and 95% of South Africa's crude oil is imported though its demand is rising (Carter & Gulati, 2014). Under these circumstances, climate change will worsen and meeting WFE needs of the country's population will be more challenging. In the water sector, drought and floods will be a common occurrence amidst temperature rises. Ziervogel et al. (2014) have already reported some these effects in South Africa by highlighting that average yearly temperatures of the country have increased by a factor of 1.5 over the past five decades and the frequency of extreme rainfall increased. Additionally, the authors noted that events of extremely cold/hot weather, reduced rainfall during autumn and shifts in rainfall intensity and seasonality were common occurrences in the past 50 years. Similarly, the Department of Environmental Affairs (2013), predicted that erratic rainfall events and temperature rises in the country were expected. In the food sector, accessibility and availability of reliable food systems is depleting due to loss of livestock, reduced crop yields, low productivity and high farming costs due to increased pest infestation and diseases. In the energy sector, increased inputs are now required to pump water at longer distances and purify low quality water. These additional energy uses compete with demands for the resource in the food and water sectors.

To address the risks that climate change poses on the WFE nexus sustainability, a deeper understanding of its impacts in the various nexus components is needed. This was the main objective of formulating the long-term adaptation scenario (LTAS) project by the DEA (2013) whose vision was to adapt and mitigate climate change. The next sub-sections discuss the impacts of climate change on the WFE nexus using case scenarios of South Africa.

EFFECTS OF CLIMATE CHANGE ON WATER SECURITY

Climate change projections in South Africa report that the water sector will be severely affected, a situation that could spread to food and energy sectors (Nhamo et al., 2018). Yearly rainfall is projected to diminish by 20% by 2080 according to Conway et al. (2015) and these changes will increase drought events. This is a concern in South Africa, which is nearing physical water scarcity and has exploited most of its unevenly distributed water resources (Carter & Gulati, 2014). Some of the effects of climate change on water security include reduced availability of drinking and agricultural water due to rising water and air temperatures, reduced oxygen concentrations in water affecting aquatic habitats, reduced water quality and its increased ability to incubate and transmit water-borne diseases and salt intrusion in water systems (DEA, 2013). Additional effects are increased rainfall intensities and flash flood events, regional overflowing of water leading to its mixing with wastewater and increased drought events in other areas. For instance, waters of Limpopo River Basin (LRB) of Southern Africa are predicted to reduce significantly by 2050 due to increased withdrawal to meet agricultural demands amidst rising temperatures (Zhu & Ringler, 2012). The rise of temperatures in the country is set to increase the demand for green water from blue water resources. According to Carter and Gulati (2014), this trend is a result of shifts from dryland agriculture to irrigation using ground-and-surface-water resources in efforts to produce more food while consuming more energy.

EFFECTS OF CLIMATE CHANGE ON FOOD SECURITY

With increased water insecurity because of climate change, the food sector will be negatively affected due to dependency on the resource. According to Carter and Gulati (2014), climate change has adversely affected wheat, maize and potato crop production in South Africa. These crops are significant food sources for the country's population. Reduction of precipitation by 10% reduced maize yield by 4% (Akpalu et al., 2004) while total value losses of this crop due to CO_2 fertilisation was R46 million and R681 million without such fertilisation modifications (Turpie et al., 2002). Johnston et al. (2012) predicted that maize and wheat yields would decrease by 25% between 2010 and 2050. With climate change, potato farmers will depend on irrigation, fertilisers and agrochemicals to improve production. Franke et al. (2013) who assessed modifications in potato farming in Limpopo amidst climate change claimed that the modifications would increase production cost, which is not sustainable economically. Van der Waals et al. (2013) also reported that excessive precipitation will reduce potato yields as the incidences of brown spot, early blight and rotting diseases will rise in South Africa.

Apart from crop productivity and food production reductions, climate change will negatively affect food quality in South Africa. According to Fujisawa et al. (2013), rising temperatures affect the quality of apples in that their size is smaller due to water stress and most fruits have sunburns due to rising temperatures. Livestock production in South Africa will reduce with climate change as animals will be subject to heat stress, their yields will be lower and their water requirements will rise amidst shortage of the resource. According to Benhin (2006), Northern Cape Province of South Africa, which is the largest livestock producer, is adversely affected by extreme weather and increased drought events hence, reduced production. Masipa (2017) associated climate change to food insecurity in rural South Africa reporting that 35% of the country's rural population is food insecure and in 2015, Free State, Kwa Zulu Natal and Eastern Cape provinces were hardest hit by food insecurity and malnutrition as a result of extreme drought.

EFFECTS OF CLIMATE CHANGE ON ENERGY SECURITY

The rising environmental temperatures resulting from climate change will affect the energy sector, as more cost and resources will be incurred in cooling. In South Africa where more than 90% of energy generation is from coal, more water will be pumped to cool coal-fired energy plants installed with wet-cooling systems (Conway et al., 2015). Eskom, the country's main power utility is said to consume 20% of freshwater resources in the country at coal power plants for cooling (Conway et al., 2015). These negative impacts on the water sector are further exacerbated by reducing quantities and competing demands for the resource while a dry-cooling system can be adopted as an alternative. Changes in rainfall patterns have adversely affected hydropower generation that meets 1.5% of the country's energy demand and trends are skewed to an even drier climate (Carter and Gulati, 2014). Indirectly, the energy sector has been affected by efforts towards climate change adaptation in South Africa. Increased irrigation in areas that practiced dryland agriculture previously increase energy costs of pumping and supplying water while a higher temperature increase energy demands associated with air conditioning. Coalmine expansion in South Africa is under criticism because of its competing demand for water needed by the food sector considering that demands are rising with climate change (Mpandeli et al., 2018). The country therefore needs to diversify its power sources and shift from CO_2 emitting coal-dependent sources.

ROADMAP TO WFE NEXUS IMPLEMENTATION IN SOUTH AFRICA

Considering the identified interconnections of WFE sectors and their linkage to climate change in the case of South Africa, the need to develop models cultured towards sustainable growth is indispensable. A three-tier road map that will identify the magnitude of the WFE nexus challenges, optimize the nexus management and strengthen collaboration networks between South Africa and other countries has been proposed by this study and is illustrated in Figure 3.

Figure 3. The framework to implement the WFE nexus in South Africa

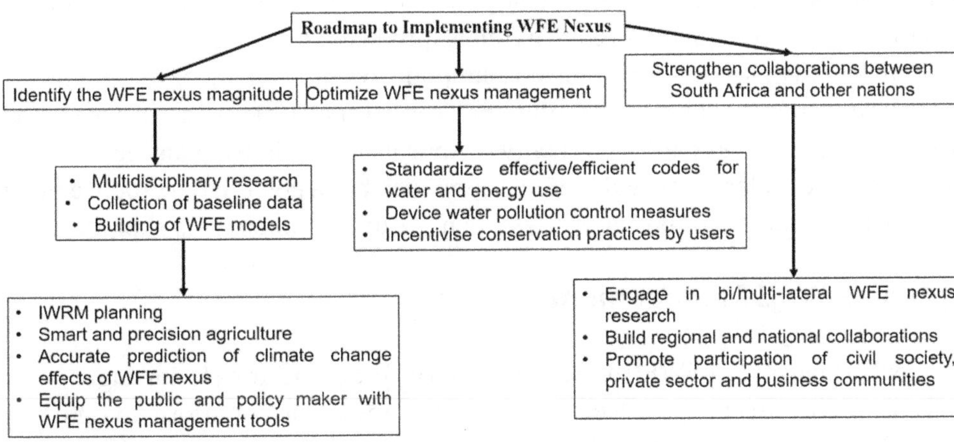

Identifying Issues in the WFE Nexus

Baseline data paucity remains a major challenge in integrating WFE nexus components in South Africa. In particular, determination of crop water requirement and amount of energy used in the water sector remains largely unknown and although data exists, it is available in piece bits across many agencies that do not cooperate (Mpandeli et al., 2018; Nhamo et al. 2018). To alleviate this challenge, research networks that collect, analyse and store data on the WFE nexus regionally are a necessity. Mohtar et al. (2015) supported these suggestions towards a platform for trade-off analysis of the nexus components. In this context, financial costs, carbon emissions, energy imports, water, land, energy and financial requirements will be weighed against their outputs and towards integrated and sustainable accounting. With adequate data and its analysis, existent gaps in linking WFE will be bridged through development of

WFE models. To be successful, such models should adopt IWRM and planning, evaluate the use of smart and precision agriculture techniques, predict climate change effects on the WFE components accurately and inform the public and policy makers on efficient approaches and tools to adopt to climate change shocks on the nexus (Pardoe et al., 2018).

Some of the country's notable action plans to collect data has been in efforts to gather information on energy-water data regarding consumption rates of the resource during coal power generation. Similarly, lifecycle-water consumption analysis in alternative hydropower generation, food production, processing and supply is underway. A case example is the introduction of mini-hydropower plants with a capacity to produce a maximum of 1-megawatt (MW) power in Ethekwini municipality of South Africa (Water Research Commission, 2018). In addition to analysis on water for coal, the country is investigating on the quantities of the resource used for fuel (natural gas or biofuels) production. Collecting such baseline data will identify the risks of existent consumption trends in reference to WFE nexus and devise sustainable alternative solutions. Nhamo et al. (2018) for instance noted that Southern Africa's coal-based energy sector consumes higher quantities of water while the use of solar, wind and biofuels combined has a potential to produce more than 20, 000 terawatts of power yet, less than 1% of this potential is exploited.

Optimise WFE Management

Efficient management of WFE demand is a priority in South Africa following existent analysis of climate change effects on these resources. Current efforts to optimize the WFE nexus are promoting water and energy use efficiency through policies. South Africa and its neighbours have strengthened their regional policies towards WFE security with an urgency to manage climate change effects in their SADC[1] regional agriculture policy (RAP) (Nhamo et al., 2018). In addition, they are pooling their energy resources together to plan on its sustainable use and distribution in a grouping dabbed the Southern Africa Power Pool (SAPP)[2] (Nhamo et al., 2018). Currently, South Africa is developing codes and standards on efficient use of water and energy to promote effective use of available resources amidst scarcity due to climate change. In Middle Breede basin of South Africa, the water budget model has been adopted towards improved WFE nexus as highlighted by Seeliger et al. (2018). In the budget, water for food is supplied using gravity-fed pipes that do not use energy unlike the conventional canal irrigation system. The authors noted that the reconfigured model is using less water to produce more food and the energy costs of production have significantly decreased.

To optimize the WFE nexus, the country is putting concerted efforts to minimize pollution in the three sectors. Some measures in place include raising the energy-water footprint and renewed efforts to reduce environmental pollution. The country is committed implement the polluter pays principle to ensure water quality is not deteriorated by activities in agriculture and energy sectors (Water Research Commission, 2018). The country is campaigning and funding to adopt renewable energy and green building[3] to enhance water and energy efficiency as Watson (2013) noted. These measures will improve resource productivity and adapt to the negative environmental effects of climate change. Nhamo et al. (2018) also noted that South Africa is optimising the WFE nexus management through intensified public awareness towards recognition and integration of resource conservation practices in the country's socio-economy and people's livelihoods. Mpandeli et al. (2018) noted that public awareness on WFE nexus management could be enhanced by implementing climate change adaptation efforts in the nexus context, integrating stakeholders of WFE sectors in climate change management planning and through strengthened policies and institutions. The country is using incentives for consumers and industries to promote end-use conservation of WFE resources. Some examples in the energy sector include the exemption of proceeds for adopters of clean development mechanisms[4], research and development (R&D) allowances for energy efficiency-based projects, accelerated depreciation and energy savings allowances to users of renewable energy (Dippenaar, 2018).

Strengthen Collaborations of South Africa With Other Countries

South Africa is one of Africa's largest producer and its WFE sectors have a great impact on the ecology and economy of the region. To cement its WFE nexus management amidst climate change, the country must devise collaborative approaches locally, regionally, with the private sector and other countries to offset the current challenges in implementing the nexus. This is possible by engaging in multi-lateral research on the nexus in the context of climate change. Nhamo et al. (2018) documented that South Africa is cooperating with regional countries under SADC to assess the opportunities of WFE nexus considering that these countries share resources. The authors noted that such multi-lateral initiatives will promote coordinated planning regionally, cascaded policy adoption on nexus management, interregional development, interdisciplinary cooperation and synergies towards integrated efficiency in resource use. Ziervogel et al. (2014) also noted that South Africa is in partnership with international organisations such as the World Wildlife Fund (WWF) and regional academicians to collect and analyse data on the management of WFE nexus towards climate change adaptation.

These partnerships are expanding the scale and discipline of focus to quantify scenarios through modelling and devise solutions through multi-sectoral management of the WFE sectors. Initiatives focusing on mapping the transboundary resources of the region are underway and make it easy for policymakers to plan and visualize the consumption patterns of the resources towards efficient energy use and food production. In a regional strategic action plan iv (RSAP IV)[5], South Africa and other SADC nations committed to adapt to climate change by using their transboundary water resources efficiently to promote sustainable agriculture and energy provision (Nhamo et al., 2018). Through joint collaborations with the Global Water Partnership, World Bank and the European Union, South Africa is expanding its WFE nexus outlook to leverage opportunities of integrated resource management with other economies despite the climate change challenges (Gulati et al., 2014; Water Research Commission, 2018).

RECOMMENDATIONS AND THE WAY FORWARD

There is evidence that the WFE nexus approach to resource management promotes sustainable development even with the environmental challenges posed by climate change (Gulati et al., 2014; Nhamo et al., 2018; Mpandeli et al., 2018). Furthermore, evidence shows that interrelating WFE sectors is a growing interest for researchers, institutions and policy formulators in a bid to devise realistic climate change adaptation and mitigation measures (Smajgl et al., 2015; Cai et al., 2018; Zhang et al., 2018). However, successful implementation of the nexus must address its existent contentious issues, bottlenecks, conditions and opportunities through structured dialogues, which result to investment identification, pilot demonstrations and realistic action plans. According to Markantonis et al. (2019), such outlooks will assist decision-makers to formulate implementation plans geared towards WFE nexus management for socio-economic development with sound environmental management policies and institutions. The following sub-sections discuss the recommendations proposed as a way forward to successful implementation of the nexus in South Africa.

Enforcement of Water Pricing

Pricing of water is an under looked economic issue that has great potential to influence the WFE nexus particularly in the food sector, which is the largest water consumer. As an economic instrument, water pricing should be designed, conceived and implemented in a way that promotes efficiency in all WFE nexus aspects. A number of pricing strategies based on polluter pays, full cost pricing and willingness to pay principles are in place to balance existent demands to available supplies in South Africa (Cole et al., 2018). Their implementation though intricate, offers an opportunity for the WFE nexus to consider the full costing of resources economically.

Such costs expand their scope from abstractions to treatment, environmental, resource and ecosystem preservation costs as water resources are exploited. Water pricing evaluates specific uses of the resource and reflect on opportunity costs incurred during its development. These economic decisions are crucial support factors during policy making on resource management and will enhance incentivised conservation for better WFE nexus management.

Introduction of Integrated Assessment Approaches

To optimize the management of the WFE nexus, there is a need for South Africa to assess its economic gains, do a cost benefit, and cost effectiveness analysis in addition. Furthermore, the country can adopt choice experiments, benefit transfer and contingent valuation among other preference methods to make optimal decisions on WFE nexus management (Carter & Gulati, 2014).

Introduction of Market Instruments

To supplement economic instruments such as subsidies, property rights and taxes, South Africa can introduce market instruments that allow effective and efficient allocation of resources. Such initiatives require a clear understanding of involved stakeholders in the WFE nexus, technological scope and market structures. For instance, the introduction of taxes to water resources promote its efficient management even in the food and energy sectors through the introduction of technologies to use the resource sparingly. In this regard, Rogers et al. (2002) noted that the high economic and market value associated with water resources in reference to the other nexus components necessitates its prioritized management.

Increased Investment in the WFE Nexus

Financing the WFE nexus is another potential consideration towards its better management. The public sector that steers up long-term perspectives of the nexus should provide adequate funding to support such visions. Additionally, the private sector has a role to play by investing in planning and R&D stages of nexus implementation. Public-private sector investment provide synergy in effective management of resources and if fostered, they can promote cost effectiveness and timely delivery of WFE management plans. Markantonis et al. (2019) highlighted examples where the public-private sectors can jointly finance towards nexus optimization. These examples included improved access to WFE resources, enhanced storage of water resources to mitigate climate change, development of alternative energy sources that are considered greener, smart irrigation for efficient energy and water use and the use of diverse water and power supply systems to reach many users. These investments can be replicated in South Africa but require multi-sectoral and interdisciplinary knowledge and information exchange.

Effective Data Collection, Analysis and Management

With prior knowledge that gaps exists in understanding the WFE nexus as a result of data unavailability in South Africa, there is need to invest in collecting data and analysing it to make scientific references and decisions on resource management. This recommendation is more relevant with the erratic climate change events that can be averted through prediction. With accurate data, comparisons of the economy in the nexus can be done to inform on existent gaps and alternative improvement measures to undertake. Technologies used in energy and food production can be maintained and improved if accurate data is provided. Scientific and research institutions should steer up data collection with the involvement of interest parties. The collection of micro data to estimate price elasticity and demand of WFE resources is crucial in decision and policymaking (Markantonis et al., 2019).

Stakeholder Cooperation for Participatory Decision-making

A solid WFE nexus should be pillared on dialogue, which is realized through public-private partnership and participation of stakeholders of interest. Nexus dialogues start at micro-regions and expand to macro-environments using the success stories of the former. For instance, success stories of WFE nexus management in Limpopo river basin (Zhu & Ringler, 2012) and Breede catchment (Seeliger et al., 2018) can be replicated to other regions of South Africa. Similarly, the initiatives to manage the nexus at national levels can be transferred to regions such as SADC in the case of South Africa. Such cooperative tendencies however demand the building of trust among the involved parties.

Institutional Restructuring

Building a WFE nexus framework should be pillared in working institutions that partner to share experiences, build capacity and trust through stakeholder participation. According to Mpandeli et al. (2018), institutional soundness in running the nexus does not mean building new institutions but effectively managing the existent ones in an interlinked manner. With effective management, institutions integrate sustainable ownership that advocates for innovation to solve the WFE nexus challenges including climate change. In addition, policy implementation is easier as participatory decision-making is priority. To this end, South African institutions involved in the WFE nexus management should be restructured to accommodate the precepts of effective management of resources, their evaluation and monitoring.

CONCLUSION

This chapter highlights the importance of the WFE nexus in socio-economic development and human well-being. It gives an expanded description of the nexus as interrelations and processes in the water-food-energy sectors using examples. The WFE nexus is vulnerable to climate variability and change. Particular attention is given to water resources whose fluctuations affect the food and energy sectors. The management of the nexus should consider the impacts of climate change and enhance integrated resources management and planning. This chapter proposes a three-tier road map toward better WFE nexus management in South Africa, which will incorporate data collection and analysis, optimisation of the nexus and collaboration of the country with local, regional and international stakeholders.

REFERENCES

Akpalu, W., Hassan, R., & Ringler, C. (2009). *Climate variability and maize yield in South Africa: Results from GME and MELE methods.* IFPRI Discussion Paper No. 843. International Food Policy Research Institute.

Beck, M. B., & Walker, R. V. (2013). On water security, sustainability, and the water-food-energy- climate nexus. *Frontiers of Environmental Science & Engineering*, *7*(5), 626–639. doi:10.100711783-013-0548-6

Benhin, K. J. (2006). Climate Change and South African Agriculture: Impacts and Adaptation Options. Centre for Environmental Economics and Policy in Africa (CEEPA).

Benson, D., Gain, A. K., & Rouillard, J. J. (2015). Water governance in a comparative perspective: From IWRM to a 'nexus' approach? *Water Alternatives*, *8*(1), 756–773.

Bogardi, J. J., Dudgeon, D., Lawford, R., Flinkerbusch, E., Meyn, A., Pahl-Wostl, C., Vielhauer, K., & Vörösmarty, C. (2012). Water security for a planet under pressure: Interconnected challenges of a changing world call for sustainable solutions. *Current Opinion in Environmental Sustainability*, *4*(1), 35–43. doi:10.1016/j.cosust.2011.12.002

Bohensky, E., Connell, D., & Taylor, B. (2009). *Experiences with Integrated River Basin Management, International and Murray Darling Basin: Lessons for Northern Australia. Norther Australia Land and Water Science Review*. CSIRO.

Bracken, N., Macknick, J., Tovar-Hastings, A., Komor, P., Gerritsen, M., & Mehta, S. (2015). *Concentrating Solar Power and Water Issues in the U.S. Southwest*. Joint Institute for Strategic Energy Analysis. doi:10.2172/1176743

Cai, X., Wallington, K., Shafiee-Jood, M., & Marston, L. (2018). Understanding and managing the food-energy-water nexus – opportunities for water resources research. *Advances in Water Resources*, *11*, 259–273. doi:10.1016/j.advwatres.2017.11.014

Carter, S., & Gulati, M. (2014). *Climate change, the Food Energy Water Nexus and food security in South Africa. In Understanding the Food Energy Water Nexus*. WWF-SA.

Chen, B., Han, M. Y., Peng, K., Zhou, S. L., Shao, L., Wu, X. F., Wei, W. D., Liu, S. Y., Li, Z., Li, J. S., & Chen, G. Q. (2018). Global land-water nexus: Agricultural land and freshwater use embodied in worldwide supply chains. *The Science of the Total Environment*, *613*, 931–943. doi:10.1016/j.scitotenv.2017.09.138 PMID:28946381

Clark, C. E., Horner, R. M., & Harto, C. B. (2013). Life Cycle Water Consumption for Shale Gas and Conventional Natural Gas. *Environmental Science & Technology*, *47*(20), 11829–11836. doi:10.1021/es4013855 PMID:24004382

Conway, D., Garderen, E., Deryng, D., Dorling, S., Krueger, T., Landman, W., Lankford, B., Lebek, K., Osborn, T., Ringler, C., Thurlow, J., Zhu, T., & Dalin, C. (2015). *Climate and southern Africa's water-energy-food nexus*. Available online:https://repository.up.ac.za/bitstream/handle/2263/52832/Conway_Climate_2015.pdf?sequence=1&isAllowed=y

Dale, V. H., Efroymson, R. A., & Kline, K. L. (2011). The land use climate change energy nexus. *Landscape Ecology*, *26*(6), 755–773. doi:10.100710980-011-9606-2

Department of Environmental Affairs (DEA). (2013). Long-Term Adaptation Scenarios Flagship Research Programme (LTAS) for South Africa. Pretoria, South Africa: DEA.

Dippenaar, M. (2018). The role of tax incentives in encouraging energy efficiency in the largest listed South African businesses. *Suid-Afrikaanse Tydskrif vir Ekonomiese en Bestuurswetenskappe*, *21*(1), a1723. doi:10.4102ajems.v21i1.1723

Endo, A., Tsurita, I., Burnett, K., & Orencio, P. M. (2017). A review of the current state of research on the water, energy, and food nexus. *Journal of Hydrology. Regional Studies, 11*, 20–30. doi:10.1016/j.ejrh.2015.11.010

EPA. (2015). *Mississippi River/Gulf of Mexico Watershed Nutrient Task Force 2015 Report to Congress*. U.S. Environmental Protection Agency.

FAO. (2014). *The Water-energy-food Nexus: a New Approach in Support of Food Security and Sustainable Agriculture*. Food Agric. Organ.

Franke, A. C., Haverkort, A. J., & Steyn, J. M. (2013). Climate change and potato production in contrasting South African Agro-Ecosystems 2. Assessing risks and opportunities of adaptation strategies. *Potato Research, 56*(1), 51–66. doi:10.100711540-013-9229-x

Fujisawa, M., New, M., & Johnston, P. (2013). Is Agricultural Sector Listening to Us? African Climate and Development Initiative, University of Cape Town.

Gephart, J. A., Troell, M., Henriksson, P. J. G., Beveridge, M. C. M., Verdegem, M., Metian, M., Mateos, L. D., & Deutsch, L. (2017). The 'seafood gap' in the food-water nexus literature—Issues surrounding freshwater use in seafood production chains. *Advances in Water Resources, 110*, 505–514. doi:10.1016/j.advwatres.2017.03.025

Grafton, R. Q., Libecap, G. D., Edwards, E. C., O'Brien, R. J., & Landry, C. (2012). Comparative assessment of water markets: Insights from the Murray–Darling Basin of Australia and the Western USA. *Water Policy, 14*(2), 175–193. doi:10.2166/wp.2011.016

Groenfeldt, D. (2010). The next nexus: Environmental ethics, water management and climate change. *Water Alternatives, 3*, 575.

Gulati, M., Jacobs, I., Jooste, A., Naidoo, D., & Fakir, S. (2014). The water-energy-food security nexus: Challenges and opportunities for food security in South Africa. *Aquatic Procedia, 1*, 150–164. doi:10.1016/j.aqpro.2013.07.013

Gunkel, G. (2009). Hydropower - A Green Energy? Tropical Reservoirs and Greenhouse Gas Emissions. *CLEAN - Soil, Air. Water (Basel), 37*, 726–734.

Gupta, A. (2017). Water-energy-food (WEF) nexus and sustainable development. Geophysical Monograph Series, 221-241.

Hermann, S., Welsch, M., Segerstrom, R.E., Howells, M.I., Young, C., Alfstad, T., Rogner, H. & Steduto, P. (2012). Climate, land, energy and water (CLEW) interlinkages in Burkina Faso: an analysis of agricultural intensification and bioenergy production. *Natural Resources Forum, 36*, 245-262.

Housh, M., Yaeger, M. A., Cai, X., McIsaac, G. F., Khanna, M., Sivapalan, M., Ouyang, Y., Al-Qadi, I., & Jain, A. K. (2015). Managing Multiple Mandates: A System of Systems Model to Analyze Strategies for Producing Cellulosic Ethanol and Reducing Riverine Nitrate Loads in the Upper Mississippi River Basin. *Environmental Science & Technology*, *49*(19), 11932–11940. doi:10.1021/acs.est.5b02712 PMID:26348783

IPCC. (2013). Summary for Policymakers. In Climate Change 2013: The Physical Science Basis. Contribution of Working Group I to the Fifth Assessment Report of the Intergovernmental Panel on Climate Change. Cambridge University Press.

Jarvie, H. P., Sharpley, A. N., Flaten, D., Kleinman, P. J., Jenkins, A., & Simmons, T. (2015). The Pivotal Role of Phosphorus in a Resilient Water–Energy–Food Security Nexus. *Journal of Environmental Quality*, *44*(4), 1049–1062. doi:10.2134/jeq2015.01.0030 PMID:26437086

Johnston, P., Hachigonta, S., Sibanda, L. M., & Thomas, T. S. (2012). Southern African Agriculture and Climate Change: A Comprehensive Analysis -South Africa. International Food Policy Research Institute (IFPRI).

Keskinen, M., Guillaume, J. H. A., Kattelus, M., Porkka, M., Rasanen, T. A., & Varis, O. (2016). The water-energy-food nexus and the transboundary context: Insights from large Asian rivers. *Water-Sui*, *8*(5), w8050193. doi:10.3390/w8050193

Kunkel, K. E., Easterling, D. R., Kristovich, D. R., Gleason, B., Stoecker, L., & Smith, R. (2012). Meteorological causes of the secular variations in observed extreme precipitation events for the conterminous United States. *Journal of Hydrometeorology*, *13*(3), 1131–1141. doi:10.1175/JHM-D-11-0108.1

Li, X., Feng, K., Siu, Y. L., & Hubacek, K. (2012). Energy-water nexus of wind power in China: The balancing act between CO_2 emissions and water consumption. *Energy Policy*, *45*, 440–448. doi:10.1016/j.enpol.2012.02.054

Markantonis, V., Reynaud, A., Karabulut, A., El-Hajj, R., Altinbilek, D., Awad, I., Bruggeman, A., Constantianos, V., Mysiak, J., Lamaddalena, N., Salah, M., Monteiro, H., Pistocchi, A., & … . (2019). Can the implementation of the water-energy-food nexus support economic growth in the Mediterranean region? The current status and the way forward. *Frontiers in Environmental Science*, *7*, 1–11. doi:10.3389/fenvs.2019.00084

Masipa, T. S. (2017). The impact of climate change on food security in South Africa: Current realities and challenges ahead. *Jamba: Journal of Disaster Risk Studies*, *9*(1), a411. doi:10.4102/jamba.v9i1.411 PMID:29955344

Mekonnen, M. M., Gerbens-Leenes, P. W., & Hoekstra, A. Y. (2015). The consumptive water footprint of electricity and heat: A global assessment. *Environmental Science. Water Research & Technology*, *1*(3), 285–297. doi:10.1039/C5EW00026B

Mekonnen, M. M., & Hoekstra, A. Y. (2016). Four billion people facing severe water scarcity. *Science Advances*, *2*(2), e1500323. doi:10.1126ciadv.1500323 PMID:26933676

Mohtar, H., Assi, T., & Daher, B. (2015). *Bridging the Water and Food Gap: The Role of the Water-Energy-Food Nexus*. Working Paper. United Nations University Institute for Integrated Management of Material Fluxes and of Resources (UNU-FLORES).

Mpandeli, S., Naidoo, D., Mabhaudhi, T., Nhemachena, C., Nhamo, L., Liphadzi, S., Hlahla, S., & Modi, A. (2018). Climate change adaptation through the water-energy-food nexus in Southern Africa. *International Journal of Environmental Research and Public Health*, *15*(10), 2306. doi:10.3390/ijerph15102306 PMID:30347771

Mu, J., & Khan, S. (2009). The effect of climate change on the water and food nexus in China. *Food Security*, *1*(4), 413–430. doi:10.100712571-009-0042-y

Muller, M. (2015). The 'nexus' as a step back towards a more coherent water resource management paradigm. *Water Alternatives*, *8*(1), 675–694.

Nhamo, L., Ndlela, B., Nhemachena, C., Mabhaudhi, T., Mpandeli, S., & Matchaya, G. (2018). The water-energy-food nexus: Climate risks and opportunities in Southern Africa. *Water (Basel)*, *10*(5), 567. doi:10.3390/w10050567

Nistor, M. M., Cheval, S., Gualtieri, A. F., Dumitrescu, A., Boţan, V. E., Berni, A., Hognogi, G., Irimuş, I. A., & Porumb-Ghiurco, C. G. (2017). Crop evapotranspiration assessment under climate change in the Pannonian basin during 1991-2050. *Meteorological Applications*, *24*(1), 84–91. doi:10.1002/met.1607

Pardoe, J., Conway, D., Namaganda, E., Vincent, K., Dougill, A., & Kashaigili, J. (2018). Climate change and the water-energy-food nexus: Insights from policy and practice in Tanzania. *Climate Policy*, *18*(7), 863–877. doi:10.1080/14693062.2017.1386082

Ray, D. K., Gerber, J. S., MacDonald, G. K., & West, P. C. (2015). Climate variation explains a third of global crop yield variability. *Nature Communications*, *6*(1), 5989. doi:10.1038/ncomms6989 PMID:25609225

Rogers, P., Silva, R., & Bhatia, R. (2002). Water is an economic good: How to use prices to promote equity, efficiency, and sustainability. *Water Policy*, *4*(1), 1–17. doi:10.1016/S1366-7017(02)00004-1

Salam, A., Pandey, P., Shrestha, S., & Anal, A. K. (2017). The need for the nexus approach. In P. Abdul Salam, S. Shrestha, V. Prasad Pandey, & A. K. Anal (Eds.), *Water-Energy-Food Nexus: Principles and Practices*. John Wiley and Sons, Inc. doi:10.1002/9781119243175.ch1

Sanders, K. T., & Webber, M. E. (2012). Evaluating the energy consumed for water use in the United States. *Environmental Research Letters*, *7*(3), 34034. doi:10.1088/1748-9326/7/3/034034

Schnier, S., Cai, X., & Cao, Y. (2016). Importance of natural and anthropogenic environmental factors to fish communities of the Fox River in Illinois. *Environmental Management*, *57*(2), 389–411. doi:10.100700267-015-0611-0 PMID:26404430

Scott, C. A., & Sugg, Z. P. (2015). Global energy development and climate-induced water scarcity-Physical limits, sectoral constraints, and policy imperatives. *Energies*, *8*(8), 8211–8225. doi:10.3390/en8088211

Seeliger, L., Clercq, W., Hoffmann, W., Cullis, J., Horn, A., & Witt, M. (2018). Applying the water-energy-food nexus to farm profitability in the Middle Breede catchment, South Africa. *South African Journal of Science*, *114*(11/12), 1–10. doi:10.17159ajs.2018/5062

Shah, T. (2009). *Taming the Anarchy: Groundwater Governance in South Asia*. Routledge.

Shah, T., Scott, C., Kishore, A. & Sharma, A. (2007). Energy-irrigation nexus in South Asia: improving groundwater conservation and power sector viability. *BMJ Clinical Resources, 2*, 992.

Simpson, G., & Jewitt, G. (2019). The development of the water-energy-food nexus as a framework for achieving resource security: A review. *Frontiers in Environmental Science*, *7*(8), 1–9. doi:10.3389/fenvs.2019.00008

Simpson, T. W., Sharpley, A. N., Howarth, R. W., Paerl, H. W., & Mankin, K. R. (2008). The new gold rush: Fueling ethanol production while protecting water quality. *Journal of Environmental Quality*, *37*(2), 318–324. doi:10.2134/jeq2007.0599 PMID:18268293

Smajgl, A., Ward, J., & Pluschke, L. (2016). The water-food-energy nexus-realising a new paradigm. *Journal of Hydrology (Amsterdam)*, *533*, 533–540. doi:10.1016/j.jhydrol.2015.12.033

Smith, C. M., David, M. B., Mitchell, C., Masters, M. D., Anderson-Teixeira, K. J., Bernacchi, C. J., & DeLucia, E. H. (2013). Reduced Nitrogen losses after conversion of row crop agriculture to perennial biofuel crops. *Journal of Environmental Quality*, *42*(1), 219–228. doi:10.2134/jeq2012.0210 PMID:23673757

Song, Y., Cervarich, M., Jain, A. K., Kheshgi, H. S., Landuyt, W., & Cai, X. (2016). The interplay between bioenergy grass production and water resources in the United States of America. *Environmental Science & Technology*, *50*(6), 3010–3019. doi:10.1021/acs.est.5b05239 PMID:26866460

Suen, J. P., Eheart, J. W., Herricks, E. E., & Chang, F. J. (2009). Evaluating the potential impact of reservoir operation on fish communities. *Journal of Water Resources Planning and Management*, *135*(6), 475–483. doi:10.1061/(ASCE)0733-9496(2009)135:6(475)

Turpie, J., Winkler, H., Spalding-Fecher, R., & Midgley, G. (2002). *Economic Impacts of Climate Change in South Africa: A Preliminary Analysis of Unmitigated Damage Costs*. Southern Waters Ecological Research & Consulting & Energy & Development Research Centre, University of Cape Town.

UN. (2015). *Sustainable Development Goals*. Retrieved from: http://www.un.org/sustainabledevelopment/sustainable-development-goals/ 10.18356/90048b6a-en

Van der Waals, Kruger, Franke, Havenkort, & Steyn. (2013). Climate change and potato production in contrasting South African agro-ecosystems 3. Effects on relative development rates of selected pathogens and pests. *Journal of the European Association for Potato Research*.

Vengosh, A., Jackson, R. B., Warner, N., Darrah, T. H., & Kondash, A. (2014). A critical review of the risks to water resources from unconventional shale gas development and hydraulic fracturing in the United States. *Environmental Science & Technology*, *48*(15), 8334–8348. doi:10.1021/es405118y PMID:24606408

Walsh, M. J., Van Doren, L. G., Shete, N., Prakash, A., & Salim, U. (2018). Financial trade-offs of energy and food uses of algal biomass under stochastic conditions. *Applied Energy*, *210*, 591–603. doi:10.1016/j.apenergy.2017.08.060

Water Research Commission. (2018). *Assessing the state of the water-energy-food (WEF) nexus in South Africa*. Available Online: http://www.wrc.org.za/wp-content/uploads/mdocs/KV%20365-18.pdf

Watson, A. (2013). Reduce, reuse, recycle: green technologies and practices at work. *Beyond the Numbers, 2*, 19-31.

Webber, M. E. (2015). A puzzle for the planet. *Scientific American, 312*(2), 62–67. doi:10.1038cientificamerican0215-62

Weitz, N., Strambo, C., Kemp-Benedict, E., & Nilsson, M. (2017). Closing the governance gaps in the water-energy-food nexus: Insights from integrative governance. *Global Environmental Change, 45*, 165–173. doi:10.1016/j.gloenvcha.2017.06.006

Woo, D. K., & Kumar, P. (2016). Mean age distribution of inorganic soil-nitrogen. *Water Resources Research, 52*(7), 5516–5536. doi:10.1002/2015WR017799

Xu, Z., Yin, X., Sun, T., Cai, Y., Ding, Y., Yang, W., & Yang, Z. (2017). Labyrinths in large reservoirs: An invisible barrier to fish migration and the solution through reservoir operation. *Water Resources Research, 53*(1), 817–831. doi:10.1002/2016WR019485

Zeng, R., Cai, X., Ringler, C., & Zhu, T. (2017). Hydropower versus irrigation—An analysis of global patterns. *Environmental Research Letters, 12*(3), 34006. doi:10.1088/1748-9326/aa5f3f

Zhang, C., Chen, X., Li, Y., Ding, W., & Fu, G. (2018). Water-energy-food nexus: Concepts, questions and methodologies. *Journal of Cleaner Production, 195*, 625–639. doi:10.1016/j.jclepro.2018.05.194

Zhu, T., & Ringler, C. (2012). Climate change impacts on water availability and use in the Limpopo river basin. *Water (Basel), 4*(1), 64–86. doi:10.3390/w4010063

Ziervogel, G., New, M., Garderen, E., Midgley, G., Taylor, A., Hamann, R., Stuart-Hill, S., Myers, J., & Warburton, M. (2014). Climate change impacts and adaptation in South Africa. Royal Geographical Society, 605-620.

KEY TERMS AND DEFINITIONS

Climate Change: A state of change in the characteristics of weather such as its precipitation and temperature over a prolonged period due to natural or manmade causes.

Eco-Hydrologist: An individual who specialises in understanding ecological and water interactions.

Integrated Water Resources Management (IWRM): The practice that advocates for planned and collaborative development of land and water resources to optimise their productivity and improve socio-economic welfare sustainably without losing the integrity of essential ecosystems.

Nexus: A link or connection.

Research and Development (R&D): Activities directed towards products and processes improvements through innovation.

Sustainable Development Goals (SDGs): These are global goals, which were agreed upon by UN member states in 2015 and whose focus is to eradicate poverty, protect the environment and safeguard socio-economic welfare.

Water Pricing: The process of valorizing water through a public policy.

ENDNOTES

1 An acronym representing the Southern African development community, which is an organization that fosters integration, cooperation, and socio-economic development in member countries.

2 An organization consisting of SADC member countries with exception of Mauritius that aims at investing in sustainable energy development towards economic growth.

3 The application of resource efficient and environmentally friendly processing in all building lifecycle processes.

4 Mechanisms that promote reduced emission of greenhouse gases.

5 An initiative to enhance better management and development of water resources in the SADC region.

Chapter 9
Industry 4.0-Based Large-Scale Symbiotic Systems for Sustainable Food Security in Namibia

K. S. Sastry Musti
https://orcid.org/0000-0003-4384-7933
Namibia University of Science and Technology, Namibia

ABSTRACT

Climatic changes can cause severe food and water shortages, and desert nations such as Namibia can be challenged more than other countries for obvious reasons. Dependency on imports for food and electricity in Namibia is continuous in recent times. However, Industry 4.0-based large-scale symbiotic systems can potentially help in achieving a sustainable food security regime, as they operate under controlled conditions. Namibia is blessed with abundant sunshine and land availability, and hence, ample opportunities do exist for producing solar energy, which is used to meet the energy requirements of symbiotic systems. This chapter examines typical local operating conditions and then makes a strong case for fully automated symbiotic systems that use low-cost desalination and renewable energies.

INTRODUCTION

Since this chapter focuses on symbiotic systems for Namibia; it is important to understand the Namibian context and climatic conditions.Global warming and changing climatic conditions have impacted Namibia. Over the last decade, Namibia

DOI: 10.4018/978-1-7998-3343-7.ch009

has faced very low rainfall that resulted in drought and food shortage. Farmers have lost crops and livestock due to shortage of water and animal fodder. At present Namibia is a net importer of vegetables, rice and fish from other countries and this leads to food dependency and insecurity. According to the World Food Programme / Namibia,

"Namibia produces about 40 percent of the food it consumes and is highly dependent on imports. This means that while food is available, price fluctuations can make it difficult to access for 28 percent of Namibian families to access food. This particularly affects the 80 percent of the population who depend on markets to fulfill their food needs. Smallholder farmers also have limited access to nutritious food due to recurrent droughts and floods, low productivity and limited access to land."

One of the major challenges is that the land availability in Namibia for agricultural forming is limited due to the fact that Namibia is a desert nation. During 2014 to 2016, Namibia experienced severe food insecurities that mainly affected rural communities due to draught and very low rainfall conditions. Specifically, southern and western parts of Namibia witnessed abnormally low rain during 2015.At the same time northern Namibia experienced heavy rain fall that caused floods, washed away crops and even resulted in soil erosion.These rains started late and there was no significant follow up rainfall to support agricultural production. During 2015, local authorities have carried out a post-harvest crop assessment and found that crop production in the country was estimated to be 46 per cent below average for the 2014/15 cropping season (Namibia Food Insecurity, 2018). In 2019, Namibia has even declared state of emergency due to severe conditions of drought (Reliefweb, 2019) and in fact, the nation has to face such situation three times in six years. The problem of water shortage in Namibia is well-known and Windhoek, the capital cityis one of the severely affected places. In 2019, water shortage in the capital city became even worse as it received approximately 163,000 cubic meters of water against the required demand of 539,000 per week (Namibian, 2019).

Obviously erratic rain fall patterns due to climatic changes have impacted Namibia badly. Then semi-arid and/ or desert nature of the land makes it difficult to source ground water and hence reduced opportunities for traditional farming. The 2013-2020 climate change strategy & action plan has set out four thematic areas, viz., Food security and sustainable biological resource base; sustainable water resource base; human health and well-being; and infrastructure well-being as critical to Namibia. Among these food security and sustainable water resource base are very critical for obvious reasons. A good number of animals have died due to acute drought conditions, Ministry of Finance, had to announcetax exemption on the importation of most fodder products to afford producers the opportunity to import the bulk of the approved products in these challenging times.

Fortunately, there are solutions, thanks to advances in smart technologies for climate friendly farming. Hupenyu et al., (2019) presented the nature of soil conditions in Namibia, and reviewed various methods of climate smart agricultural methods. Large scale symbiotic systems such as aquaponic and /or aeroponic and / or hydroponic systems have been well studied recently and there have been several real world experiences and success stories have been reported. Aquaponic systems grow aquatic animals such as fish and several varieties of plants and operate under controlled setting and use very less water. Joyce et al (2019) stated that

"Hydroponics initially developed in arid regions in response to freshwater shortages, while in areas with poor soil, it was viewed as an opportunity to increase productivity with fewer fertilizer inputs. In the 1950s, recirculating aquaculture also emerged in response to similar water".

Aquaponics was also shown to be an adaptable and cost-effective technology given that farms could be situated in areas that are otherwise unsuitable for agriculture, for instance, on rooftops and on unused, derelict factory sites. A wide range of cost savings could be achieved through strategic placement of aquaponics sites to reduce land acquisition costs, and by also allowing farming closer to suburban and urban areas, thus reducing transportation costs to markets and hence also the fossil fuel and CO_2 footprints of production.

From the above, it can be seen that a) Namibia is indeed affected by climatic changes and thus suffering from food and water shortages b) There is a significant interest in symbiotic systems c) symbiotic systems can potentially solve the food security issues for semi-arid countries like Namibia d) Applying principles and concept of industry 4.0 for symbiotic systems is in the nascent stage, especially for the systems involving different configurations e) There is a need for legal framework, national policies, education, and awareness about impact of climatic changes on food and water availability.

Keeping in view of the above, this book chapter focuses on various issues dealing with symbiotic systems and their applicability to Namibia. In the first place, the book chapter takes up the present food and water shortage problems in Namibia beside its abundant natural resources including land availability and sun shine. Then successful experiences from the literature on symbiotic systems in Africa and beyond will be summarized to propose the possibility of establishing modern, large scale symbiotic systems that include aeroponic, aquaponic and hydroponic sub-systems using industry 4.0 principles to suit the typical Namibian operating conditions. Besides that, a structured literature review of large scale vertical symbiotic structures for selected vegetables, fruits, grains, and variety of fish; with typical controlled climatic conditions will be provided. Various parameters including temperature, pH, flow rates, pathogens and bacteria will be identified and industry 4.0 based mechanisms for monitoring in real-time will be discussed. Application of low-cost desalination and solar energy technologies for powering symbiotic systems will be examined.

OVERVIEW OF SYMBIOTIC SYSTEMS

Benefits of symbiotic or aquaponic systems have been widely reported. Typically, such systems circulate water and feed continuously to symbiotic community of fish, plants and bacteria through a closed loop system. Carefully monitored and controlled symbiotic systems usually provide higher yields when compared to traditional techniques of plant cultivationand/or fish production (Palm et. al., 2019). In fact, aquaponic systems have existed since 1960s and gradually evolved over the time. There are different models and configurations such as single loop, multi-loop, coupled and stand-alone systems. Palm et. al (2019) provided an overview of the historical development of acquaponic systems, various design configurations, treatment of saline and brackish water, proper choices of fish and plants and other process management issues related to coupled aquaponics especially in Europe.

Proksch and Kotzen (2019) have illustrated a few real life, large-scale aquaponic systems from USA, Canada and Europe. Specifically the concept of Aquaponics as Controlled Environment Agriculture(CEA) is explained along with its benefits. Some of those benefits include: efficient use of water efficiency, specifically in large-scale urban farming. A few studies have even quantified the inputs and outputs in typical CEA systems. Typically, conventional systems require large volumes of water, up to 375,000 liters for producing1 kilogram of fish; whereas CEA/aquaponic system requires less than 100 liters of water (Goddek et al., 2015) for the same volume of fish. Sulh et al., (2016) suggested that Double recirculating aquaponic systems(DRAPS)in controlled setting can be more effective. Their study illustrates that the joint use of one cubic meter of freshwater resulted in 46 kilograms of tomato and 1.5 kilograms of tilapia. Even the efficiency of fertilizer use was improved by 23%. There are different types of designs and configurations of large scale aquaponic (or symbiotic) systems, as several factors can influence designs and configurations of overall ecosystems (Proksch and Kotzen, 2019). However, most studies point to the fact that it is prudent to grow tilapia fish as that species has high tolerance to a wide temperature range and also to use shaded aquaculture tanksto prevent overheating and algae growth.

The range of vegetables that are currently proven very suitable to aquaponic systems are: Water spinach, lettuce, water cress, Okra, Chives, Red Amaranth, Pak-choi, Cucumbers, Sweet peppers, tomatoes and eggplants. Rakocy (2012) suggested that 57 grams of feed per day per square meter of lettuce growing area. It should be noted that there are contradicting views on coupled and decoupled systems. A case study from US indicated that raising tilapia fish will lead to net loss; and cultivating vegetables alone will yeild profits. However the study considered the local costs of water, land and energy and thus may not suit very well for other sites.

A good number of research articles around the world have reported on different symbiotic systems on wide ranging issues. A real world investigation on range of vegetables and fruits is carried out in Albania (Bakiu, 2017) and it concluded some interesting facts that favor symbiotic systems. However, the results have also pointed the importance of using regular soil. Analene et al., (2017) from Oman reported on aquaponic systems that are powered by solar photovoltaic systems. This work uses modern technologies such as Internet of Things (IoT) in laboratory scale greenhouse systems and reports the success of producing fish and vegetables all around the year. Another interesting investigation from Kenya (Dijkgraaf et al., 2019) reports scientific modeling approaches to aquaponic systems with symbiotic systems with a biomass digester. Aquaponics in the Mediterranean Region and the requirements for full featured software have been presented by Valentini (2017). Doyle et al., (2018) reported on food and water shortages issues specifically related to East Africa and sub-Saharan Africa. Their work consisted of developing a prototype climate smart system to grow fish and vegetables to suggest as a possible sustainable solution to the food insecurity problem in Africa. Interestingly, all the above and a majority of other experiences are based on funded research or regular university research. On the other hand, smart farming techniques in symbiotic systems have been reported too. Vernandheset.al (2017), Silva et al (2016), Aishwarya et al., (2018), Pasha et al (2018) and a few others have considered the use of IoT and other electronic means to automatically monitor and control the symbiotic systems.

The literature thus far focused on various aspects of symbiotic systems. A few papers have described IoT / industry 4.0 based monitoring; a few focused on system modeling, design and implementation; a significant articles focused on various plants/ fish that can be raised; a few explained challenges in operating; and a few dealt with overall energy costs and economic issues. It can be seen that the very concept of symbiotic systems is highly interdisciplinary in nature (Junge et al, 2019 et al) and that researchers may have different objectives for each of their contributions. Energy and water requirements vary significantly from one geographic location to the other due to differences in temperature, pressure and humidity etc. Even climatic conditions in different seasons do play their role. Contrary to the common belief/ thinking that large scale systems may yield more benefits; a few articles suggested that medium scale ecosystems are more economically promising when compared too large or too small scale systems. Further, most research contributions focused on all-year round operation of aquaponic/ symbiotic systems; and thus did not look into the prospects of operating them over a specific period that may be conducive for plant/ fish growth in the given specific location.

It is important to consider the problem of 'food miles' while dealing the food security problem. The concept of food miles envisages the environmental impact of food supply chain. However, food miles are not always correlated with the actual environmental impact of food production. In comparison, the percentage of total energy used in home food preparation is 26% and in food processing is 29%, far greater than transportation. To counter such criticisms, a thorough result-oriented and constructive approach should be taken by the nations facing food and water shortage problems and put appropriate policy regulation framework in place.

In general, water requirements for raising the cattle (to produce meat) are more than the requirements for a typical symbiotic system. Even the cattle require a lot of grass/feed for proper growth. Wheat grass is an excellent source of food for most animals that are raised for meat, with the exception of few animals such as pigs. Most people may be under the impression that symbiotic systems advocate more on vegetarianism as the emphasis is on plant based produce besides fish. However, symbiotic systems can also produce alternate sources of fodder and in fact, in very high quantities at lower expenses. Several experiences around the world, including Namibia have successfully demonstrated this aspect of producing higher yields of fodder with significantly lower investments and costs. This means, symbiotic systems can potentially empower farmers with higher yields of protein rich fodder at significantly lower investments and costs. This ecosystem may help resolving several issues specifically in Namibia, where scores of animals die in drought conditions and thus putting the farmers to significant financial losses.

In a way, symbiotic systems can be treated as an industrial process and hence most operations can be automated. Some of the challenges such as maintenance of temperature, pH and flow rate of the water, nutrient composition in the system, growth of fish, plants and produce – all can be monitored continuously in real time and necessary control actions can be taken whenever needed. Industry 4.0 principles and methodologies can effectively assist in automating large scale symbiotic systems. However, only a few articles are pointing to this aspect and hence is a good case for deeper investigation.

SYMBIOTIC SYSTEMS FOR NAMIBIA

Applying symbiotic systems to Namibian context can be very interesting due to the factors associated to the local setting. Land availability is very high in Namibia, though it is semi-arid and a majority of the land may not be directly suitable for conventional methods of farming. Another major problem is the availability of water for farming. It should be noted that Namibia has access to sea coast and hence low cost desalination for smart farming is a possibility in principle. In fact, Goddek, (2018) suggested in favor of using desalination techniques for aquaponics for Namibia. This will positively support an excellent diversity of food spectrum, combination of salt water and fresh water fishes. On the hindsight, for Namibian conditions, medium to large scale symbiotic systems are the best way forward to mitigate food shortage problems. For instance, vertical aeroponicsystems utilize less space and even less water when compared to conventional farming. At the same time, regular soil (that is not suitable to farming) need not be used in such vertical smart farming systems. Such systems can be established even in urban setting. There are a few smart farming projects in Namibia that reported recentlyNamibianSun (2017); (Pupkewitz, 2019). Goddek, (2017) reported a 500-1000 m^2 pilot system that is being developed jointly with the state and private sector. However, national level framework for policies and regulations are required to be in place (Goddek, 2017) to support new range of entrepreneurs in this sector. Pupkewitz, the popular business stores in Namibia even organized a workshop to the public to demonstrate how to grow organic vegetables inside sustainable aquaponic systems (Pupkewitz, 2019). All of these sources are pointing aquaponic systems as the solution for the food security problems faced by Namibia.

INDUSTRY 4.0 FOR SYMBIOTIC SYSTEMS

Fourth industrial revolution or Industry 4.0 basically combines several different layers of technologies into one. Specifically, various technologies such as sensors (that measure various parameters), edge side computing technologies (to process the data collected by the sensors onsite itself), communication technologies (to transfer the data to remote servers, mostly through wireless means) and cloud computing technologies (to extract information from the 'huge or big data' collected) and data visualization technologies (to present the information to the end the user(s) in wide ranging and interactive interfaces (Musti et al., 2020). In the case of symbiotic systems, a wide range of sensors do exist in open markets for measuring the key parameters of the ecosystem, such as temperature, relative humidity, moisture and atmospheric pressure. The good news is that the commercial costs of these sensors are only decreasing over the years. Several different sensors also exist to measure water levels, pH of the water and nutrient levels in feed stock tanks etc. When it comes to operational monitoring; water flow and material consumption can be measured with sensors as well. Then sensors also exist for measuring plant side measurements such as leaf wetness, color and texture etc. In other words, there will be several sensors exist in a typical symbiotic system however small or big. Those sensors operate in tandem and in real time to collect data; and then process the data and further communicate the processed data to the central servers. It can be seen that entire system generates a lot of data that results in 'big data' that is likely vary in nature and volumes over the time.

Table 1. Essential components that need to be studied specifically for Namibian context

Component	Sub-component	Specific Purpose/Requirement
Energy	Energy generated from Solar PV	Online Energy Offline Energy (battery storage)
	Energy required for the operation of symbiotic system	• Water circulation for plants • Water purification/ filtering • Desalination of water (if applicable) • Lighting • Space heating or Cooling • Other needs
Water	Overall water input to the system	Quantity of intake water
	Water used for plants/fish	Quantity spent for fish/ plants
	Water used for other purposes	Quantity of water spent for other purposes/ lost in evaporation
Nutrients	Externally added nutrients	Quantity of nutrients for fish and plants
Time	Time for plant growth	Time from seeding to harvest for each type of plant
	Time for fish growth	Time from seedlings to a specific size
	Time between replacement of filtering components	How long the filtering components lost? How much water they can filter?
	Production / operation time in a calendar year	Production and operating times of plants and fish

Hence, the symbiotic systems, with the IoT infrastructure and the huge volumes of data are excellent examples of 'cyber-physical systems'. And thus, they need Enterprise Information Systems to manage various aspects. Though several authors contributed towards IoT based monitoring systems; the big data based or industry 4.0

based systems have not been very well treated thus far. Significant differences exist between mere monitoring systems and industry 4.0 MEIS. The later essentially can provide insights into various key aspects related to the operation of the ecosystems, in this case the symbiotic systems. For instance, it is possible to find solutions for specific questions / aspects as outlined in table 1.

These essential components need to be studied for proper quantification in terms of measurement and overall costs. Big data that is obtained, processed and transmitted from sensors can shed light into details such as quantification and costs associated.And of course, to accomplish the quantification and cost estimation about key operating components of the symbiotic systems, industry 4.0 based EMIS needs to be properly prototyped first and then implemented.

CHALLENGES IN MANAGING SYMBIOTIC SYSTEMS

Though there are significant advantages associated with symbiotic systems, there exist a good number of challenges as well. The following sections deal with those challenges in detail.

Water Shortage – A Major Challenge

As illustrated above, water is indeed a scarce commodity in Namibia. There are laws in place in Namibia that prohibit explicit usage of water for irrigation purposes, keeping in view of water scarcity. Hence, sourcing the required amount of water that too in urban areas will be a major challenge. It is possible to produce water through desalination process, but the conventional desalination process itself can be very expensive. However, a few studies have indicated the possibilities of sourcing the water through low cost desalination methods. Besides suggestions made by Goddek and Keessman, (2018), Power System business development unit of NamPower has also investigated the possibilities in this area (Hoffmann and Dall, 2018). Their study looked into the prospects of combining desalination with concentrated solar power systems. It is to be noted that the location, Arandis(a town in Namibia that has the world's largest open-pit uranium mine and this town is also called uranium capital of the world) town is selected by the Namibian Power Utility, NamPoweras a potential candidate for establishing a solar park.Their study suggests that combined generation of water and electricity is very much viable. However, their study is not entirely specific to supplying water to any symbiotic system and that the system costs may be higher; water quality may be the best to suit mostly human consumption. To produce plant grade water through low cost desalination is somewhat challenging. For, this both reverse osmosis and filtration loops need to carefully designed. Bachoo,

(2016) illustrated several contemporary methodologies and proposed a new models for modeling the both the loops. Hence it is now, high time for further examining the potential use of low cost desalination technologies that operate with solar energy for symbiotic systems; thus making use of vast coast line that Namibia is blessed with.

Indeed, there were a few efforts that resulted in establishing small scale desalination plants solely powered by solar power. One of the notable projects is the joint development initiative between Finland's University of Turku and University of Namibia in setting up 100% solar-powered desalination system that can produce 3000 liters of fresh water in an hour. Prospects of deploying/ using such units for meeting the requirements of properly sized symbiotic systems and the related economic feasibility issues need to be studied.

Water Contamination Challenges

There are several sources that contribute to water contamination. Though there are several bacterium and/or pathogens exist; a few are discussed more than others. E. Coli and Pythium are the two prominent contaminants that were dealt by many researchers. Pythium is a type of parasite that comes under the broad classification of fungi. Most Pythium species thrive on plant ecosystems, which means symbiotic systems can be vulnerable with Pythium infestation. Pythium can spread itself by moving water, small water puddles, wet moving equipment (lawnmowers) and even with animals. Water logged soils with pH higher than; and/or over-fertilized (typically more nitrogen and lower calcium levels) do support the growth of Pythium infestation that generally leads to root rot in most plant systems.

Escherichia coli (E. Coli for short) is one of the several variant strains the bacterium Escherichia coli. Most variants live in the intestines of animals and humans who may be healthy, without any apparent symptoms. Though majority of these variants can be harmless, some of them are proven to be very powerful toxins that can cause severe illness in humans. E. coli is commonly found in water bodies that have been contaminated with feces from infected humans or animals. There may be several ways for water bodies to get contaminated with E. Coli. Some of the possibilities include: good water sources getting mixed up with water from sewerage system through leakages; contaminated flood waters, water from shallow, private wells etc.

The two major problems viz., Pythium or E. Coli contamination have been flagged by several authors. However, it is important to note that these problems do exist in traditional farming also. A few (less number of) studies may indirectly suggest that rigidly protected/ controlled environments such as greenfarms or symbiotic systems suffer more from Pythium and/ E. Coli, when compared to open farming systems; due to their inherent closed loop operating conditions. Beside the above two major contaminating aspects, there are a few other challenges as well in the form of –

Algae, Antinomies, Nematodes, Meloidogynes, Heterodera. All these aspects will impact quality of produce in varying degrees depending on severity of infestation.

INDUSTRY 4.0 BASED TECHNOLOGIES FOR DETECTING AND MITIGATING THE WATER CONTAMINATION

Interestingly the world IoT, with recent advances in sensor technologies; has its own impact in terms timely detection of soil born deceases. Climent et al., (2017) proposed a new low-cost, portable electronic nose instrument with the name MOOSY4. The novelty of this contribution is that, this instrument takes care of the full cycle of data – from measurement to processing and to pushing the information to the user interface. This device is essentially designed towards Sulphur based water contamination. Sherchan et al., (2018) have highlighted the aspects of bacteria's abilities in revoking their repair mechanisms in response to ultra-violet treatment of the water. Their work evaluated the performance of three different sensors for detecting water born bacteria. The work examined water from different sources and from laboratory to field levels. Their work points towards the successful use of low cost sensors, though off-line total in identifying the microbial load, including non-culturable microbes. Kim, (2015) suggested a remote monitoring and detection of harmful bacteria using aIoT based sensory system that uses a webcam for visual feed. Then the sensor technology itself is leaping forward in different directions. MIT's research in this area resulted in field realization of the concept 'bacteria-on-a-chip'; which is considered as a ground breaking event in the area of synthetic biology. This research looks into the possibilities of combining new range of sensors that are made out of living cells with ultra-low-power electronic components. Such sensors expected to travel along the flow of medium (typically in human stomach) and primarily look for the presence of bacteria that may react to the existing living cells inside the sensor.

Nutrient film technique (NFT), Ozone treatment, Hydrogen Peroxide, ultra-violet treatment (prominently) etc have been used besides natural remedies such as heating up the water (thus killing the bacteria to the possible extent), several stages of filtering (to trap micro level bacteria and pathogens) and open air aerating of the water (to infuse oxygen) have been used to deal with various harmful bacteria and pathogens. The combination of natural remedies specifically water heating, aerating and filtering need to be deeply investigated to combat the menace of water contamination. Energy requirements for heating and aerating the water can be derived from solar energy during most parts of the day in Namibia due to the local climatic conditions. However, it should be noted that the problem of water contamination is not really unique to greenhouses and/or symbiotic systems; indeed this exists in the regular open farming as well. In other words, the symbiotic systems cannot be viewed with a negative outlook due to this problem. Rather, it is the time to explore and investigate the means of combating the water contamination through appropriate research by taking the advantage of recent developments in both sensory technologies and prevention methodologies.

THE FOOD-MILES CHALLENGE

The 'food-miles' is another challenge that is raised by many researchers. Proksch and Kotzen, (2019) provided clarification on the issues related to transportation or food miles. In case of Singapore, nearly the entire food needs to be imported and thus it makes sense to opt for urban CEA to cut the transportation chains and thus mitigating the food miles issues. The strategy cannot be applied to Spain, where supply chains are much shorter, between the farms to the city. Even cities like Boston and Stockholm the benefits due to urban CEA are insignificant, when compared to the benefits from transporting the food through urban transport mechanisms. In case of Namibia, a majority of the food needs to be imported from different neighboring, but significantly from South Africa. Though on the hindsight, the situation of Namibia may favor urban CEA, rigorous studies are required to estimate the impact of food miles and then benefits and comparative studies should be undertaken.

OPERATIONAL CHALLENGES IN SYMBIOTIC SYSTEMS

Managing the temperature inside the symbiotic systems is a major challenge. Either to heat and or to cool the large space, significant amount of energy (typically electricity) will be required. However, there are different ways to manage the challenge of maintaining the temperature within the required range. First and foremost solution is to produce solar energy within the symbiotic system. Roof-tops can be used to capture sunlight and to produce the required energy and even beyond.

In the case of Namibia, due to its abundant availability of sunshine and land availability (of course not in urban areas), it is much easy to add solar panels to produce energy. It should be noted that overall energy prices per kilowatt in case of solar energy segment are decreasing rapidly over the last few years. Even the battery storage technologies are becoming cheaper. These factors will naturally result in cheaper electricity, albeit there will be initial expenses to set up the solar plant. Even it is possible to store heat energy in large water tanks that are partially submerged in the ground. New energy regulatory framework supports modified single buyer model, wherein financially lucrative options are available in renewable energy and trading the same. This aspect of Namibia now makes it an attractive investment destination in the energy area.

Once electrical energy is available, at cheaper prices then it is possible to invest in water flow processing and control mechanisms and space heating or cooling processes. Besides the option of solar energy based capacity addition, evaporative cooling and fog cooling are well known strategies for effective temperature control. However, evaporative cooling is very effective in dry climates but requires high water use, which may be a limitation tofarms in arid areas of the world. On the other hand, fog cooling is more friendly as far as water usage is concerned. This technique basically sprays misty water with overhead sprinklers/misters until the temperature reaches within acceptable range. Though fog cooling increases the relative humidity of a growing space, saves a lot of water (Proksch et al., 2019).

EDUCATIONAL AND SKILLS-TRAINING NEEDS FOR SYMBIOTIC SYSTEMS

Primary and secondary schools do teach biological sciences to an extent. It is important to teach the basics of plant and animal ecosystems at the fundamental stage. One of the major problems in Africa and by extension in Namibia is relatively higher school dropout rates. To make the schools interesting, action and contributory based activities are important. Many schools across the world have undertaken raising plants (popularly known as plant walls or wall gardens) as an activity. The outcome of this activity (or project) will result in a big wall laden with various small sized plants. In most cases, disposed plastic bottles are used for holding the plants.

Under this activity, students are typically asked to do the desktop research to view different types and configurations of wall gardens. They are asked to collect used (waste) plastic bottles and then put a plant in it. In secondary schools students can even be asked to design and implement the water lines with PVC pipes. Such projects basically make the student community to involve more in displaying their talents, as it leads to an experience of witnessing a natural growth of something they themselves created. This in turn may lead to mitigate school dropout rate to an extent; however this suggestion needs to be verified through action based research, specifically in Namibian context. However, several benefits can be seen on the hindsight as this connects humans to the natural ecosystems.

At tertiary/ university level, a lot more elaborate efforts should be in place, specifically in terms of action research in the areas of simulation, design and analysis of various aspects related to symbiotic systems; besides raising just wall gardens. One may wonder why university or tertiary level students only should take up such research (and why not schools?). As stated above, advanced research into symbiotic systems do require – software simulators (either commercial or open source), real, full scale prototypes of symbiotic systems, IoT equipment, energy apparatus to power up the prototypes etc may be most likely to be available only in the university systems (and not in the schools). Tertiary or university level students are skilled enough to operate such systems or even may be willing to learn and acquire the skills required for such projects. Aspects of safety can be better understood and followed by tertiary level students, though they may require specific training in handling the apparatus. In engineering schools, even students may be asked to develop software and prototype/ tools or systems to assist and/or promote symbiotic systems.

EFFORTS IN NAMIBIA TOWARDS SYMBIOTIC SYSTEMS

This section illustrates a few projects and activities that were undertaken in Namibia and are very much related to the current discussion of large scale symbiotic systems.

There are a good number of organizations and firms in Namibia that support good causes of sustainable and ecofriendly farming. Another example is a non-profit organization that trains youth in setting up rainwater harvesting systems and nitrogen fixing tress systems. Another organization, Namibia Future Farming (NFF) has a more vibrant setting as it operates more than ten aquaponic sites across Windhoek. NFF has installed a few aquaponic systems in different schools in the capital city of Windhoek, Namibia. Fish, fruits and vegetables are being produced in these systems. NFF claims (Travel news Namibia, 2016) that these aquaponic systems use only 10% of the water that is normally required for regular, soil-based farming. These systems recycle the water on continuous basis in a continuous, closed loop and thus avoid significant loss of water, with the exception of losses due to evaporation and transportation. Plants are used as natural filters for the sediments in the water. Interesting point here is that Travel News Namibia (2016) states that

"Aquaponics is ideal for Namibia, as the climate and environmental conditions do not limit the operation of these systems. Not only can aquaponics be beneficial to agriculture, but it can also be a solution to the problem of malnutrition in the country"

The significant aspect is that, this project has received a liberal funding of N$ 2 million from the Finnish embassy and other donors. The Dagbreek area is located in urban setting and is facing water shortages. The students who stay at hostel consume the fruits and vegetables and the excess is sold in the markets. This project serves as proof of concept and is a testimony for the sustainable urban farming.

Agritech Namibia presented a complete cycle of fodder development project models (Meatco, 2019). Under these models, it is possible to grow fodder just under ten days, that too without use of soil. The model grows fodder either from barely or maize and cites the successful experiences of other countries including Brazil, Australia and Japan. The project model uses a substrate instead of regular soil and fed by controlled flow of water, mixed with typical nutrients. The typical yield in this regime are estimated around 400 kilograms of fodder for 40 kilograms of seeds. Another important project is “Eloolo Permaculture Initiative”, which advocates “working with the nature, as opposed to working against the nature. This is a group of environmentalists in food systems and specialist in projects that deal with plants, animals, buildings and infrastructure. Under its umbrella, a garden was put up at a primary school as a community project. They even published a handbook with guidelines that is simple to follow.

IMPACT OF GLOBAL AND BEST EXPERIENCES AND RESPONSE

The Brazilian city, Belo Horizonte is well-known for its initiatives for beating hunger. The Brazilian national policy and strategy on zero-hunger is well implemented in the Belo Horizonte city with the help of more than twenty (20) different programs. The city has achieved the success through a) careful and protracted supervision of all programs by specialized teams drawn from various concerned departments b) multi-stakeholder policy involving public, private and government c) legislation and regulations in place to support food security regime at the national level. Belo Horizonte is able to reduce child mortality, reduce malnutrition in children and also in adults, provide higher and stable income to farmers. No wonder, the city has won the “Future Policy Award” from World Future Council. This success and experience of the Brazilian city not only serves as a model for many parts of the world; but also can potentially reduce the food miles; and thus in mitigating the greenhouse gases.World Future Council, the City of Windhoek, the City of Belo Horizonte and the Food and Agricultural Organization of the United Nations (FAO) have organized a workshop in 2014 in Windhoek. Following which, memorandum of understanding was signed to work together with specific focus on urban agriculture and municipal food banks. However, it should be noted that city of Windhoek itself faces severe water shortages frequently. At the same time, water demand of the city is continuously growing, but the supply sources are crippling due to adverse impacts of climatic changes. Hence, it is not recommended to use current water resources for symbiotic systems. Significant efforts should be made towards adding solar powered desalination plants near the coastline and then establish the symbiotic systems close to those units. This requires location identification and cost benefit studies. Alternately, additional pipeline infrastructure needs to be added to transport water to urban areas that are far from the coastline. However such initiative may involve higher capital investments, but should be considered as part of long term planning. To enable the good initiatives in large scale symbiotic systems, appropriate legislation, careful monitoring and other factors as seen in the case of Belo Horizonte; need to be in place. However, Windhoek and by extension, all the other cities need to take appropriate initiatives in time to move the nation forward in terms of realizing food security. There are excellent examples from different parts of the world to follow.

It can be seen from the above that a good number of projects are already in action and few more are in pipeline presently in Namibia. The problems with raising costs of supplementary fodder products can be effectively resolved through hydroponics.

PROSPECTS OF ECO-TOURISM WITH SYMBIOTIC SYSTEMS

A few authors have pointed out the significant interest of the public in visiting the symbiotic systems and the related prospects of green tourism. Modern tourists value environmental friendly eco-systems, such as greenhouse systems. Several holiday destinations in the world actively advertise about their massive, lush green gardens as tourist attractions. Large symbiotic systems can provide such strength as well, if designed carefully and maintained properly. Within the country, consumer satisfaction and perception studies need to be conducted for the sale of aquaponic products. Recent studies suggest that there is a higher public support to symbiotic systems.Major reasons for such perceptions include - increased awareness about climatic changes and related impact on availability of food; curiosity of the source of the food that they consume etc. In case of Namibia, a majority of tourism is about visiting sand dunes, vast coast lines and desert conditions; besides historical places and monuments. Introduction of large scale symbiotic systems will provide the tourists with a completely contrasting experience within the same country. In other words, tourists will have opportunities to visit greener symbiotic systems to arid desert conditions in one go.

WAY FORWARD

From the above sections, several points can be summarized and they are, not in any specific order of merit: a) low cost, solar powered desalination plants (that can produce water and electrical power) need to be designed specifically for medium to large scale symbiotic systems. This alone requires significant efforts from engineers, industry and entrepreneurs. In the absence of such systems, symbiotic systems may deplete water, a resource that Namibia is carefully conversing at present. b) Real-time monitoring and control of symbiotic systems indeed require state-of-the art Industry 4.0 techniques. Though there are several systems exist, symbiotic systems will have different requirements such as monitoring the eco-systems for possible pathogens and contaminations. c) Big data harvested out of symbiotic systems within Namibia, over a period of time can result in several useful findings. For instance, such findings can help in designing and operating a symbiotic system, elsewhere in Namibia d) Investigations into number of symbiotic systems, their sizing/ scaling, selecting a proper configuration and even geo-locating within Namibia, while taking into various resources, costs, needs and requirements of food and food miles; will be of significant interest. e) Schools and universities may have to identify their own role and responsibilities in promoting the awareness in this area; while the policy makers and regulators may have to look into the legal framework to encourage stakeholders towards establishing symbiotic systems to resolve food insecurity issues.

CONCLUSION

Symbiotic systems have significant environmental benefits that include – lower water consumption; no or less use of fertilizers, pesticides; reduced impact of food miles, sustainable food security. These benefits can be achieved along with other financial benefits such as higher yields for the same area cultivated; better quality of the produce and lesser labor/ operating costs. Industry 4.0 based, EMIS systems help in better managing the modern symbiotic systems. The natural climatic conditions of Namibia, though may not favor traditional farming, it is possible to achieve sustainable food security through symbiotic systems.

REFERENCES

Aishwarya, K. S., Harish, M., Prathibhashree, S., & Panimozhi, K. (2018). Survey On Automated Aquponics Based Gardening Approaches. *2018 Second International Conference on Inventive Communication and Computational Technologies (ICICCT)*, 1377–1381. 10.1109/ICICCT.2018.8473155

Pappa, Kapsis, Mente, & Berillis. (2017). Aquaponics Software in Greece. *Journal of Fisheries*, *11*(3), 1–4.

Bachoo & Sastry. (2016). Tensor product model for a Reverse Osmosis system. *IEEE ANDESCON*, *1-4*. doi:10.1109/ANDESCON.2016.7836195

Bakiu, Tafaj, & Taci. (2017). First Study about Aquaponic Systems in Albania. *Journal of Marine Biology and Aquaculture Research, 1*(1).

Climent, Pelegri-Sebastia, Sogorb, Talens, & Chilo. (2017). Development of the MOOSY4 eNoseIoT for Sulphur-Based VOC Water Pollution Detection. *Journal of Sensors, 17*(8). Advance online publication. doi:10.339017081917

De Silva, P. C. P., & De Silva, P. C. A. (2016). Ipanera: An Industry 4. 0 based Architecture for Distributed Soil-less Food Production Systems. *2016 Manufacturing & Industrial Engineering Symposium (MIES)*, 1–5. https://doi.org/10.1109/MIES.2016.7780266

Dijkgraaf, K. H., Goddek, S., & Keesman, K. J. (2019). Modeling innovative aquaponics farming in Kenya. *Aquaculture International*, (May). Advance online publication. doi:10.100710499-019-00397-z

Doyle, L., Oliver, L., & Whitworth, C. (2020). Design of a Climate Smart Farming System in East Africa. *2018 IEEE Global Humanitarian Technology Conference (GHTC)*, 1–6.

Era, N. (2016). *Age of aquaponics dawns on Namibia*. Available online: https://neweralive.na/posts/age-aquaponics-dawns-namibia

Goddek & Keesman. (2017). *Aquaponics - climate smart solutions to enhance food security in Namibia*. Wageningen University and Research.

Goddek, S., & Keesman, K. J. (2018). The necessity of desalination technology for designing and sizing multi-loop aquaponics systems. *Desalination*, *428*, 76–85. doi:10.1016/j.desal.2017.11.024

Hoffmann, J. E., & Dall, E. P. (2018). Integrating desalination with concentrating solar thermal power: A Namibian case study. *Renewable Energy*, *115*, 423–432. doi:10.1016/j.renene.2017.08.060

Joyce, A., Goddek, S., Kotzen, B., & Wuertz, S. (2019). Aquaponics: Closing the Cycle on Limited Water, Land and Nutrient Resources. In S. Goddek, A. Joyce, B. Kotzen, & G. Burnell (Eds.), *Aquaponics Food Production Systems*. Springer. doi:10.1007/978-3-030-15943-6_2

Junge, R., Bulc, T. G., & Anseeuw, D. (2019). Aquaponics as an Educational Tool. In Aquaponics as an Educational Tool. https://doi.org/ doi:10.1007/978-3-030-15943-6

Keesman, K. J., Körner, O., Wagner, K., Urban, J., Karimanzira, D., Rauschenbach, T., & Goddek, S. (2019). *Aquaponics Systems Modelling*. doi:10.1007/978-3-030-15943-6_11

Kyukwang, K., & Myung, H. (2015). Sensor Node for Remote Monitoring of Waterborne Disease-Causing Bacteria. *Journal of Sensors, 15*(3), 10569–10579. doi:10.3390150510569 PMID:25951340

Meatco, N. (2019). *Hydroponics Fodder Production: a glimmer of hope for farming*. Available online: https://www.meatco.com.na/news/559/Hydroponics-Fodder-Production-a-glimmer-of-hope-for-farming/

Ministry of Environment and Tourism. (2013). *National Climate Change Strategy & Action Plan 2013 – 2020*. Available online Namibia Food insecurity https://reliefweb.int/report/namibia/namibia-food-insecurity-emergency-plan-action-final-report-operation-n-mdrna009

Mupambwa, Hausiku, Dakarai, Nciizah, & Dube. (2019). The unique Namib desert-coastal region and its opportunities for climatesmart agriculture: A review. *Cogent Food & Agriculture, 5*(1)., https://doi.org/ doi:10.1080/23311932.2019.1645258

Nagayo, M., Vega, & Rodrigo. (2017). An Automated Solar-Powered Aquaponics System towards Agricultural Sustainability in the Sultanate of Oman. *IEEE International Conf. on Smart Grid and Smart Cities*, 1–8

Namibian. (2019). *Windhoek declares water crisis*. https://www.namibian.com.na/188436/archive-read/Windhoek-declares-water-crisis

NamibianSun. (2017). *Aquatics the answer*. Available online, https://www.namibiansun.com/news/aquaponics-the-answer

Palm, H. W., Knaus, U., Appelbaum, S., Strauch, S. M., & Kotzen, B. (2019). Coupled Aquaponics Systems. In S. Goddek, A. Joyce, B. Kotzen, & G. Burnell (Eds.), *Aquaponics Food Production Systems*. Springer. doi:10.1007/978-3-030-15943-6_7

Pasha, A. K., Ramdhani, M. A., & Mulyana, E. (2018). System Design of Controlling and Monitoring on Aquaponic Based on Internet of Things. *2018 4th International Conference on Wireless and Telematics (ICWT)*, 1–5.

Proksch, G., Ianchenko, A., & Kotzen, B. (2019). *Aquaponics in the Built Environment*. Aquaponics in the Built Environment. doi:10.1007/978-3-030-15943-6_21

Pupkewitz. (2019). *PupkewitzMegaBuild Garden Talk: Introducing Aquaponics*. Available on line, https://www.megabuildonline.com/articles/media-centre/news/46

Reliefweb. (2019). *President declares state of emergency over drought*. https://reliefweb.int/report/namibia/president-declares-state-emergency-over-drought

SastryMusti. (2020). *Industry 4.0-Based Enterprise Information System for Demand-Side Management and Energy Efficiency*. Novel Approaches to Information Systems Design. doi:10.4018/978-1-7998-2975-1.ch007

Scattini, N., & Maj, S. P. (2017). *Aquaponics – A Process Control Approach*. https://doi.org/ doi:10.5539/mas.v11n11p43

Sherchan, Miles, Ikner, Yu, Snyder, & Ian. (2018). Pepper Near Real-Time Detection of E. coli in Reclaimed Water. *Journal of Sensors*, *18*(8). Advance online publication. doi:10.339018072303

Stouvenakers, G., Dapprich, P., & Massart, S. (2019). Plant Pathogens and Control Strategies in Aquaponics. Plant Pathogens and Control Strategies in Aquaponics. https://doi.org/ doi:10.1007/978-3-030-15943-6

Suhl, J., Dannehl, D., Kloas, W., Baganz, D., Jobs, S., Scheibe, G., & Schmidt, U. (2016). Advanced aquaponics: Evaluation of intensive tomato production inaquaponics vs. conventional hydroponics. *Agricultural Water Management*, *178*, 335–344. doi:10.1016/j.agwat.2016.10.013

Travel News Namibia. (2016). *Aquaponics; the way forward for Namibian agriculture?* Available online, http://www.travelnewsnamibia.com/news/aquaponics-way-forward-namibian-agriculture/

Vernandhes, W., Salahuddin, N. S., Kowanda, A., & Sari, S. P. (2017). Smart Aquaponic with Monitoring and Control System Based On IoT. *Second International Conference on Informatics and Computing (ICIC)*, Jayapura, Indonesia. 10.1109/IAC.2017.8280590

World Food Programme / Namibia. (n.d.). https://www1.wfp.org/countries/namibia

World Future Council. (2016). *Growing food in Windhoek, Urban agriculture policy and practice*. Available online: https://www.worldfuturecouncil.org/wp-content/uploads/2017/03/Food-Handbook-final-web.pdf

KEY TERMS AND DEFINITIONS

Aquaponics: An ecosystem that combines conventional aquaculture (water born animals such as fish and snails) and plants with or without soil in a symbiotic environment.

Big Data: Is a field that deals with analysis, information extraction and visualization from very large data sets that are too complex to be processed by traditional systems.

Climate Change: Set of changes that occurs due to long and unexpected changes, such as global warming.

Industry 4.0: It is a part of the fourth industrial revolution that combines the modern concepts such as IoT, cloud computing, big data, and information systems.

Internet of Things (IoT): Is a system of sensors and computing devices that communicate wirelessly to transfer the data and information to central servers without human intervention.

Symbiotic Systems: An ecosystem in which two or more different biological organisms such as fish and plants, coexist through a closely controlled environment with mutual dependence.

Chapter 10
Assessing Gender Equality in Climate Change Advocacy Campaign for Sustainable Agricultural Food Security in Uganda:
Gender Equality in Climate Information Services for Agriculture in Africa

Wilson Okaka
Kyambogo University, Uganda

ABSTRACT

This chapter examines climate change and variability emergency disaster risks on agricultural food security of the local communities in Africa with a focus on gender equality lens in Uganda. Ugandan women contribute up to 75% of domestic food production and yet they are often overburdened with reproduction, household management, gender-specific discrimination, and adverse climate change effects like agricultural droughts, flash flooding, violent windstorms, or water stress. To ensure sustainable food security in the face of climate change vulnerability risks, the role of women is vital. Communication strategy to promote local climate information service (CIS) delivery system has been developed by the local government district planners in the park areas, but there is a lack of capacity to raise public awareness of the gender equality for the empowerment of women and girls for sustainable food security through agriculture production in Uganda for enhanced livelihood assets.

DOI: 10.4018/978-1-7998-3343-7.ch010

INTRODUCTION

This chapter discusses the need for an effective gender equality communication campaign in creating, raising, developing, monitoring, and sustaining public awareness of gender equality and equity mainstreaming in climate change information services for sustainable development goals (SDGs) in Africa with a focus on Uganda. According to the United Nations Development Programme (UNDP) and Uganda government (2005 & 2010), gender equality promotes gender-specific equal enjoyment by both women and men or boys and girls of socially valued goods, rights, opportunities, benefits, or resources with a focus on all spheres of sustainable social, economic, environment, legal, and political development. It is imperative that public awareness of all the practical and strategic gender equality issues are ruthlessly created, raised, developed, and sustained at every doorstep across Africa. Awareness of gender equality must reign deep, high, wide, or broad across-cultures and all demographics.

Evidence of good practice proves that mainstreaming the gender equality communication campaign promotes sustainable community awareness, education, information, entertainment, and empowerment (Okaka, 2013). Policy advocacy communication campaign that infuses gender equality in climate smart-agriculture is, highly recommended because the strategy enhances local community based adaptation with a focus on gender equality in small-holder agriculture. This, in turn, empowers women and girls to produce more food for cash and sustainable livelihoods (GU, 2015). At the same time, it promotes women's human rights, eliminates the so-called gender role which limits women's land ownership for sustainable agriculture and household food security as envisaged by the 2030 SDGs (UN, 2015). Likewise, the Ugandan national climate change policy promotes awareness raising and the role of women in implementing national and decentralized adaptation and mitigation priority areas like agriculture, food, and nutrition security; land use, socio-economic infrastructure, biodiversity, forests, tourism, settlements, and water in tandem with the East African Community (EAC) regional policy that stresses climate change adaptation or mitigation of emissions (EAC, 2011& GU, 2015).

In addition, mainstreaming gender equality communication campaign data are usually collected in order to measure environmental performance, measuring environmental performance is important for a number of reasons, namely: to provide feed-back on system behavior and policy performance; to improve chances of successful adaptations; to improve implementation; and to increase accountability (NEMA, 2003, 2014). A range of communication options should be considered, including: those largely oriented to print, those that can be delivered through radio and television, internet-based reporting, and use of alternative communication tools such as posters, discussions, theatre and songs. Data collection based on gender specificity is a critical indicator of climate charge development management in Africa and in Uganda, for example, data are usually collected the National Meteorological Authority to measure climate charge performance should be desegregated to reflect specific gender inputs ethically (GU, 2015).

GENDER EQUALITY IN CLIMATE CHANGE EMERGENCY

A gender equality issue involves three main elements: gender discrimination, gender gaps and gender oppression. Gender issues affect access to and control over resources, benefits and opportunities within every sector. Gender equality issues affect the success and sustainability of development plans and programmes within all sectors. In Africa, key gender issues are: family life, employment, decision making, education, health, and agriculture. Public communications campaigns should be guided by informed ethical, theoretical and gender concerns. The key targets and indicators of SDG 5 on gender equality and empowerment of all women and girls are expected to speed up progress in the African regional climate change adaptation and mitigation policies. These obviously assume gender parity as elaborated on in the UN SDG 5 and the relevant targets (UN, 2015).

According to the salient provisions of the SDG 5 targets on gender equality for the empowerment of all women and girls in Uganda the country needs to eliminate all forms of discrimination against women and girls, stop all forms of violence against them everywhere, enhance full participation of women in all spheres of opportunities like leadership and decision-making at all levels of social, political, economic, and environment development (UN, 2015). In contrast, with their male counterparts in the country, Ugandan women normally suffer triple jeopardy in the frequent and catastrophic climate-induced hazards: like torrents of flash floods, hydrological and agricultural droughts, scorching heat waves, landslides, wind storms a spate of lightning, wild bush, and forest fires (GU, 2015). Most of the main human vulnerabilities and livelihood impacts in the areas are (GU, 2015): reduced agricultural production, water shortage, fuelwood crisis, groundwater depletion, increased disease and conflicts, male dominate out migrations, public health problems, intensive food security, crop failures, loss of critical natural biodiversity eco-services and forest areas, land conflicts, loss of land, degradation, and land-use changes, In their responses, the local communities in the areas have been involved in the following adaptation and mitigation projects:

- Climate change and development planning project
- Community tree growing project
- Community water and sanitation project
- Drought adaptation project

- Indigenous knowledge (IK) and natural resources management project
- Land degradation management project
- Strengthening meteorological services
- Vectors, pests and disease control project
- Water for production project

Today, it has also emerged that some of the needs or lessons learned from the effective climate change information interventions call for a variety of coordinated advocacy campaign strategiess (Okaka, 2012): At the same time, the UN Economic Commission for Africa (UNECA), the African Union Commission (AUC), the African Development Bank (AfDB), and the African Climate Policy Centre (ACPC) have combined their efforts to build local, institutional, national, sub-regional, and regional capacities in climate information gathering, dissemination for adaptation, and mitigation practices to curb the challenges of climate change emergency risks and vulnerabilities with a focus on gender equality for women empowerment and key interventions have been launched(Kofi, 2014).

The regional strengthening of climate information services on the continent will enhance the development of the vital resources like water, land use, energy, food security, and more opportunities for the majority of the African youth and women who have the potential to maximize Africa's agricultural production value chain for regional competitiveness (Denton, 2014). Likewise, unfettered dissemination of usable climate change information service to the local small-holder farmers will raise awareness levels on the opportunities and benefits of the role of gender equality advocacy in empowering the women farmers (EAC, 2011& GU, 2915).

The preliminary findings of a Ugandan national study on communicating climate change issues in the country revealed that many stakeholders (32%), reported that there was low access to climate change information; 20% said information availability was very low, and only 12% of them had high access to climate information services in the country(GU, 2015). At the same time, it has been reported that the current vulnerability status of social and economic systems depend largely on the quality and quantity or standard of our economic and institutional infrastructure in the climate change emergency (IPCC & UNFCCC, 2014). In addition, the promotion of equal participation of men and women in climate change emergency adaptation priorities should focus on freshwater for all purposes, biodiversity and efficient natural resources management conservation, and infrastructure building; institutional capacity arrangements; effective weather and climate monitoring systems for floods or droughts; rehabilitation of watersheds, and construction of new reservoir capacity (EAC, 2011&GU, 2015). The national adaptation priority for Uganda and other East African states targets agriculture, especially regarding the changes, crop varieties, efficient water resources management, irrigation farming systems, smart-agriculture good practice, and changes in planting or tillage (EAC, 2011&GU, 2015).

The current known benefits of limiting greenhouse gas emissions and enhancing sinks are: the climate change damages and adaptation costs are curbed and the economic and environmental benefits of the relevant policy innovation for climate change response are enhanced. In Uganda, the key objective for achieving sustainable development is contingent upon reviving meaningful inclusive growth and development countrywide (UNDP, 2005 One of the practical ways of attaining inclusive development and growth is through sharing of information. However, a recent national survey conducted in Uganda found that 48% said they experienced low levels of climate change information services, 44% reported very low access to climate change information sharing, and worse still, up to 54% of the people asked said experienced low access to climate change information for planning and decision making as opposed to 10% of stakeholders who had very high access (GU, 2015). The national (MNR, 2000) environment policy for Uganda stresses the essential need to address the implications of gender roles in developing environment and natural resource management policy. The objective is to integrate gender concerns in environmental policy planning, decision making, and implementing at all levels to ensure sustainable social and economic development in Uganda and all the other 53 African Union states.

The common adage is that sustainable development is focused on people at the center. This calls for the infusion of key ethical and gender issues in the following pillars of the human development paradigm based on productivity, equity, sustainability, cooperation, and security. People must be empowered to increase their inputs/outputs for gainful income generation and remuneration employment. Economic growth is a subset of the human development model based on gender mainstreaming in decision-making. Uganda's vulnerability and adaptation assessments indicate that the poor, most being women and children, are the most vulnerable to climate change impacts. The greater vulnerability of women is mostly due to gender inequality (GoU, 2013).

PROSPECTS FOR PRACTICAL PROPOSED SOLUTIONS

To begin with, discrimination against women is any distinction, exclusions or restrictions made based on sex, which has the effect or purpose of impairing both the spirit and letter of the Convention in their own countries (CEDAW, 1999). Discrimination is one of the key gender and ethical issues listed in the article. The convention obligates the states to, institute the appropriate constitution, policy, legal, institutional, and administrative reforms or initiatives in order to implement the letter and spirit of the document. States should address the development and advancement of women, acceleration of equality between men and women, sex role and stereotypes, suppression of the exploitation of women, political and public life. Its perpetuity grossly undermines SDG 5 and SDG 13 progress in Africa.

To clarity, gender equality media awareness campaigns are very critical for effective climate policy, research, conventional, and legal implementation; they are vital for policy and decision-makers themselves; the campaigns provide information and knowledge for the empowerment of the intended people; they are important for policy implementers and key stakeholders as they act as guides, and the professionals or the campaigners rely on them as they provide more insight and raw materials for better campaigns strategies.

This time around, the entire SDGs must be guided by gender in all the units of data collection (component by component). Time is up for costly gender blunders. The 2030 SDGs are a series of targets and indicators which were approved by the UN in 2015. Climate coinage services, goods, and processes are the pivot of all the 17 SDGs (.UN, 2015). The proposed SDGs by 192 UN member states contain 17 goals, 169 targets, and 304 indicators covering a wide range of sustainable development issues (.UN, 2015). These include: ending poverty and hunger, improving health and education, making our cities and human settlements sustainable, gender equality, combating climate change, protecting oceans, and forests (.UN, 2015&GU, 2015).

Regarding gender disparity, the UNCEDAW (GU, 1999) and the African Charter on Women's Rights obligate states to provide for: international representation and participation of women; national laws; education; employment; equality and participation of women; national laws; equality and access to appropriate health care; economic and social benefits; rural women; equality before the law and in a civil matter; and equality in marriage and family life. Most African states have eagerly responded to the CEDAW but public awareness of them is still too low or nonexistent (Okaka, 2006; 2016). Since community media such local radio stations are very popular with most individuals, local communities, and households, they can be very useful, if they are well programmed and managed to inform the communities about the good practice of gender equality in community-based climate-smart agriculture production (BNNRC, 2012).

Furthermore, engendered community radio can play a specific and crucial role in encouraging women leadership and participation in climate-smart agriculture for sustainable food nutrition security for socio-economic household and community welfare in the south-western Ugandan in particular and across the entire country (Basnet, 2012&GU, 2015). Gender equality in climate agriculture promotes women to achieve their fundamental human rights, social, economic, ecological, and political participation for sustainable food security in households or local communities (AU 2O15).

To achieve effective gender equality advocacy communication campaigns, it is prudent to lift or ease the common obstacles that most girls and women face on a regular basis at all levels of the society (GU 2007). In all the adjacent districts to the Lake Mburo National Park, climate-induced agricultural drought has persistently eroded food nutrition. In many instances, the local communities had relied on relief emergency food supplies for the socio-economic welfare of the districts and subsequently, the perennial food insecurity is characterized by inadequate accessibility, availability, utilization, and stability of food including humanitarian emergency food relief (IDDP 2017). In this case, the combined use of multimedia communication channels will provide the maximum climate information access to target audience behavior change (Okaka & Nagasha 2016).

CONCLUSION AND RECOMMENDATIONS

The current status of climate change and vulnerability impact issues, loss, and damage in Africa demand very urgent, prudent, and robust attention to gender equality for women empowerment mainstreaming in all development areas like natural resources, soil and land use, wildlife, biodiversity, health, the earth in the future, population, land use, and industrial activity.

Furthermore, the continent should adopt plausible economic variables, the political indicators for resilience development in Africa should practice gender equality mainstreaming and women in all spheres of policy focus such as the degree of national integration. For example, a Kenyan study on the policy-makers' sources of information; ranked the following sources in their descending order of importance: mass media both the electronic and print (led by radio); special groups (NGOs); special media like policy briefs, brochures, pamphlets, newsletters); and internet websites with social media becoming more popular.

It is also imperative that communities are empowered and that both men and women participate meaningfully in planning, testing, and rolling out adaptation and mitigation activities in rural and urban areas. We need to include gender and climate change in education curriculum or training programs. All climate change policies and activities must be gender-sensitive, and the capacity of stakeholders at all levels be strengthened to promote gender-sensitive approaches to climate change adaptive capacity for practical adaptation, mitigation, and resilience (GU, 2013).

Besides, it is now known that the media are the most effective and efficient channels in affecting the vital attitudinal and knowledge change for a public health management system for all. Multimedia and inter-personal communications campaign would yield better outtakes in creating, raising, developing, and sustaining high awareness for empowerment in Africa. Effective communication campaigns can create, raise, develop, monitor, and sustain public awareness of gender equality and equity mainstreaming in climate change information services for SDGs in Africa and other poor developing countries across the globe.

Already, there are vital gender inequality and equity information, policy, research, knowledge, and practice gaps across Africa due to poor or wrong: attitudes, behavior, planning, perceptions, policies, management, budgeting, leadership, education, communication, or practice. To summarize, engendered cost-effective policies based on indigenous knowledge system (IKS), economic incentives, and coordinated instruments; will curb mitigation and adaptation costs. A prudent gender equality policy in climate change advocacy communication campaigns promotes the critical role of women in agriculture production and value chain for sustainable food security in households, communities, and nationally as well as sustainable global food nutrition security status as well as to achieve Uganda national SDGs' progress. Currently, adequate access to extreme weather events that are induced by climate change information services is still very low and so is the status of food security in the country as a whole. Likewise, cultural backwardness in the areas requires more awareness of the huge opportunities and benefits of the massive contributions of women and girls in creating and sustaining Africa's global competitiveness in food and nutrition security and in food trade at the community, national, and global levels.

REFERENCES

ACPC Ministry of Water, Lands, and Environment. (2001). *Wetland Sector Strategic Plan 2001-2010.* Kampala: Government of Uganda.

Day, B. A., & Monroe, C. M. (1999). Environmental education and communication for sustainable world. In *Handbook for International Practitioners.* GreenCom.

Denton, F. (2014). *Does climate – smart agriculture hold the key to Africa banishing hunger by 2025? CCDA-LIVE (Wed. 8 Oct. 2014).* ACPC/UNECA.

European Union. (2000). *Caring for Our Future: Action for Europe's Environment* (3rd ed.). Luxembourg: EU.

Eyotaru, O. (2013). East African Integration Hindered by Information Gaps – Minister. Daily Monitor Newspaper, p. 1.

Fatima, D. (2013). *Africa on the rise: can the opportunities from climate change spring the continent to transformative development?* An inaugural speech at the 3rd annual climate change and development from 21-23 October 2013 at United Nations conference centre, Addis Ababa, Ethiopia.

Government of Uganda. (1999). *A simplified version of the United Nations Convention on the elimination of all forms of discrimination against women*. Uganda Government.

Government of Uganda. (2015). *Climate change policy 2013. Ministry of water and environment*. Government of Uganda.

IPCC. (2014). *IPCC Synthesis of Scientific-Technical Information relevant to interpreting Article 2 of the UN Framework Convention on Climate Change*. IPCC.

Kofi, A. D. (2014). CCDA- Live. New initiative will help finance efforts to harness information to fight climate change. *Daily news*, *10*, 1.

Lopes, C. (2013). *Can climate change spring Africa's transformative development? A Statement by the UN Undersecretary-General and ECA Executive Secretary on 21 October 2013,on third annual conference on climate change and development in Africa from 21-23 October 2013 at UN conference centre*. Addis Ababa: UNECA.

National Environment Management Agency. (2002). *State of the environment report for Uganda 2002*. Ministry of Water, Lands and Environment. Kampala: Government of Uganda.

National Environment Management Authority. (2000/2001). *State of environment report 2000/2001*. Government of Uganda. Kampala: NEMA.

National Environment Management Authority (NEMA). (2003). *Guidelines for District State of Environment Report (DSOER)*. Preparation in Uganda [Final Report]. Kampala: NEMA.

Okaka, W. (2002). An environmental education program: Uganda Polytechnic Kyambogo. Applied Environmental Education and Communication, 1(1), 45-52.

UNDP. (2005). *Uganda human development report 2005. Linking environment to development: A deliberate choice*. UNDP.

UNECA / ACPC. (2014). *Fourth Conference On Climate Change and Development in Africa Summary and Recommendations*. Marrakech, Morocco: UNECA / ACPC.

Wafula, D. (2012). *Where do policy-makers get information? Science Communication Training* Workshop *for Bio-Innovate Projects*. BIPCEA.

Wilson, O. (2007). *The Role of Media Communications in Developing Tourism Policy and Cross-Cultural Communication for Peace, Security for Sustainable* Tourism Industry in Africa. A Paper at the 4th International Institute of Peace through Tourism African Conference on Peace through Tourism at Educators' Forum, Kampala.

Wilson, O. (2010). Developing Regional Communications Campaigns Strategy For Environment and Natural Resources Management Policy Awareness for The East African Community. *Research Journal of Environmental and Earth Sciences*, *2*(2), 106–111.

Wilson, O. (2013). *The dawn of community radio in Bangladesh*. Lambert Academic publishers.

KEY TERMS AND DEFINITIONS

Agriculture: Is the traditional or scientific farming methods that involve the production of crops or plants, animals, or livestock, as well as fisheries for subsistence or basic livelihood by a small-holder farmer or for large scale investment for big cash or trade by a commercial farmer.

Climate: Is the predictable average record or information of state or pattern of every day's weather or prevailing condition of a country, state, or region over a period of at least 30 years as the seasons, such as, for example, the bi-modal climatic change of dry and wet seasons as in Uganda.

Climate Change: Is the quantitative statistical record of a long term adjustment of regular annual two climatic wet and wet seasons, in Uganda, which occurs in districts or nationally over a long period of over 30 years as a result of natural and human drivers with a focus on human activities.

Climate Variability: Is any unusual adjustment, alteration, deviation, or irregular change from the normal or means or average annual climate record or season of a state or a place over several weeks or months from the usual pattern as experienced by a state or a place within the country.

Climatic Change: Is the usual pattern of change from the normal or mean or average annual climate record or season of a state, country, district, or a place; to the next recorded pattern as experienced by the community, district, sub-region, county, state or a place within the state from dry season to wet season as experienced in the case of Ugandan bi-modal climate patterns today.

Community: Refers to a specific cluster of human settlement in urban (city, municipality, town, trading center), peri-urban (suburb), or rural (village) locations that are often inhabited by related or unrelated people with similar or different socio-economic demographic backgrounds.

Food Security: Is the status of access to sustainable quality and quantity of the personally desired type of predictable supply of nutritious food from anywhere, daily and annually for the present and future needs at the household, community, district, county, national, regional, and global levels without any level of social, economic, environmental, legal, or political barriers or restrictions.

Gender Equality: Is the natural attitude and practice of sharing benefits, resources, opportunities, or challenges on the fair or equal assessment of abilities or potential of men, women, boys, or girls in all social, economic, environment, legal, administrative, technical, scientific, and political fields.

Gender Roles: Are the archaic, illegal, unconstitutional cultural norms, attitudes, or practices, which discriminate against women or girls on the basis of their biological sex, to deny them access to equal opportunities, or benefits of social, economic, environmental, and political rights.

Household: Is a social infrastructure unit of human settlement, habitation, or accommodation, which normally hosts or houses individuals, relatives or non-relatives, or members of a nucleus family that may be headed by men, women, or children, as practically experienced in Uganda today.

Livelihood Asset: Is a basic need that supports a normal social, economic, and ecological resource that are essential for human life at individual, household or family, and local community level.

Small-Holder Farmer: Is a small-scale peasant (poor farmers) who practices subsistence farming or agriculture on a small plot of land using manual labor based on rudimentary technology mainly for household food supply and for small cash at the early days of good harvest periods.

Weather: Is the current persistent or intermittent prevailing status of a local, regional, or national short-term atmospheric conditions of the time or hours, day (s), or week that is characterized by temperature, vapor, drizzle, rainfall, visibility, sunshine, cloud, wind, snowing, or humidity.

Section 3
Climate Change and Aquaculture

Chapter 11
Effects of Climate Change and Anthropogenic Activities on Algivorous Cichlid Fish in Lake Tanganyika

Renalda N. Munubi
Sokoine University of Agriculture, Tanzania

Hieromin A. Lamtane
Sokoine University of Agriculture, Tanzania

ABSTRACT

Over the last century, water temperatures in Lake Tanganyika have risen due to climate change, which increased thermal stratification and reduced the magnitude of nutrient availability. A rise in temperature increases the C:N:P ratio resulting in a poor algal diet. In addition, lake littoral habitat is experiencing increased sediment load due to deforestation of the watershed caused by anthropogenic activities. Sediments cover benthic algae and reduce its nutritional value, consequently affecting the foraging behavior, distribution, and growth performance of algivorous fish. Algae and algivorous fish are an important link in the lake food chain; therefore, if the rise in temperature will continue as predicted, then this may have a cascading effect for the rest of the community in the food chain including human being. This, in turn, may contribute to food insecurity at local and regional levels. To counteract this adaptation and mitigation measures such as environmental monitoring systems and creating new opportunities should be considered.

DOI: 10.4018/978-1-7998-3343-7.ch011

INTRODUCTION

Lake Tanganyika is the second deepest lake in the world, holding approximately 1/6 of the world's freshwater (Wetzel, 2001). It lies in the East African Rift Valley and is shared by four countries with the shoreline of about 1,800 kms, among these 9percent is in Burundi, 43percentis in the Democratic Republic of the Congo, 37percent is in Tanzania, and 11percent is found in Zambia (Fig. 1). The lake supports the most productive freshwater pelagic fishery, with remarkable cichlids fish diversity with estimated 250 morphologically and behaviourally diverse fish species that mainly inhabit the lake's littoral habitats (Snoeks, 2000). Most of the fish in the rocky littoral zone of Lake Tanganyika are algivorous cichlids that graze on the attached algae (periphyton), which are found on the rocky shore habitats of the littoral zone (Takeuchi, Ochi, Kohda, Sinyinza& Hori, 2010). In general, Lake Tanganyika ecosystems represent biodiversity hotspots of endangerment (Dudgeon et al., 2006) because of their critical contribution to global biodiversity.

Lake Tanganyika is a significant source of food and livelihood to millions of people dwelling inside and outside of its basin. The basin is rich in minerals and fertile soils. The main economic activities of the 10 million people living in the lake's catchment area are small-scale agricultural activities (Nkotagu, 2008) including farming of crops such as maize, tobacco, rice, sugarcane, coffee, beans, groundnuts, cassava; and keeping of livestock such as cattle and goats for their livelihood. The lake is also used asa means of transportation of goods and generation of hydroelectric power, communication, and a source of potable water for the sustenance of human, livestock, and agricultural activities (Coulter, 1991). It is also a source of revenue from fishing and ecotourism in the riparian countries bordering the lake.

The fisheries of Lake Tanganyika are by far the most important source of animal protein for human consumption along the lake. The lake supplies between 25-40 percent of the protein needs of the local population in the four riparian countries (Mölsä, Reynolds, Coenen, &Lindqvist, 1999). About 45,000 people are directly involved in the lake fisheries which constitutes between 55-90percent of the commercial fishery and 80-99percentof the traditional artisanal fishery (Rufli, 2001).

Globally, greenhouse gas emission has increased leading to atmospheric concentration of gases such as carbon dioxide, methane, and nitrous oxide. The effects of these gases are likely to be the dominant cause of the observed global warming (Burnett et al., 2011). About 60percent of these emissions have been stored on land, marine, and freshwater systems. This might have increased the pH of both marine and fresh water ecosystems. Moreover, lake water pH is under the (stronger) influence of land and sediment and thus, the effect of atmospheric CO_2may play a relatively minor role. One of the most immediate and obvious effects of global warming is the increase in temperatures around the world. The average global temperature has

Figure 1. Geographical location of Lake Tanganyika

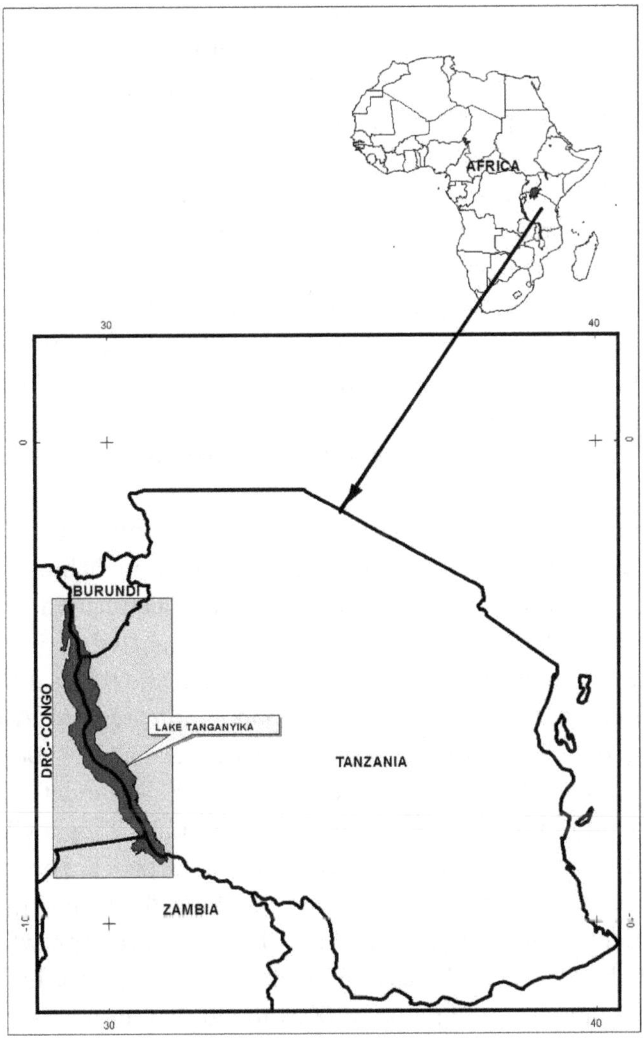

increased by about 1.4 degrees Fahrenheit (0.8 degrees Celsius) over the past 100 years, according to the National Oceanic and Atmospheric Administration (IPCC, 2018, NOAA,2019)

Researchers have shown that the lake is also currently confronted by degradation of its watershed (Alin et al., 2002) and global climate change (Kraemer et al., 2015; O'Reilly, Alin, Plisnier, Cohen & McKee, 2003; Verburg, Hecky & Kling, 2003). Climate change is altering hydrological cycles across atmospheric, terrestrial, and aquatic components of the ecosystems, causing an increase in ambient water temperature and carbon dioxide concentration. These changes are expected to lead to the alteration in the nutrient availability (Prado & Da Silva, 2017). Like other large water bodies worldwide, Lake Tanganyika fishes are suffering from the negative impacts of climate change (Alin& Cohen, 2003). This is because over the last century, the temperatures of Lake Tanganyika's surface and deep-water have been rising(O'Reilly, Alin, Plisnier, Cohen & McKee, 2003) resulting into increased thermal stratification and stability of the water column (Kraemer et al., 2015), which, in turn, reduced frequency and magnitude of availability of nutrients that are essential for primary production. Threats already have an impact on the functions of the lake resulting to reduced nutrients availability (Corman et al., 2010) and primary productivity (O'Reilly, Alin, Plisnier, Cohen & McKee, 2003). Since the food web of Lake Tanganyika is based on algal productivity (Sumaila, Cheung, Lam, Pauly& Herrick,2011). the decline in primary productivity may affect both near shore and near pelagic fishery productivity (Stenuite et al., 2007; Donohue & Molinos, 2009).

Apart from changes in climatic variables such as precipitation, temperature, pH, and lake levels, currently, much of Lake Tanganyika's littoral habitat is experiencing increased sediments load due to deforestation of the watershed caused by anthropogenic activities (Alin et al.,1999). Sedimentation has been reported as one of the factors, which affect the health of Lake Tanganyika ecosystem and this is due to increasing human economic activities along the lake and population pressure (Alin et al., 2002; Dobiesz et al., 2010). According to Donohue, Duck, & Irvine, (2003), the highest sediment load originates from the catchment of the rivers. Sediments have been found to affect biodiversity in large water bodies including Lake Tanganyika. A study by Donohue and Molinos (2008) showed that, sediments which are deposited in water bodies may cover benthic algae, potentially reducing their nutritional value and consequently affecting the foraging behaviour, distribution, and growth performance of algivorous fish (Munubi, McIntyre & Vadeboncoeur, 2018). In another study, Alin et al. (1999) and McIntyre et al. (2005) found that an increase of sediments in lakes resulted into a decline of benthic algivores at shallower depth. In addition, sedimentation reduces penetration of light into the water, depriving the plants of the light needed for photosynthesis and reduces habitat complexity by filling crevices and other sheltered areas in rocks (Donohue & Irvine, 2004) that are important as hiding, breeding and feeding sites for algivorous fish.

Describing the relationship between climate change and algivores fish may be critical in understanding the processes by which environmental changes can affect the distribution, behaviour, growth, and reproduction of algivorous cichlids such as *Trophiusspp*. In this chapter, the authors used algivorous cichlids as an example, to illustrate how climate change can alter nutrients stoichiometry, algal food quality, fish feeding behaviour, growth, and distribution. Authors also discussed the consequences of climate change on food security and socio-economic development, mitigation and adaptation strategies.

FEEDING HABITS OF TANGANYIKA ALGIVOROUS CICHLID FISH

Lake Tanganyika presents one of the most peculiar tropical freshwater ecosystems with an evolution of an endemic fish fauna, dominated by cichlids. Five tribes of the family Cichlidae (Tropheini, Lamprologini, Ectodini, Eretmodini, and Tilapini) have acquired several ecomorphs that are closely related to feeding habits in herbivory such as grazing, browsing, scraping, biting, and scooping (Hori, Gashagaza, Nshombo&Kawanabe, 1993). Most of the species belonging to these tribes are aufwuchs-eaters, which feed on a thin bio-cover of algae, microorganisms, sponges, small crustaceans, and detritus (Yamaoka, 1983) and they have colonized different habitats of the rocky shore of the lake.

The differences in feeding habits was classified based on the shape of the teeth and movement of the jaw while feeding. For example, grazers such as *Petrochromis*spp are aufwuchs eaters which combs loose unicellular attached algae along the rocky shore of the lake (Yamaoka, 1997), they contain gill rakers for filtration of small algae, whereas browser such as *Tropheus*spp bite and tear off pieces of filamentous algae using the upper and lower teeth (Mbomba, 1983; Yamaoka, 1983). Scrapers rub epiphyton from rock surfaces by moving teeth along the substrate or rock (Yamaoka, Hori & Kuratani. 1986). In addition, these sympatric species are able to coexist on account of their slightly different distinct specialized trophic morphologies in their jaw structures and intestine lengths(Wagner et al., 2009),physiological abilities, such as secretion of digestive enzymes (Sturmbauer, Mark & Dallinger, 1992), and feeding habits (Yamaoka, 1983)

Algivorous cichlid fish are primary consumers that form an important link in the food chain by providing nutrients and energy to higher organisms in the trophic levels. However, the rise in temperatures and CO_2 concentration due to anthropogenic activities in Lake Tanganyika has the potential of altering both the quality and the quantity of algae (periphyton). Periphytonare tiny plants attached on the rocky shore of Lake Tanganyika, their growth depends on the availability of nutrients such as C:N:P. They are an important link in the food chain in the lake because algivorous fish feeds on them. If algal are harmed, by any stressor, then this may have a cascading effect for the rest of the community in the food chain including human beings. Research conducted by Descy et al. (2005) and Bergamino et al. (2010),showed that changes in phytoplankton composition and abundance in Lake Tanganyika are partly contributed by climate change.

CLIMATE CHANGE AND CARBON:NITROGEN:PHOSPHORUS RATIO

Aquatic environments will experience direct effects of temperature change at a slower rate due to the relatively high specific heat capacity of water (Ogungbe, Alabi, & Ometan, 2015). Indirect effects of climate change in aquatic systems include changes in the rates of photosynthesis and variability in the availability of nutrients, which include C, N, and P (Verburg & Hecky, 2009; Corman et al., 2010). In recent decades, anthropogenic activities have altered the cycle of nitrogen (N), phosphorus (P), and carbon (C) in plants including algae. This may be due to an increase in excessive application of fertilizers, an increase in temperature that reduces availability of nutrients (Kraemer et al., 2015), and an increase in CO_2 emissions, which rapidly limits the increase of plant production (Gleadow, Evans, McCaffery & Cavagnaro, 2009).In order for plants to grow and survive, they need to adapt to these changes by changing metabolic activities at different levels of organisms such as molecular, biochemical, anatomical, and morphological (Eggert, 2012).

Temperatures of Lake Tanganyika's surface and deep-water have risen (O'Reilly, Alin, Plisnier, Cohen & McKee, 2003; Verburg, Hecky & Kling, 2003) and wind velocities in the Lake Tanganyika watershed have declined by 30percentsince the late 1970's (Corman et al.,2010; O'Reilly, Alin, Plisnier, Cohen & McKee, 2003). The increased air and surface water temperatures and the decreased wind velocity have increased thermal stratification and stability of the water column (Kraemer et al., 2015) resulting in the reduced frequency and magnitude of nutrient upwelling and, hence, reduced the mixing of nutrients (O'Reilly, Alin, Plisnier, Cohen & McKee, 2003; Verburg, Hecky & Kling, 2003). The reduction in nutrient mixing decreases the availability of nutrients that are essential for primary production (O'Reilly, Alin, Plisnier, Cohen & McKee, 2003) from the deep to the surface water nutrient reservoir making nitrogen (N) and phosphorus (P) less readily available for uptake by algae (Schindler & Eby, 1997). The decrease in N and P may increase the ratio of carbon to nitrogen to phosphorus (C: N: P) in the water and algal mats. A high C:N:P ratio is associated with poor quality algal food (Table 1) and is another indirect effect of climate change (Lichtenbelt, 1992).Thus, climate change affects both nutrient concentrations and C:N:P stoichiometry of organisms and ecosystems.

Plants (algae) fix CO_2 during photosynthesis, as the rates of photosynthesis increases, the amount of C and the overall ratio of C:N:P increases (Thompson, Gamage, Hirotsu, Martin, & Seneweera, 2017).Plants need other elements for growth and the raised CO_2does not directly make these mineral elements more available and may even decrease the uptake of some of these elements for example, Nitrogen and phosphorus (Cotrufo, Ineson & Scott, 1998). Therefore, the ability of plants to respond to elevated $CO_2$with increased photosynthesis and growth promotes the conditions of high carbon to low mineral availability (Sardans, Rivas-Ubach & Peñuelas, 2012). A high C:N:P ratio of plant tissues is associated with poor quality algal food for herbivores fish and increased microbial decomposition (Enriquez, Duarte &Sand-Jensen,1993;Jefferies & Maron, 1997).

The ability of plants to respond to elevated $CO_2$with increased photosynthesis may be limited under conditions of low availability of other nutrients. Several studies (e.g. Dijkstra et al., 2012; Deng et al., 2015) have reported a decrease in plant N:P ratios in response to elevated CO_2.Nitrogen (N) and/or phosphorus (P) are important for photosynthetic processes, new cell construction, metabolism, and synthesis of protein(Chapin, Matson & Vitousek, 2011). Though CO_2 can increase the growth of plants, the plants may become less nutritious for algivorous fish as the sources of energy, due to high C:N:P ratios. Therefore, elevated atmospheric CO_2will have major implications for plant physiology and growth. Higher C:N:P ratio as a result of increased atmospheric CO_2implies poor quality diet, this may affect both nearshore and pelagic fisheries. This is supported by the findings reported by Munubi, McIntyre & Vadeboncoeur, (2018, Fig. 2) that fish densities correlate positively with algal primary production and the amount of nutrients available in the algal diet (% P and %N).

In Lake Tanganyika, food web is based primarily on algal productivity (Sumaila, Cheung, Lam, Pauly & Herrick 2011). An increase in temperature may also increase photosynthesis rate of aquatic plants. However, studies by O'Reilly, Alin, Plisnier, Cohen & McKee, (2003) showed that primary production in Lake Tanganyika has decreased because of warming events. A decrease in primary production (biomass) of algae limits food availability to herbivores in aquatic systems, and consequently lowers fish growth and abundance (Munubi, McIntyre & Vadeboncoeur, 2018). Low food abundance (primary productivity) and poor food quality (high C:N:P) compounded with an increased demand for food due to an increase in metabolic rate (Kim, 2014) and energy requirements (Mueller and Diamond, 2001) could exert too much stress for herbivorous fish to overcome.

CLIMATE CHANGE AND FISH FORAGING BEHAVIOUR

The effects of elevated water temperature on the Earth's ecosystems are expected to decrease the availability and abundance of algal food resources. Studies by Bergamino et al. (2007) and Ndebele et al. (2010) revealed that, as temperature increases, the production of palatable algal species such as chlorophytes and diatoms decreases; while the production of toxic, nitrogen-fixing cyanobacteria, which are poor quality and less palatable food, increases (Wagner and Adrian, 2009). If this continues, it might lead to reduced primary consumers' (e.g. zooplankton)

Figure 2. Relationship between fish swimming activities and algal quality. As the algal quality increases the swimming activities deceases
(adopted from Munubi, 2015)

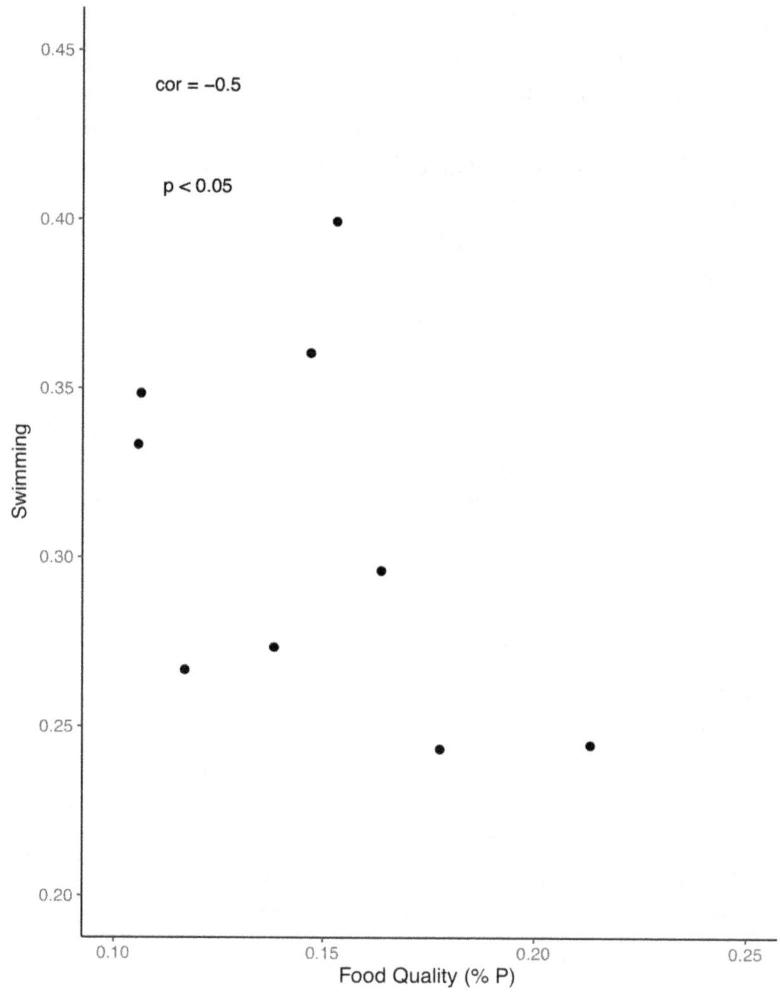

production and a consequent decline in fish stocks (Ndebele et al., 2010; Cohen et al., 2016), which are secondary consumers in the food chain. Therefore, in order to maintain high growth rates, high reproduction, and high survival rates, flexibility in feeding behaviour is an important adaptive attribute of the foraging fishes such as algivorous(Dill, 1983), and effective foraging strategies must take into account this heterogeneous distribution of food.

Fish should respond to low levels of food availability by altering their behaviour in ways that ensure higher feeding rates, larger feeding territories, and broader diets. For the algivorous fish when fed on poor quality diet, two biological responses to climate change must occur: i)changes in behavioural and ii) morphological modifications (Evans and Moustakas, 2018). In general, a study by Wright and Westoby (2003) revealed that, algivorous respond to poor quality food by (i) increasing the feeding rates to compensate for the poor food quality and by (ii) feeding for a longer period on food of inferior quality containing high C: N: P ratio. However, a study by Munubi (2015), did not show any compensation activities when fish were exposed to poor quality diet, instead the study revealed that fish in sites with poor quality diet, reduced their foraging efficiency by decreasing their feeding rates and time spent feeding per patch while increasing their swimming activities (Fig. 2). Munubi (2015) attributed the increase in swimming activities (recorded in the afternoon) to be the result of searching for food. However, this increase in swimming activity may increase metabolic costs or energy expenditure as revealed by Schreck, (2010), which may be reflected in an increase in oxygen consumption (Boscolo-Galazzo, Crichton, Barker& Pearson 2018). Subsequently, the solubility of oxygen in water tends to decrease with increase in temperature (Walczyńska & Sobczyk 2017), hence, aquatic organisms are more likely to be affected as temperature increases due to climate change. In addition, Munubi (2015) also found that the fish at sites with poor food quality (low %P) spent less time feeding in each session and algivorous fish feed selectively on diatoms, which are palatable (Bergamino et al.,2007) and rich in essential fatty acids (EFA) (Jónasdóttir, 2019). In addition, rise in temperature increases the C:N:P and brings about rapid rate of maturity of forages. As forages mature, there is an increase in cell wall contents, which, in turn, may decrease the digestibility of the dry matter of the forages and, hence, increases the digestibility costs of the plant (Minson & McLeod, 1970). The high costs of digestion, foraging activity, and basal metabolism associated with increase in temperature must be met by high-energy intake.

The studies conducted by Radomski and Goeman (2001) and Cornelissen (2011) revealed that climate change caused by anthropogenic activities in Lake Tanganyika will likely alter both algal quality and the quantity, and hence, affect both near shore and pelagic fisheries which depends on primary productivity of algae. Munubi, (2015) compared the foraging behaviours between sites with high quality and low-quality diets in Lake Tanganyika and found that algivorous fish rely on a combination of both change in behavioural (change in feeding behaviour) and morphology (increase in gut length) to adapt to low quality diet. Flexibility allows these primary consumers to adapt to and exploit nutrients and energy from the poor algal diet. The study results also showed a positive correlation between sediment accumulation on the periphyton and both gut length and stomach fullness. An increase in relative intestinal length may facilitate a more complete extraction of nutrients and energy from the contaminated food. Sediment deposited on top of the algae covers and contaminates algae, and thus depletes the quality and quantity. This may increase the digestion costs (Navarro, Iglesias, & Ortega, 1992) to herbivores and consequently, producing a negative cascading growth effect through the food chain.

Climate change and anthropogenic activities may decrease both nutritional quality and the quantity of feed available to the fish. This may negatively affect the growth rate of herbivores since the amount of energy and nutrients acquired per unit time are determined by both the time allocated for foraging activities (Abrams, 1984) and the quality and quantity of food. However, algivores including *T. brichardi* can modify their feeding behaviour and gut morphology to counteract the effects of food quality contaminated with sediments. A study by Munubi, McIntyre & Vadeboncoeur, (2018) revealed that fish at sites with high quantity of sediments and organic matter have longer intestine length.

CLIMATE CHANGE AND FISH METABOLISM

Cichlid fish are ectothermic animals, which mean that their body temperature depends on the temperature of the surroundings (McNab, 2002). Therefore, a rise in water temperature associated with climate change will likely increase stress to tropical ectotherms that have narrow thermal windows relative to temperate ectotherms(Sunday, Bates, & Dulvy, 2011).Studies have shown a positive relationship between the rise in temperature of the surrounding environment and metabolic rate of fish when the temperature is within the acceptable range. Climate change is predicted to increase ambient temperatures of aquatic systems and decrease the quality of food available to algivorous fish.

Kim, (2014) conducted an experiment to evaluate the effect of the combination of increased temperature from 26 ^{0}C to 32 ^{0}C and decreased food quality on metabolism and growth of algivorous cichlid (*tropheusduboisi*). The experiment was conducted using algae collected from Lake Tanganyika as poor quality food with high C:P ratio and high quality food(Flake food) with low C:P ratio. The results revealed that the rise in temperature from 26^{0}C to 29^{0}C increased the growth and metabolic rate of cichlids, but declined grow that higher temperatures of 32 ^{0}C, suggesting that between 26 and 29^{0}Cthe biochemical reaction rates increase steadily with temperature of up to optimum (Hochachka & Somero, 2002). Based on these results, the author recommended the optimal temperature for growth was above 26 ^{0}C and below 32^{0}C. Within optimum temperatures, cichlid fish can move fast and consume a lot of food to fuel metabolic processes. However, an experiment conducted by Kim, (2014) showed also that temperature of above 32 ^{0}C is detrimental to the growth of *T. duboisi* because the fish did to grow. At this temperature, fish growth rate decreases regardless of whether they were fed poor or high-quality food. These results suggest that the combined effects of higher temperature > 32 ^{0}C and lower food quality are too much for the fish to overcome; hence, they began to lose biomass.

All fish species are characterized by an ideal range of temperature in which they show their optimum growth (Person-Le, Buchet, Vincent, Le & Quemener, 2006;Oyugi, Cucherous set, Baker, &Britton, 2012). The effect of the magnitude of the temperature increase depends on the species-specific pejus temperature, and whether that animal will experience those temperatures depends on the environment in which it is located. The study by Kim (2014) indicated that fish that fed on lower quality algal diet grew by 50percentless compared to the fish that fed on high quality algal food at all temperatures. Therefore, the strong effect of poor quality food on growth rate and relative growth rate is likely to be even more critical in Lake Tanganyika where the algal food is of poor quality. A study by Deng et al. (2014) revealed that nutrients are more important than water temperature on distribution and abundance of phytoplankton communities. An experimental model conducted by O'Connor, Gilbert & Brown, (2011)revealed that both phytoplankton and zooplankton population will decrease as temperature increases. The rise in temperatures may decrease the number of herbivores, due to the fact that being ectothermic, their physiological activities such as consumption and metabolic demands are highly sensitive to temperature and increases with rising temperatures (Lee, Ban, Ikeda, & Matsuishi, 2003). However, experimental studies suggest that as temperature increases, nutrient enrichment and primary productivity may also be limited due to less inorganic nutrients (Power, 1983; O'Reilly, Alin, Plisnier, Cohen & McKee, 2003; Verburg, Hecky & Kling, 2003), hence poor diet which may support a smaller population of herbivores (Schoo, Aberle, Malzahn & Boersma, 2012). Therefore, continued warming could lead to further decline of herbivores and ultimately lead to less food for other fish species in the food chain.

One of the key factors that limit accuracy in predictions is that climate change affects positively and negatively the species performances, and there is an interaction of species within and between the trophic level and these species do not necessarily react to climate change in a similar way (Van der Putten et al., 2004; Schweiger, Settele, Kudrna, Klotz & Kühn, 2008). Therefore, fish will simultaneously be exposed to the indirect and direct effects of climate change, lower food abundance, and lower food quality compounded with an increased demand for food resulting from an increase in metabolic rate. This may indirect control other fish species in the food chain due to the fact that the mineral nutrient contents of the secondary consumers as prey for higher trophic levels depend on the biochemical composition, and hence, the quality of the algal food available to them (Schoo, Aberle, Malzahn & Boersma, 2012), thus herbivores feeding on poor quality food (high C: P) become in turn, low quality food for other organisms in the food web.

Evidence suggests that increase in temperature resulting from climate change will result in changes in recruitment success (Shoji, Toshito, Mizuno, Kamimura, Hori, & Hirakawa, 2011) because of the decrease in production or survival of the pelagic eggs or larvae owing to changes in the quality/quantity of nursery habitats. Fishes at early stages are usually more sensitive to climate change stressors than are fishes at later stages, and therefore increase vulnerability (Harrod, Ramirez, Valbo–Jørgensen & Funge-Smith, 2018).Considering the positive relationship between ambient temperature and fish growth rates, it is important to examine how changing temperatures will affect the energy demands of fish. Studies(e.g. Rosenfeld *et al.,* 2015) revealed that higher temperatures increase metabolic rates hence high-energy demands; but there is a threshold where energy demands cannot be met and the growth and function of the animal begins to decline. For example, at temperatures above the threshold, proteins in the body may begin to denature and metabolic pathways begin to break down. The point at which basal metabolism begins to decline with increasing temperature is the pejus temperature, and for Tropheus spp. it was suggested to be at temperature between 26 to 29 °C (Kim, 2014). At higher temperature, fish in Lake Tanganyika will likely experience increased food demand (Lemoine, Burkepile & Parker, 2014) and decreased supply and quality of food, therefore, they may not be able to compensate. As the lake temperature increases metabolic rates will increase, requiring the fish living in Lake Tanganyika to either eat more of the low-quality algae or decrease growth rates (Kim, 2014), for example research by Lemoine, (2015) shows that rise in temperatures increases the metabolic demand of N, and herbivores, compensates the increased demand by significantly increasing consumption rates of low-quality diets. When a fish reaches a critically high temperature, physiological disorganization occurs, leading to loss of equilibrium (Lutterschmidt and Hutchison, 1997).It is most likely that changing algal quality, foraging behaviour and metabolism of fish due to climate change will definitely alter fish distribution in a lake.

CLIMATE CHANGE AND FISH DISTRIBUTION

Rise in temperature will modify nutrients stoichiometry which is important in production and distribution of plant species community and primary production (Prado and De Silva, 2017). Changes in plant species distribution may influence positively or negatively fish population. These climate-driven biotic changes will likely vary between and within ecosystems because research showed that species responses to climate change are not consistent; some species are apparently able to respond more quickly than others do (Lehodey et al., 2006). Research shows that species at lower trophic levels (plants including algae) respond more quickly to a change in climate than species at higher levels (fish). Negative changes may drive organisms (including algivorous fish) to extinction, but the magnitude of the effects depends on whether the species are able to alter their geographic distribution or modify their digestive physiology in response to climate change. All living organisms have a tolerance level for temperature range. A rise in temperature beyond the threshold level can change the way a living organism develops, because it changes the metabolic rates of plants or animals. For example, studies (e.g. Beaulier et al., 2013; Savadovaet al., 2018)revealed that higher temperatures could cause replacement of some plant species by more arid-type plants or vice versa.

The negative effect of the rise in temperature is dependent upon both the magnitude of environmental temperature increase and the physiology of the animal (Hop & Graham, 1995). Different review of published data supports the hypothesis that global warming results in a shift in abundance and distribution (in patterns of occurrence with latitude and depth) of fish species and algal species. This means that fish will be challenged with new environments in which they have to compete and persist (Donaton, Durham, Cerrato, Schwerzmann & Thorne, 2018). In these situations, one of the mechanisms will be for the fish species to exhibit clear changes in migration patterns related to climate-induced changes in natural food availability (Munubi, McIntyre & Vadeboncoeur, 2018) and productivity (Power, 1983). The changes in the distribution and abundance of fish species may be the result of the movement made when searching for suitable ecosystems (Kwon et al., 2012; Weijerman, Lindeboom, & Zuur, 2005) such as availability of high quality food and good water quality. In the case of algivorous fish in Lake Tanganyika, Munubi *et al.,* (2018) observed that high fish densities were found in areas with high food quality (Fig. 3) and that fish densities correlates positively with algal quality (% P). According to Munubi, McIntyre & Vadeboncoeur, (2018) sites with high algal quality had higher fish density and negatively correlate with C: P ratio while sites with lower algal quality (high C: P ratio) had lower fish density (Fig. 4). In this case, environmental response of plants to climate change has a negative influence on food production globally with future losses that will reduce food security in the future.

CLIMATE CHANGE, FOOD SECURITY, AND SOCIAL ECONOMIC DEVELOPMENT

In this section, the authors discuss the association between climate change, food security, and social economic development. Fish provides nutritious food, and fishing generates income and employment. Trade in fishery products contributes to poverty reduction and national economic growth in all countries sharing the Lake Tanganyika. According to Chu, Mandrak & Minns, (2005) and Ficke, Myrick, &

Figure 3. Relationship between fish density and algal phosphorus (%). As the algal quality increases, so does the fish density
(adopted from Munubi, McIntyre & Vadeboncoeur, 2018)

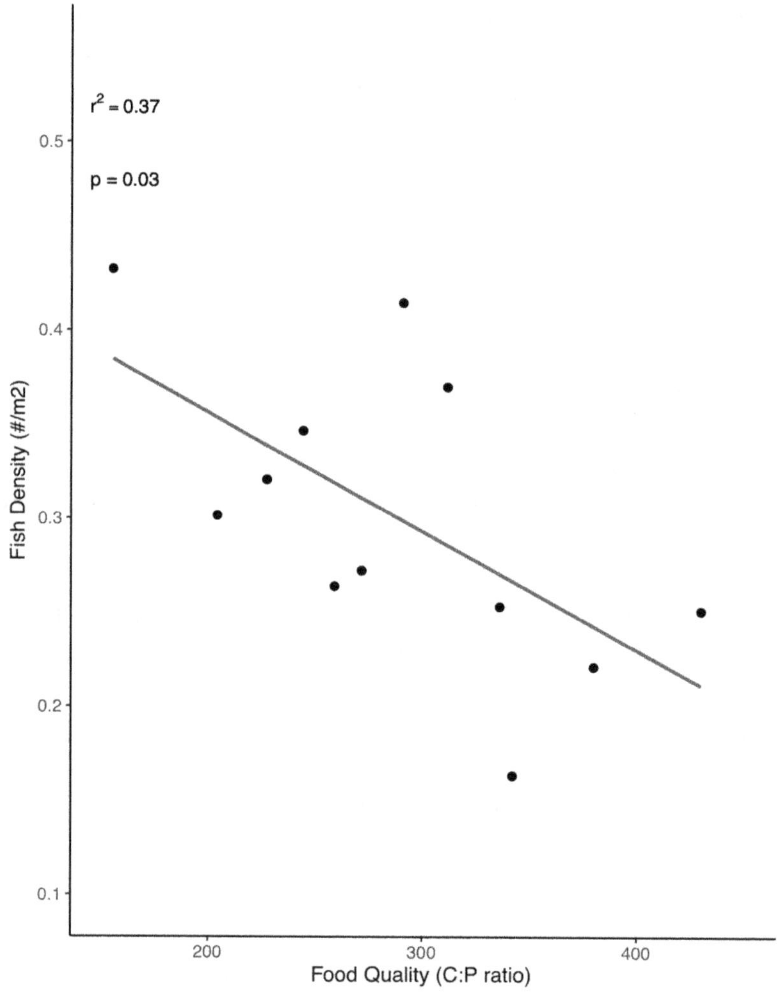

Hansen, (2007), climate change might be responsible for changing the timing and amount of precipitation that could affect spawning, dispersal and growth of fish and eventually affect the fish yields. The effect can also be indirectly affecting the primary producers (lower trophic level), which include phytoplankton or periphyton (Brgamino *et al.*, 2010). In Lake Tanganyika, higher water temperatures may reduce the availability of algivorous fish stocks by changing the abundance of food available (O'Reilly, Alin, Plisnier, Cohen, & McKee, 2003) and finally affect other fishery species in the food web (Schoo, Aberle, Malzahn, Boersma, 2012) that depends

Figure 4. Relationship between fish density and algal C: P ratio. Climate change will increase the ratio; hence, this may decrease the density of fish
(Adopted from Munubi, 2015)

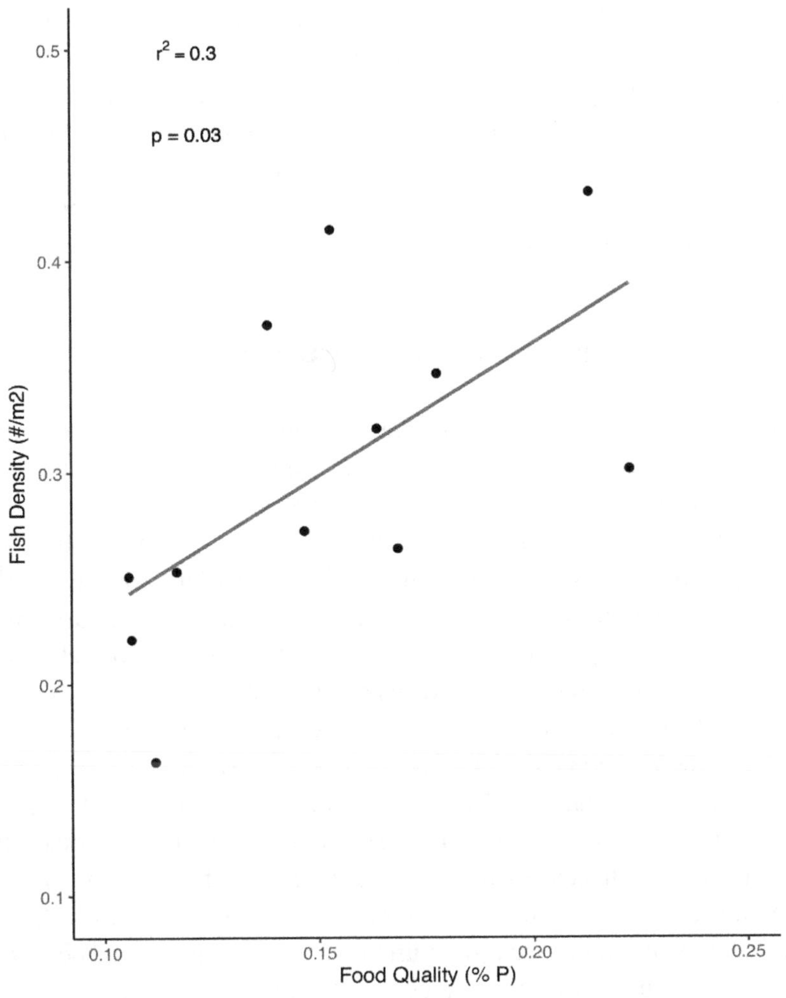

on algivorous fish as a source of energy. Again, looking at the bottom up cascade, climate change may affect the amount of nutrients available to plankton and therefore reducing yield in planktivorous fish. Therefore, climate changes may affect fisheries directly by decreasing fish stocks because secondary consumers depend on the availability and quality of primary productivity or indirectly by influencing fish prices or the cost of goods and services required by fishers.

It is envisaged that climate change will be affecting communities that depend on the lake for their livelihoods at three levels namely, subsistence, artisanal, and ornamental fish harvesters. The subsistence harvesters are doing fishing as their cultural activity; the artisanal are getting money and spread the earnings throughout the community, and contribute to the regional economies. The ornamental harvesters are also getting money through local trade and export. The three levels of harvesters will be affected, as the fish habitat will change and necessitating a change of the fishing ground, fishing gears, and vessels in order to access a particular targeted fish stock and thereby spending more money, something which would affect their economies. Furthermore, there will be an indirect effect like a wide spread of vector-born infection such as malaria. Other effects on community will be poor nutrition, social dislocation, and migration.

ADAPTATION AND MITIGATION MEASURES

Possible adaption for people surrounding the lake might be looking for more livelihood opportunities, human capital, and higher education to enhance adaptive capacity. Aquaculture has been considered as one of the livelihood opportunities, which are created by climate change. In addition, climate-smart agriculture may increase agricultural productivity and income, hence, building resilience through adaption to and mitigation of the impact of climate change. Other suggested mitigation measures include credit availability, technological innovation, ecosystem based approach to fisheries management, and training and ensuring buffer zones are accessible to fish (He *et al.*, 2018). It is known that fish population can be increasing mobility and access to the buffer zones and therefore be more adaptive as they can respond rapidly to climate change. Environmental monitoring including water level measurement may assist rapid action and prevention once communicated through local networks. Furthermore, mitigation measures can be an intervention of removing other stressors that exacerbate the climate change for example, reforestation to reduce sedimentation in the lake catchment. This is because sediment loads may clog gravel beds impeding fish such as algivorous cichlids from spawning, feeding, and survival hence reducing recruitment rates.

CONCLUSION

Like other lakes in the world, Lake Tanganyika is suffering from the impact of climate change. It has been observed that temperature as one of the climatic variables and other anthropologic activities such as deforestation affects the performance of both primary and secondary producers in the lake. The increase in water temperatures leads to increased thermal stratification and stability of the water column. This in turn, reduces the availability of nutrients that are essential for primary production from the deep-water reservoir to the surface. When faced with poor quality food, algivorous fish (e. g *T. brichardi*) have to change their behaviour to compensate for the poor food quality by reducing their foraging efficiency, decreasing their feeding rates and time spent feeding per patch while increasing their swimming activities, which may increase the metabolic costs or energy expenditure. It is envisaged that as the temperature rises and sediments loading continues as predicted, there would be large shifts in food chains with severe consequences to global food security and conservation of species. It is recommended that conservation measures should be considered through adaptation and concrete adaptations measures including creating buffers zones and opening new opportunities including integrated aquaculture.

REFERENCES

Abrams, P. (1984). Foraging Time Optimization and Interactions in Food Webs. *American Naturalist*, *124*(1), 80–96. doi:10.1086/284253

Alin, S. R., Cohen, A., Bills, R., Gashagaza, M. M., Michel, E., Tiercelin, J. J., & Ntakimazi, G. (1999). Effects of landscape disturbance on animal communities in Lake Tanganyika, East Africa. *Conservation Biology*, *13*(5), 1017–1033. doi:10.1046/j.1523-1739.1999.96476.x

Alin, S. R., & Cohen, A. S. (2003). *The surface level of the lake fluctuates at a variety of time and depth scales in response to both climatic and geomorphic factors*. Academic Press.

Alin, S. R., O'Reilly, C. M., Cohen, A. S., Dettman, D. L., Palacios-Fest, M. R., & McKee, B. A. (2002). Effects of land-use change on aquatic biodiversity: A review from the paleorecord at Lake Tanganyika, East Africa. *Geology*, *30*(12), 1143–1146. doi:10.1130/0091-7613(2002)030<1143:EOLUCO>2.0.CO;2

Bergamino, N., Horion, S., Stenuite, S., Cornet, Y., Loiselle, S., Plisnier, P. D., & Descy, J. P. (2010). Spatio-temporal dynamics of phytoplankton and primary production in Lake Tanganyika using a MODIS based bio-optical time series. *Remote Sensing of Environment*, *114*(4), 772–780. doi:10.1016/j.rse.2009.11.013

Bergamino, N., Loisellea, S. A., Cozar, A., Dattiloa, A. M., Bracchini, L., & Rossi, C. (2007). Examining the dynamics of phytoplankton biomass in Lake Tanganyika using empirical orthogonal functions. *Ecological Modelling*, *204*(1-2), 156–162. doi:10.1016/j.ecolmodel.2006.12.031

Boscolo-Galazzo, F., Crichton, K., Barker, S., & Pearson, P. (2018). Temperature dependency of metabolic rates in the upper ocean: A positive feedback to global climate change? *Global and Planetary Change*, *170*, 201–212. Advance online publication. doi:10.1016/j.gloplacha.2018.08.017

Burnett, A. P., Soreghan, M. J., Schols, C. A., & Brown, E. T. (2011). Tropical east African climate change and its relation to global climate: A record from Lake Tanganyika, tropical east Africa, over the past 90+kyr. *Palaeogeography, Palaeoclimatology, Palaeoecology*, *303*(1-4), 155–167. doi:10.1016/j.palaeo.2010.02.011

Carpenter, S., Kitchell, J., & Hodgson, J. (1985). Cascading Trophic Interactions and Lake Productivity. *Bioscience*, *35*(10), 634–639. doi:10.2307/1309989

Chapin, F. S., Matson, P. A., & Vitousek, P. M. (2011). Landscape Heterogeneity and Ecosystem Dynamics. In *Principles of Terrestrial Ecosystem Ecology*. Springer. doi:10.1007/978-1-4419-9504-9_13

Chu, C., Mandrak, N. E., & Minns, C. K. (2005). Potential impacts of climate change on the distributions of several common rare freshwater fishes in Canada. *Diversity & Distributions*, *11*(4), 299–310. doi:10.1111/j.1366-9516.2005.00153.x

Cohen, A. S., Gergurich, E. L., Kraemer, B. M., McGlue, M. M., McIntyre, P. B., Russell, J. M., & Swarzenski, P. W. (2016). Climate warming reduces fish production and benthic habitat in Lake Tanganyika, one of the most biodiverse freshwater ecosystems. *Proceedings of the National Academy of Sciences of the United States of America*, *113*(34), 9563–9568. doi:10.1073/pnas.1603237113 PMID:27503877

Corman, J. R., McIntyre, P. B., Kujoba, B., Mbemba, W., Fink, D., Wheeler, C. W., Gans, C., Michel, E., & Flecker, A. S. (2010). Upwelling couples chemical and biological dynamics across the littoral and pelagic zones of Lake Tanganyika, East Africa. *Limnology and Oceanography*, *55*(1), 214–234. doi:10.4319/lo.2010.55.1.0214

Cornelissen, T. (2011). Climate change and its effects on terrestrial insects and herbivory patterns. *Neotropical Entomology*, *40*(2), 155–163. doi:10.1590/S1519-566X2011000200001 PMID:21584394

Cotrufo, M. F., Ineson, P., & Scott, A. (1998). Elevated CO2 reduces the nitrogen concentration of plant tissues. *Global Change Biology*, *4*(1), 43–54. doi:10.1046/j.1365-2486.1998.00101.x

Coulter, G. W., & Spigel, R. H. (1991). Hydrodynamics, p. 49–75. In G. W. Coulter (Ed.), Lake Tanganyika and its life. Oxford Univ. Press.

De Sassi, C., & Tylianakis, J. M. (2012). Climate Change Disproportionately Increases Herbivore over Plant or Parasitoid Biomass. *PLoS One*, *7*(7), e40557. doi:10.1371/journal.pone.0040557 PMID:22815763

Deng, Q., Hui, D. F., Luo, Y. Q., Elser, J., Wang, Y. P., Loladze, I., Zhang, Q., & Dennis, S. (2014). Down-regulation of tissue N:P ratios in terrestrial plants by elevated CO_2. *Ecology*, *96*(12), 3354–3362. doi:10.1890/15-0217.1 PMID:26909440

Descy, J. P., Hardy, M. A., Stenuite, S., Perlot, S., Leporco, B., Kimerei, I., Sekadende, B., Mwaitega, S. R., & Sinyenza, D. (2005). Phytoplankton pigments and community composition in Lake Tanganyika. *Freshwater Ecology*, *50*(4), 668–684. doi:10.1111/j.1365-2427.2005.01358.x

Dijkstra, F. A., Pendall, E., Morgan, J. A., Blumenthal, D. M., Carrillo, Y., LeCain, D. R., Follett, R. F., & Williams, D. G. (2012). Climate change alters stoichiometry of phosphorus and nitrogen in a semiarid grassland. *The New Phytologist*, *196*(3), 807–815. doi:10.1111/j.1469-8137.2012.04349.x PMID:23005343

Dill, L. M. (1983). Adaptive flexibility in the foraging behavior of fishes. *Canadian Journal of Fisheries and Aquatic Sciences*, *40*(4), 398–408. doi:10.1139/f83-058

Dobiesz, N. E., Hecky, R. E., Johnson, T. B., Sarvala, J., Dettmers, J. M., Lehtiniemi, M., Rudstam, L. G., Madenjian, C. P., & Witte, F. (2010). Metrics of ecosystem status for large aquatic systems – A global comparison. *Journal of Great Lakes Research*, *36*(1), 123–138. doi:10.1016/j.jglr.2009.11.003

Donaton, J., Durham, K., Cerrato, R., Schwerzmann, J., & Thorne, L. H. (2018). Long-term changes in loggerhead sea turtle diet indicate shifts in the benthic community associated with warming temperatures. *Estuarine, Coastal and Shelf Science*. Advance online publication. doi:10.1016/j.ecss

Donoghue, I., & Irvine, K. (2003). *Effects of sediment particle size composition on survivorship of benthic invertebrates from Lake Tanganyika, Africa.* ArchivfuerHydrobiologie.

Donohue, I., & Garcia Molinos, J. (2009). Impacts of increased sediment loads on the ecology of lakes. *Biological Reviews of the Cambridge Philosophical Society*, *84*(4), 517–531. doi:10.1111/j.1469-185X.2009.00081.x PMID:19485985

Donohue, I., & Molinos, G. J. (2009). Impacts of increased sediment loads on the ecology of lakes. *Biological Reviews of the Cambridge Philosophical Society*, *84*(4), 517–531. doi:10.1111/j.1469-185X.2009.00081.x PMID:19485985

Donohue, T., Duck, R. W., & Irvine, K. (2003). Land use, sediment loads and dispersal pathways from two catchments at the southern end of Lake Tanganyika, Africa: Implications for lake management. *Environmental Geology*, *44*(4), 448–455. doi:10.100700254-003-0779-0

Dudgeon, D., Arthington, A. H., Gessner, M. O., Kawabata, Z., Knowler, D., Lévêque, C., Naiman, R. J., Prieur-Richard, A.-H., Soto, D., Stiassny, M. L. J., & Sullivan, C. A. (2006). Freshwater biodiversity: Importance, threats, status and conservation challenges. *Biological Reviews of the Cambridge Philosophical Society*, *81*(2), 163–182. doi:10.1017/S1464793105006950 PMID:16336747

Eggert, A. (2012). Seaweed responses to temperature. In Seaweed Biology. Springer-Verlag. doi:10.1007/978-3-642-28451-9_3

Enriquez, S., Duarte, C. M., & Sand-Jensen, K. (1993). Patterns in decomposition rates among photosynthetic organisms: the importance of detritus C:N:P content. *Oecologia*, *94*(4), 457–471. doi:10.1007/BF00566960 PMID:28313985

Evans, M. R., & Moustakas, A. (2018). Plasticity in foraging behaviour as a possible response to climate change. *Ecological Informatics*, *47*, 61–66. doi:10.1016/j.ecoinf.2017.08.001

Ficke, A. D., Myrick, C. A., & Hansen, L. J. (2007). Potential impacts of global climate change on freshwater fisheries. *Reviews in Fish Biology and Fisheries*, *17*(4), 581–613. doi:10.100711160-007-9059-5

Gleadow, R. M., Evans, J. R., McCaffery, S., & Cavagnaro, T. R. (2009). Growth and nutritive value of cassava (ManihotesculentaCranz.) are reduced when grown in elevated CO2. *Plant Biology*, *11*, 76–82. doi:10.1111/j.1438-8677.2009.00238.x PMID:19778371

Harrod, C., Ramirez, A., Valbo-Jørgensen, J., & Funge-Smith, S. (2018). How climate change impacts inland fisheries. In Impact of climate change in Fisheries and Aquaculture: Synthesis of current knowledge adaptation and mitigation options (pp. 375 –392). FAO Fisheries and Aquaculture Technical Paper No. 627.

He, P., Davy, D., Sciortino, J., Malcolm, C., Beveridge, M., Arnason, R., & Gudmundsson, A. (2018). Countering climate change: measures and tools to reduce energy use and greenhouse gas emission in fisheries and aquaculture. In Impact of climate change in Fisheries and Aquaculture: Synthesis of current knowledge adaptation and mitigation options (pp. 585-610). FAO Fisheries and Aquaculture Technical Paper No. 627.

Hochachka, P. W., & Somero, G. N. (2002). *Biochemical adaptation: mechanism and process in physiological evolution*. Oxford University Press.

Hop, H., & Graham, M. (1995). Respiration of juvenile Arctic cod (*Boreogadussaida*): Effects of acclimation, temperature and food intake. *Polar Biology, 15*(5), 359–367. doi:10.1007/BF00238487

Hori, M., Gashagaza, M. M., Nshombo, M., & Kawanabe, H. (1993). Littoral fish communities in Lake Tanganyika: Irreplaceable diversity supported by intricate interactions among species. *Conservation Biology, 1993*(7), 657–666. doi:10.1046/j.1523-1739.1993.07030657.x

IPCC. (2018). Summary for Policymakers. In *Global Warming of 1.5°C. An IPCC Special Report on the impacts of global warming of 1.5°C above pre-industrial levels and related global greenhouse gas emission pathways, in the context of strengthening the global response to the threat of climate change, sustainable development, and efforts to eradicate poverty*. World Meteorological Organization.

Jefferies, R. L., & Maron, J. L. (1997). The embarrassment of riches: Atmospheric deposition of nitrogen and community and ecosystem processes. *Trends in Ecology & Evolution, 12*(2), 74–78. doi:10.1016/S0169-5347(96)20125-9 PMID:21237983

Jónasdóttir, S. H. (2019). Fatty Acid Profiles and Production in Marine Phytoplankton. *Marine Drugs, 17*(3), 151. doi:10.3390/md17030151 PMID:30836652

Kim, L. (2014). *Effect of increased temperature and decreased food quality on metabolism and growth of an algivorous cichlid, Tropheusduboisi and effect of food habit on the field metabolism of African Cichlids* (Dissertation). Retrieved from https://etd.ohiolink.edu/

Kraemer, B. M., Hook, S., Huttula, T., Kotilainen, P., O'Reilly, C. M., Peltonen, A., Plisnier, P.-D., Sarvala, J., Tamatamah, R., Vadeboncoeur, Y., Wehrli, B., & McIntyre, P. B. (2015). Century-Long Warming Trends in the Upper Water Column of Lake Tanganyika. *PLoS One, 10*(7), e0134537. doi:10.1371/journal.pone.0132490 PMID:26225567

Kwon, Y. S., Li, F., Chung, N., Bae, M. J., Hwang, S. J., Byoen, M. S., Park, Y. S., & Park, Y.-S. (2012). Response of fish communities to various environmental variables across multiple spatial scales. *International Journal of Environmental Research and Public Health, 9*(10), 3629–3653. doi:10.3390/ijerph9103629 PMID:23202766

Lee, H. W., Ban, S., Ikeda, T., & Matsuishi, T. (2003). Effect of temperature on development, growth and reproduction in the marine copepod Pseudocalanusnewmani at satiating food condition. *Journal of Plankton Research, 25*(3), 261–271. doi:10.1093/plankt/25.3.261

Lehodey, P., Alheit, J., Barange, M., Baumgartner, T., Beaugrand, G., Drinkwater, K., & Roy, C. (2006). Climate variability, fish, and fisheries. *Journal of Climate, 19*(20), 5009–5030. doi:10.1175/JCLI3898.1

Lemoine, N. (2015). *The Effects of Climate Warming on Plant-Herbivore Interactions. Electronic Dissertation.* Florida International University. Doi:10.25148/etd.FIDC000056

Lemoine, N. P., Burkepile, D. E., & Parker, J. D. (2014). Variable effects of temperature on insect herbivory. *PeerJ, 2*, e376. doi:10.7717/peerj.376 PMID:24860701

Lichtenbelt, W. D. (1992). *Digestion in an Ectothermic Herbivore, the Green Iguana (Iguanaiguana): Effect of Food Composition and Body Temperature*. Academic Press.

Lutterschmidt, W. I., & Hutchison, V. H. (1997). The critical thermal maximum: Data to support the onset of spasms as the definitive end point. *Canadian Journal of Zoology, 75*(10), 1553–1560. doi:10.1139/z97-782

Mbomba, N. B. (1983). Comparative morphology of the feeding apparatus in cichlidian algal feeders of Lake Tanganyika. *African Study Monographs, 3*, 1–23.

McIntyre, P. B., Michel, E., France, K., Rivers, A., Hakizimana, P., & Cohen, A. S. (2005). Individual- and Assemblage-Level Effects of Anthropogenic Sedimentation on Snails in Lake Tanganyika. *Conservation Biology, 19*(1), 171–181. doi:10.1111/j.1523-1739.2005.00456.x

McNab, B. K. (2002). *The Physiological Ecology of Vertebrates: a View from Energetics.* Cornell University Press.

Minson, D. J., & McLeod, M. N. (1970). The digestibility of temperate and tropical grasses. *Proc. 11th Int. Grassl. Congr.*, 719-722.

Mölsä, H., Reynolds, J. E., Coenen, E. J., & Lindqvist, O. V. (1999). Fisheries research towards resource management on Lake Tanganyika. *Hydrobiologia, 407*, 1–24. doi:10.1023/A:1003712708969

Mueller, P., & Diamond, J. (2001). Metabolic rate and environmental productivity: Well-provisioned animals evolved to run and idle fast. *Proceedings of the National Academy of Sciences of the United States of America, 98*(22), 12550–12554. doi:10.1073/pnas.221456698 PMID:11606744

Munubi, R. (2015). *Algal quality controls the distribution, behavior and growth of algivorous cichlids in Lake Tanganyika* (Electronic Thesis or Dissertation). Retrieved from https://etd.ohiolink.edu/

Munubi, R. N., McIntyre, P. B., & Vadeboncoeur, Y. (2018). Do grazers respond to or control food quality? Cross-scale analysis of algivorous fish in littoral Lake Tanganyika. *Oecologia, 188*(3), 889–900. doi:10.100700442-018-4240-1 PMID:30099606

Navarro, E., Iglesias, J. I. P., & Ortega, M. M. (1992). Natural sediment as a food source for the cockle Cerastodermaedule (L.): Effect of variable particle concentration on feeding, digestion and the scope for growth. *Journal of Experimental Marine Biology and Ecology, 156*(1), 69–87. doi:10.1016/0022-0981(92)90017-5

Ndebele-Murisa, M. R., Musil, C. F., & Raitt, L. (2010). A review of phytoplankton dynamics in tropical African lakes. *South African Journal of Science, 106*(1-2), 13–18. doi:10.4102ajs.v106i1/2.64

Nkotagu, H. H. (2008). Lake Tanganyika ecosystem management strategies. *Aquatic Ecosystem Health & Management, 11*(1), 36–41. doi:10.1080/14634980801891373

NOAA National Centers for Environmental Information. (2019). *State of the Climate: Global Climate Report for May 2019*. www.ncdc.noaa.gov/sotc/global/201905

O'Connor, M. I., Gilbert, B., & Brown, C. J. (2011). Theoretical predictions for how temperature affects the dynamics of interacting herbivores and plants. *American Naturalist, 178*(5), 626–638. doi:10.1086/662171 PMID:22030732

O'Reilly, C. M., Alin, S. R., Plisnier, P. D., Cohen, A. S., & McKee, B. A. (2003). Climate change decreases aquatic ecosystem productivity of Lake Tanganyika, Africa. *Nature, 424*(6950), 766–768. doi:10.1038/nature01833 PMID:12917682

Ogungbe, A. S., Alabi, C. O. A., & Ometan, O. O. (2015). Comparison of Heat Retention in Fresh Water and Salt Water Samples. *European Journal of Academic Essays, 2*(1), 10–14.

Oyugi, D. O., Cucherousset, J., Baker, D. J., & Britton, J. R. (2012). Effects of temperature on the foraging and growth rate of juvenile common carp, Cyprinuscarpio. *Journal of Thermal Biology, 37*(1), 89–94. doi:10.1016/j.jtherbio.2011.11.005

Person-Le, R. J., Buchet, V., Vincent, B., Le, D. H., & Quemener, L. (2006). Effects of temperature on the growth of Pollack (*Pollachiuspollachius*) juveniles. *Aquaculture (Amsterdam, Netherlands), 251*(2-4), 340–345. doi:10.1016/j.aquaculture.2005.06.029

Prado, R. & Da Silva, G. (2017). *Ecological Response to Global Change: Changes in C:N:P Stoichiometry in Environmental Adaptations of Plants*. doi:10.5772/intechopen.69246

Radomski, P., & Goeman, T. J. (2001). Consequences of human lakeshore development on emergent and floating-leaf vegetation abundance. *North American Journal of Fisheries Management*, *21*(1), 46–61. doi:10.1577/1548-8675(2001)021<0046:COHLDO>2.0.CO;2

Rosenfeld, J. S., Van Leeuwen, T. E., Richards, J. G., & Allen, D. (2015). Relationship between growth and standard metabolic rate: Measurement artefacts and implications for habitat use and life history adaptation in salmonids. *Journal of Animal Ecology*, *84*(1), 4–20. doi:10.1111/1365-2656.12260 PMID:24930825

Rufli, H. (2001). Seasonal abundance of zooplankton and the planktivorous fish, Stolothrissa, in Tanzanian waters of Lake Tanganyika. *The great lakes of the world (GLOW): food-web, health & integrity*, 97.

Sardans, J., Rivas-Ubach, A., & Peñuelas, J. (2012). The C: N: P stoichiometry of organisms and ecosystems in a changing world: a review and perspectives. *Perspectives in Plant Ecology, Evolution and Systematics*, *14*(1), 33–47. doi:10.1016/j.ppees.2011.08.002

Savadova, K., Mazur-Marzec, H., Karosienė, J., Kasperovičienė, J., Vitonytė, I., Toruńska-Sitarz, A., & Koreivienė, J. (2018). Effect of increased temperature on native and alien nuisance cyanobacteria from temperate lakes: An experimental approach. *Toxins*, *10*(11), 445. doi:10.3390/toxins10110445 PMID:30380769

Schindler, D. E., & Eby, L. A. (1997). Stoichiometry of fishes and their prey: Implications for nutrient recycling. *Ecology*, *78*(6), 1816–1831. doi:10.1890/0012-9658(1997)078[1816:SOFATP]2.0.CO;2

Schoo, K. L., Aberle, N., Malzahn, A. M., & Boersma, M. (2012). Food Quality Affects Secondary Consumers Even at Low Quantities: An Experimental Test with Larval European Lobster. *PLoS One*, *7*(3), e33550. doi:10.1371/journal.pone.0033550 PMID:22442696

Schreck, C. B. (2010). Stress and fish reproduction: The roles of allostasis and hormesis. *General and Comparative Endocrinology*, *165*(3), 549–556. doi:10.1016/j.ygcen.2009.07.004 PMID:19596332

Schweiger, O., Settele, J., Kudrna, O., Klotz, S., & Kühn, I. (2008). Climate change can cause spatial mismatch of trophically interacting species. *Ecology*, *89*(12), 3472–3479. doi:10.1890/07-1748.1 PMID:19137952

Shoji, J., Toshito, S., Mizuno, K., Kamimura, Y., Hori, M., & Hirakawa, K. (2011). Possible effects of global warming on fish recruitment: Shifts in spawning season and latitudinal distribution can alter growth of fish early life stages through changes in daylength. *ICES Journal of Marine Science*, *68*(6), 1165–1169. doi:10.1093/icesjms/fsr059

Snoeks, J. (2000). How well known is the ichthyodiversity of the large East African lakes? *Advances in Ecological Research*, *31*, 17–38. doi:10.1016/S0065-2504(00)31005-4

Stenuite, S., Pirlot, S., Hardy, M. A., Sarmento, H., Tarbe, A. L., Leporcq, B., & Descy, J. P. (2007). Phytoplankton production and growth rate in Lake Tanganyika: Evidence of adecline in primary productivity in recent decades. *Freshwater Biology*, *52*(11), 2226–2239. doi:10.1111/j.1365-2427.2007.01829.x

Sturmbauer, C., Mark, W., &Dallinger, R. (1992). Ecophysiology of aufwuchs-eating cichlids in Lake Tanganyika: niche separation by trophic specialization. *Environ Biol Fishes, 35*, 283-290.

Sumaila, U. R., Cheung, W. W. L., Lam, V. W. Y., Pauly, D., & Herrick, S. (2011). Climate change impacts on the biophysics and economics of world fisheries. *Nature Climate Change*, *1*(9), 449–456. doi:10.1038/nclimate1301

Sunday, J., Bates, A., & Dulvy, N. (2011). Global analysis of thermal tolerance and latitude in ectotherms. *Proceedings. Biological Sciences*, *278*(1713), 1823–1830. doi:10.1098/rspb.2010.1295 PMID:21106582

Takeuchi, Y., Ochi, H., Kohda, M., Sinyinza, D., & Hori, M. (2010). 20-year census of a rocky littoral fish community in Lake Tanganyika. *Ecology Freshwater Fish*, *19*(2), 239–248. doi:10.1111/j.1600-0633.2010.00408.x

Thompson, M., Gamage, D., Hirotsu, N., Martin, A., & Seneweera, S. (2017). Effects of elevated carbon dioxide on photosynthesis and carbon partitioning: A perspective on root sugar sensing and hormonal crosstalk. *Frontiers in Physiology*, *8*, 578. doi:10.3389/fphys.2017.00578 PMID:28848452

Van der Putten, W. H., de Ruiter, P. C., Bezemer, T. M., Harvey, J. A., Wassen, M., & Wolters, V. (2004). Trophic interactions in a changing world. *Basic and Applied Ecology*, *5*(6), 487–494. doi:10.1016/j.baae.2004.09.003

Verburg, P., Hecky, R. E., & Kling, H. (2003). Ecological consequences of a century of warming in Lake Tanganyika. *Science*, *301*(5632), 505–507. doi:10.1126cience.1084846 PMID:12829786

Wagner, C., & Adrian, R. (2009). Cyanobacteria dominance: Quantifying the effects of climate change. *Limnology and Oceanography*, *54*(6part2), 2460–2468. doi:10.4319/lo.2009.54.6_part_2.2460

Wagner, C. E., McIntyre, P. B., Buels, K. S., Gilbert, D. M., & Michel, E. (2009). Diet predicts intestine length in Lake Tanganyika's cichlid fishes. *Functional Ecology*, *23*(6), 1122–1131. doi:10.1111/j.1365-2435.2009.01589.x

Walczyńska, A., & Sobczyk, Ł. (2017). The underestimated role of temperature-oxygen relationship in large-scale studies on size-to-temperature response. *Ecology and Evolution*, *7*(18), 7434–7441. doi:10.1002/ece3.3263 PMID:28944028

Weijerman, M., Lindeboom, H., & Zuur, A. F. (2005). Regime shifts in marine ecosystems of the North Sea and Wadden Sea. *Marine Ecology Progress Series*, *298*, 21–39. doi:10.3354/meps298021

Wetzel, R. G. (2001). *Limnology: Lake andRiver Ecosystems* (3rd ed.). AcademicPress.

Wright, I. J., & Westoby, M. (2003). Nutrient concentration, resorption and lifespan: Leaf traits of Australian sclerophyll species. *Functional Ecology*, *17*(1), 10–19. doi:10.1046/j.1365-2435.2003.00694.x

Yamaoka, K. (1983). Morphology and feeding behavior of five species of genus Petrochromis (Teleostei: Cichlidae). *PhysiolEcolJpn*, *19*(1), 57–75.

Yamaoka, K. (1987). Comparative osteology of the jaw of algal-feeding cichlids (Pisces, Teleostei) from Lake Tanganyika. *Kochi Daigaku Kaiyo Seibutsu Kyoiku Kenkyu Senta Kenkyu Hokoku*, *9*, 87–137.

Yamaoka, K., Hori, M., & Kuratani, S. (1986). Ecomorphology of feeding in 'goby-like' cichlid fishes in Lake Tanganyika. *Physiol. Ecol. Jap.*, *23*, 17–29.

KEY TERMS AND DEFINITIONS

Adaptations: Efforts taken to assist people or ecosystems to cope with effects of climatic variations.

Algivorous Cichlids: Fish species belonging to the Cichlidae family feeding on algae attached either on rocky or muddy habitats in freshwater ecosystem.

Anthropogenic Activities: Economic and social human activities which can be either agricultural or industrial.

Biodiversity: Is the abundance, distribution, and adaptation of different living organisms on Earth.

Climate Change: Variation of climatic variables such as temperature, pH, sea water level rise with periods of time exceeding thirty years.

Climate-Smart Agriculture: Farming organisms that can resist some of the drivers of climate change (e.g., rise in temperature, unpredictable rainfall and change in pH).

Mitigation: The act of finding solutions to alleviate or solve a problem.

Compilation of References

Aabd, N. A., Msanda, F., & El Mousadik, A. (2014). Evaluation of variability in argan oil content through different environments and preselection of elite genotypes. *Euphytica, 195*(2), 157–167. doi:10.100710681-013-0969-0

Abrams, P. (1984). Foraging Time Optimization and Interactions in Food Webs. *American Naturalist, 124*(1), 80–96. doi:10.1086/284253

ACPC Ministry of Water, Lands, and Environment. (2001). *Wetland Sector Strategic Plan 2001-2010.* Kampala: Government of Uganda.

Agostinelli, C. (2002). Robust model selection in regression via weighted likelihood methodology. *Statistics & Probability Letters, 56*(3), 289–300. doi:10.1016/S0167-7152(01)00193-6

Agovino, M., Casaccia, M., Ciommi, M., Ferrara, M., & Marchesano, K. (2019). Agriculture, climate change and sustainability: The case of EU-28. *Ecological Indicators, 105*, 525–543. doi:10.1016/j.ecolind.2018.04.064

Ahmadian, N., Demattê, J. A. M., Xu, D., Borg, E., & Zölitz, R. (2016). *A New Concept of Soil Line Retrieval from Landsat 8 Images for Estimating Plant Biophysical Parameters.* Remote Sensing Article. doi:10.3390/rs8090738

Ainsworth, C., Pitcher, T., & Rotinsulu, C. (2008). Evidence of fishery depletions and shifting cognitive baselines in Eastern Indonesia. *Biological Conservation, 141*(3), 848–859. doi:10.1016/j.biocon.2008.01.006

Aishwarya, K. S., Harish, M., Prathibhashree, S., & Panimozhi, K. (2018). Survey On Automated Aquponics Based Gardening Approaches. *2018 Second International Conference on Inventive Communication and Computational Technologies (ICICCT)*, 1377–1381. 10.1109/ICICCT.2018.8473155

Ait El Mokhtar, M., Anli, M., Ben Laouane, R., Boutasknit, A., Boutaj, H., Draoui, A., Zarik, L., & Fakhech, A. (2019a). Food Security and Climate Change. In *Handbook of Research on Global Environmental Changes and Human Health* (pp. 53–73). IGI Global. doi:10.4018/978-1-5225-7775-1.ch004

Ait El Mokhtar, M., Ben Laouane, R., Anli, M., Boutasknit, A., Fakhech, A., Wahbi, S., & Meddich, A. (2019b). Climate Change and Its Impacts on Oases Ecosystem in Morocco. In *Climate Change and Its Impact on Ecosystem Services and Biodiversity in Arid and Semi-Arid Zones* (pp. 217–245). IGI Global. doi:10.4018/978-1-5225-7387-6.ch012

Ait El Mokhtar, M., Ben-Laouane, R., Anli, M., Boutasknit, A., Wahbi, S., & Meddich, A. (2019c). Use of mycorrhizal fungi in improving tolerance of the date palm (*Phoenix dactylifera* L.) seedlings to salt stress. *Scientia Horticulturae*, *253*, 429–438. doi:10.1016/j.scienta.2019.04.066

Ait Hmida, A. (2003). Systèmes de production et stratégies des agriculteurs dans les oasis de la région d'Errachidia au Maroc. *New Medit*, *2*, 37–43.

Ait Houssa, A., Drissi, S., Asehraou, A., Asfers, A., Oubaki, L., & Chraibi, H. (2017). Changements climatiques au Maroc: Quels systèmes de culture et quelles biotechnologies pour s'y adapter? *Revue Marocaine des Sciences Agronomiques et Vétérinaires*, *5*(3).

Akaike, H. (1974). ANew Look at the Statistical Model Identification. *Institute of Electrical and Electronics Engineers*, *19*, 716–723.

Akao, Y. & Mazur, G. H. (2003). The leading edge in QFD: Past, present, and future. *International Journal of Quality and Reliability Management, 20*(1), 20-35.

Akesbi, N. (2012). Une nouvelle stratégie pour l'agriculture marocaine: Le Plan Maroc Vert. *New Medit*, *11*, 12–23.

Akpalu, W., Hassan, R., & Ringler, C. (2009). *Climate variability and maize yield in South Africa: Results from GME and MELE methods.* IFPRI Discussion Paper No. 843. International Food Policy Research Institute.

Alados, C. L., & Aich, A. (2008). Stress assessment of argan (Argania spinosa (L.) Skeels) in response to land uses across an aridity gradient: Translational asymmetry and branch fractal dimension. *Journal of Arid Environments*, *72*(4), 338–349. doi:10.1016/j.jaridenv.2007.06.015

Alcalá, F. J., Martínez-Valderrama, J., Robles-Marín, P., Guerrera, F., Martín-Martín, M., Raffaelli, G., de León, J. T., & Asebriy, L. (2015). A hydrological–economic model for sustainable groundwater use in sparse-data drylands: Application to the Amtoudi Oasis in southern Morocco, northern Sahara. *The Science of the Total Environment*, *537*, 309–322. doi:10.1016/j.scitotenv.2015.07.062 PMID:26282765

Alin, S. R., & Cohen, A. S. (2003). *The surface level of the lake fluctuates at a variety of time and depth scales in response to both climatic and geomorphic factors*. Academic Press.

Alin, S. R., Cohen, A., Bills, R., Gashagaza, M. M., Michel, E., Tiercelin, J. J., & Ntakimazi, G. (1999). Effects of landscape disturbance on animal communities in Lake Tanganyika, East Africa. *Conservation Biology*, *13*(5), 1017–1033. doi:10.1046/j.1523-1739.1999.96476.x

Alin, S. R., O'Reilly, C. M., Cohen, A. S., Dettman, D. L., Palacios-Fest, M. R., & McKee, B. A. (2002). Effects of land-use change on aquatic biodiversity: A review from the paleorecord at Lake Tanganyika, East Africa. *Geology*, *30*(12), 1143–1146. doi:10.1130/0091-7613(2002)030<1143:EOLUCO>2.0.CO;2

Ali, S., Rizwan, M., Qayyum, M. F., Ok, Y. S., Ibrahim, M., Riaz, M., Arif, M. S., Hafeez, F., Al-Wabel, M. I., & Shahzad, A. N. (2017). Biochar soil amendment on alleviation of drought and salt stress in plants: A critical review. *Environmental Science and Pollution Research International*, *24*(14), 12700–12712. doi:10.100711356-017-8904-x PMID:28374202

Almeida-Garcıa, F., & Chahine, S. (2015). Morocco, tourism. In Encyclopedia of Tourism. Academic Press.

Alongi, D. M. (2014). Carbon sequestration in mangrove forests. *Carbon Management*, *3*(3), 313–322. doi:10.4155/cmt.12.20

American Forest and Paper Association Resources Board (AFPARB). (1993). *Sustainable Forestry Principles and Implementation Guidelines. Ecosystem Management: A new Approach to Federal Forest Management and Planning*. American Forest and Paper Association Resources Board.

Amiri, F. M., & Khadivar, A. (2017). A fuzzy expert system for diagnosis and treatment of musculoskeletal disorders in wrist. *Technical Gazette*, *24*(Suppl.1), 147–155.

Amthor, J. S. (1997). Plant respiratory responses to elevated carbon dioxide partial pressure. Advances in Carbon Dioxide Effects Research. In *Proceedings of a Symposium*. American Society of Agronomy. 10.2134/asaspecpub61.c2

An, D., Xi, B., Ren, J., Wang, Y., Jia, X., He, C., & Li, Z. (2017). Sustainability assessment of groundwater remediation technologies based on multi-criteria decision making method. *Resources, Conservation and Recycling*, *119*, 36–46. doi:10.1016/j.resconrec.2016.08.002

Andriantiatsaholiniaina, L. A., Kouikoglou, V. S., & Phillis, Y. A. (2004). Evaluating strategies for sustainable development: Fuzzy logic reasoning and sensitivity analysis. *Ecological Economics*, *48*(2), 149–172. doi:10.1016/j.ecolecon.2003.08.009

Anonymous. (1998). Himalayan glacier backing off 1998. *Science*, *281*(5381), 1277. doi:10.11 26cience.281.5381.1277a

Antonelli Paul, M., Fraser Lauchlan, H., & Gardner Wendy, C. (2018). Long term carbon sequestration potential of biosolids-amended copper and molybdenum mine tailings following mine site reclamation. *Ecological Engineering*, *117*, 38–49. doi:10.1016/j.ecoleng.2018.04.001

Araujo, E., & Miyahira, S. A. (2009). Unidimensional Fuzzy Pain Intensity Scale. *IEEE International Conferences on Fuzzy Systems*, 185-190. 10.1109/FUZZY.2009.5277285

Arctic Council. (2013). *Arctic Resilience Interim Report 2013*. Stockholm Environmental Institute and Stockhol Resilience Centre.

Arkin, F. R. P. (1980). *On the Relationship between Satellite-Observed Cloud Cover and Precipitation*. Academic Press.

Arkin, P. A. (1979). *The Relationship between Fractional Coverage of High Cloud and Rainfall Accumulations during GATE over the B-Scale Array*. doi:10.1175/1520-0493(1979)107<1382:TRBFCO>2.0.CO;2

Arodudu, O., Helming, K., Wiggering, H., & Voinov, A. (2017). Towards a more holistic sustainability assessment framework for agro-bioenergy systems-A review. *Environmental Impact Assessment Review*, *62*, 61–75. doi:10.1016/j.eiar.2016.07.008

Arrus, R., & Rousset, N. (2007). *L'agriculture du Maghreb au défi du changement climatique: Quelles stratégies d'adaptation face à la raréfaction des ressources hydriques?* Academic Press.

Assem, N., El Hafid, L., & Lamchouri, F. (2017). Impact of water stress on the leaf tissues of two Moroccan varieties of durum wheat (Triticum durum Desf). *Journal of Materials*, *8*(10), 3583–3487.

Augé, R. M. (2001). Water relations, drought and vesicular-arbuscular mycorrhizal symbiosis. *Mycorrhiza*, *11*(1), 3–42. doi:10.1007005720100097

Augé, R. M., Toler, H. D., & Saxton, A. M. (2015). Arbuscular mycorrhizal symbiosis alters stomatal conductance of host plants more under drought than under amply watered conditions: A meta-analysis. *Mycorrhiza*, *25*(1), 13–24. doi:10.100700572-014-0585-4 PMID:24831020

Aung, N. N., Myo, P. P., & Moe, H. Z. (2014). Field performance evaluation of a power reaper for rice Harvesting. *International Journal of Scientific Engineering and Technology Research*, *12*(3), 2631–2636.

Azim, K., Komenane, S., & Soudi, B. (2017). Agro-environmental assessment of composting plants in Southwestern of Morocco (Souss-Massa Region). *International Journal of Recycling of Organic Waste in Agriculture*, *6*(2), 107–115. doi:10.100740093-017-0157-7

Bachoo & Sastry. (2016). Tensor product model for a Reverse Osmosis system. *IEEE ANDESCON*, *1-4*. doi:10.1109/ANDESCON.2016.7836195

Badraoui, M. (2014). The Green Morocco Plan: an innovative strategy of agricultural development. *Arab environment: Food security. Annual report of the Arab forum for environment and development*, 34-35.

Bailey, M., & Sumaila, U. (2008). Towards ecosystem-based management in the birds head functional seascape of Papua, Indonesia: the economic sub-project. In M. Bailey & T. J. Pitcher (Eds.), *Ecological and economic analyses of marine ecosystems in the Bird's Head seascape, Papua, Indonesia: II. Fisheries Centre Research Reports 16(1)*. Fisheries Centre, University of British Columbia.

Bai, X., & Schandl, H. (2010). *Urban Ecology and Industrial Ecology: The Routledge Handbook of Urban Ecology*. Routledge Publications.

Bakiu, Tafaj, & Taci. (2017). First Study about Aquaponic Systems in Albania. *Journal of Marine Biology and Aquaculture Research, 1*(1).

Bala Krishna Prasad, M., & Ramanathan, A. L. (2008). Sedimentary nutrient dynamics in a tropical estuarine mangrove ecosystem. *Estuarine, Coastal and Shelf Science, 80*(1), 60–66. doi:10.1016/j.ecss.2008.07.004

Balestrini, R., Lumini, E., Borriello, R., & Bianciotto, V. (2015). Plant-soil biota interactions. In Soil Microbiology, Ecology and Biochemistry. London: Academic Press. doi:10.1016/B978-0-12-415955-6.00011-6

Ballav, M. R., & Taranisen, P. (2011). Soil respiration and microbial population in a tropical deciduous forest soil of Orissa, India. *Flora, 206*(12), 1040–1044. doi:10.1016/j.flora.2011.07.009

Barbara, D., Kowalchuk, G., & Van, J. (2008). Climate change goes underground: effects of elevated atmospheric CO2 on microbial community structure and activities in the rhizosphere. Springer-Verlag.

Baret, F., & Guyot, G. (1991). Potentials and Limits of Vegetation Indices for LAI and APAR Assessment. *INRA Bioclimatologie. Montfavet, France, 173*, 161–173.

Baret, F., Guyot, G., & Major, D. J. (1989). Tsavi: A vegetation index which minimizes soil b r i g w e s s effects on w and. *INRA Bioclimatologie. Montfavet, France, 2*, 1355–1358.

Batts, G. R., Morison, J. I. L., Ellis, R. H., Hadley, P., & Wheeler, T. R. (1997). Effects of CO_2 and temperature on growth and yield of crops of winter wheat over several seasons. *European Journal of Agronomy, 7*(1-3), 43–52. doi:10.1016/S1161-0301(97)00022-1

Beck, M. B., & Walker, R. V. (2013). On water security, sustainability, and the water-food-energy- climate nexus. *Frontiers of Environmental Science & Engineering, 7*(5), 626–639. doi:10.100711783-013-0548-6

Belay, H., Ayalew, S., & Hailu, D. (2020). Hydrological performance evaluation of multiple satellite precipitation products in the upper Blue Nile basin, Ethiopia. *Journal of Hydrology: Regional Studies, 27*(August), 100664. doi:10.1016/j.ejrh.2020.100664

Bell, S., & Morse, S. (2008). *Sustainability indicators: measuring the immeasurable?* Earthscan.

Belmakki, M., Bartali, E. H., Xiaoru, H., & Bjerre, A. B. (2015). Identification and characterization of organic waste in Morocco, an important step towards the valorization of waste. *Revue Marocaine des Sciences Agronomiques et Vétérinaires, 3*(3), 37–45.

Ben Laouane, R., Meddich, A., Bechtaoui, N., Oufdou, K., & Wahbi, S. (2019). Effects of Arbuscular Mycorrhizal Fungi and Rhizobia Symbiosis on the Tolerance of Medicago Sativa to Salt Stress. *Gesunde Pflanzen, 71*(2), 135–146. doi:10.100710343-019-00461-x

Ben Salem, A., Messouli, M., & Yacoubi-Khebiza, M. (2011). *Developing an oasis-based water management tool: Ecohydrologic approach and weap software for a large arid catchment in Morocco*. Academic Press.

Beneduzi, A., Ambrosini, A., & Passaglia, L. M. (2012). Plant growth-promoting rhizobacteria (PGPR): Their potential as antagonists and biocontrol agents. *Genetics and Molecular Biology*, *35*(4), 1044–1051. doi:10.1590/S1415-47572012000600020 PMID:23411488

Benhin, K. J. (2006). Climate Change and South African Agriculture: Impacts and Adaptation Options. Centre for Environmental Economics and Policy in Africa (CEEPA).

Benmohammadi, A., Benmohammadi, L., Ballais, J. L., & Riser, J. (2000). Analysis of anthropic and natural interrelations: Their impact on the recrudescence of the phenomena of sand deposition and desertification in south east Morocco (Drâa valley and Ziz valley). *Sécheresse*, *11*(4), 297–308.

Benson, D., Gain, A. K., & Rouillard, J. J. (2015). Water governance in a comparative perspective: From IWRM to a 'nexus' approach? *Water Alternatives*, *8*(1), 756–773.

Bergamino, N., Horion, S., Stenuite, S., Cornet, Y., Loiselle, S., Plisnier, P. D., & Descy, J. P. (2010). Spatio-temporal dynamics of phytoplankton and primary production in Lake Tanganyika using a MODIS based bio-optical time series. *Remote Sensing of Environment*, *114*(4), 772–780. doi:10.1016/j.rse.2009.11.013

Bergamino, N., Loisellea, S. A., Cozar, A., Dattiloa, A. M., Bracchini, L., & Rossi, C. (2007). Examining the dynamics of phytoplankton biomass in Lake Tanganyika using empirical orthogonal functions. *Ecological Modelling*, *204*(1-2), 156–162. doi:10.1016/j.ecolmodel.2006.12.031

Berkes, F. (2016). Sustainability policy considerations for ecosystem management in Central and Eastern Europe. *Ecosystem Health and Sustainability*, *2*(8), e01234. doi:10.1002/ehs2.1234

Bhakta, J. N., LahiriSusmita, Pittman J, K., & Jana B. B., 2015. Carbon dioxide sequestration in wastewater by a consortium of elevated carbon dioxide-tolerant microalgae. *Journal of CO_2 Utilization, 10*, 105-112.

Bhattacharyya, P., Neogi, S., Roy, K. S., Dash, P. K., Nayak, A. K., & Mohapatra, T. (2014). Tropical low land rice ecosystem is a net carbon sink. *Agriculture, Ecosystems & Environment*, *189*, 127–135. doi:10.1016/j.agee.2014.03.013

Bhattacharyya, P., Roy, K. S., Neogi, S., Adhya, T. K., Rao, K. S., & Manna, M. C. (2012a). Effects of rice straw and nitrogen fertilization on greenhouse gas emissions and carbon storage in tropical flooded soil planted with rice. *Soil & Tillage Research*, *124*, 119–130. doi:10.1016/j.still.2012.05.015

Bhattacharyya, P., Roy, K. S., Neogi, S., Adhya, T. K., Rao, K. S., Manna, M. C., & Nayak, A. K. (2013). Influence of elevated carbon dioxide and temperature on belowground carbon allocation and enzyme activities in tropical flooded soil planted with rice. *Environmental Monitoring and Assessment*, *185*(10), 8659–8671. doi:10.100710661-013-3202-7 PMID:23612768

Bhattacharyya, P., Roy, K. S., Neogi, S., Chakravorti, S. P., Behera, K. S., Das, K. M., Bardhan, S., & Rao, K. S. (2012). Effect of long-term application of organic amendment on C storage in relation to global warming potential and biological activities in tropical flooded soil planted to rice. *Nutrient Cycling in Agroecosystems*, *94*(2-3), 273–285. doi:10.100710705-012-9540-y

Biggs, R., Schluter, M., Biggs, D., Bohensky, E., BurnSilver, S., Cundill, G., Dakos, V., Daw, T. M., Evans, L. S., Kotschy, K., Leitch, A. M., Meek, C., Quinlan, A., Raudsepp-Hearne, C., Robards, M. D., Schoon, M. L., Schultz, L., & West, P. C. (2012). Toward principles for enhancing the resilience of ecosystem services. *Annual Review of Environment and Resources*, *37*(1), 421–448. doi:10.1146/annurev-environ-051211-123836

Birge, H., Allen, C., Garmestani, A., & Pope, K. (2016). Adaptive management for ecosystem services. *Journal of Environmental Management*, *183*, 343–352. doi:10.1016/j.jenvman.2016.07.054 PMID:27460215

Bogardi, J. J., Dudgeon, D., Lawford, R., Flinkerbusch, E., Meyn, A., Pahl-Wostl, C., Vielhauer, K., & Vörösmarty, C. (2012). Water security for a planet under pressure: Interconnected challenges of a changing world call for sustainable solutions. *Current Opinion in Environmental Sustainability*, *4*(1), 35–43. doi:10.1016/j.cosust.2011.12.002

Bohensky, E., Connell, D., & Taylor, B. (2009). *Experiences with Integrated River Basin Management, International and Murray Darling Basin: Lessons for Northern Australia. Norther Australia Land and Water Science Review*. CSIRO.

Bolund, P., & Hunhammar, S. (1999). Ecosystem Services in Urban Areas. *Ecological Economics*, *29*(2), 293–301. doi:10.1016/S0921-8009(99)00013-0

Bonan, G. (2008). *Ecological climatology: Concepts and application*. Cambridge University Press. doi:10.1017/CBO9780511805530

Bor, M., Özdemir, F., & Türkan, I. (2003). The effect of salt stress on lipid peroxidation and antioxidants in leaves of sugar beet Beta vulgaris L. and wild beet Beta maritima L. *Plant Science*, *164*(1), 77–84. doi:10.1016/S0168-9452(02)00338-2

Born, K., Fink, A. H., & Paeth, H. (2008). Dry and wet periods in the northwestern Maghreb for present day and future climate conditions. *Meteorologische Zeitschrift (Berlin)*, *17*(5), 533–551. doi:10.1127/0941-2948/2008/0313

Boscolo-Galazzo, F., Crichton, K., Barker, S., & Pearson, P. (2018). Temperature dependency of metabolic rates in the upper ocean: A positive feedback to global climate change? *Global and Planetary Change*, *170*, 201–212. Advance online publication. doi:10.1016/j.gloplacha.2018.08.017

Bottani, E., Gentilotti, M. C., & Rinaldi, M. (2017). A Fuzzy Logic-Based Tool for the Assessment of Corporate Sustainability: A Case Study in the Food Machinery Industry. *Sustainability*, *9*(4), 583. doi:10.3390u9040583

Botts, E. A., Erasmus, B. F. N., & Alexander, G. J. (2015). Observed range dynamics of South African amphibians under conditions of global change. *Austral Ecology*, *40*(3), 309–317. doi:10.1111/aec.12215

Boullion, S., Borges, A., Moya, E., Diele, K., Dittmar, T., Duke, N., Kristenesen, E., Lee, S., Marchand, C., Middelburg, J., Monroy, V., Smith, T., & Twilley, R. (2008). Mangrove production and carbon sink: A revision of global budget estimates. *Global Biogeochemical Cycles*, *22*(2), 2. doi:10.1029/2007GB003052

Bousquet, P. (2000). Regional changes in carbon dioxide fluxes of land and oceans since 1980. *Science*, *290*(5495), 1342–1346. doi:10.1126cience.290.5495.1342 PMID:11082059

Bouzemouri, B. (2007). Problems of conservation and development of the argan tree. In The argan tree: lever for human development of rural Morocco symposium internationalorganisé by the association Ibn Al Baytar at the Faculty of sciences Rabat.

Boyle, R. (1992). A General History of the Air. Academic Press.

Bracken, N., Macknick, J., Tovar-Hastings, A., Komor, P., Gerritsen, M., & Mehta, S. (2015). *Concentrating Solar Power and Water Issues in the U.S. Southwest*. Joint Institute for Strategic Energy Analysis. doi:10.2172/1176743

Brandl, C., Mertens, A., & Schlick, C. M. (2017a). Ergonomic analysis of working postures using OWAS in semi-trailer assembly, applying an individual sampling strategy. *Int. J. Occup. Saf. Ergon, JOSE*, 1-8. doi:10.1080/10803548.2016.1191224

Brandl, C., Mertens, A., & Schlick, C. M. (2017b). Effect of sampling interval on the reliability of ergonomic analysis using the Ovako working posture analysing system (OWAS). *International Journal of Industrial Ergonomics*, *57*, 68–73. doi:10.1016/j.ergon.2016.11.013

Brar, B. S., Singh, K., Dheri, G. S., & Kumar, B. (2013). Carbon sequestration and soil carbon pools in a rice–wheat cropping system: Effect of long-term use of inorganic fertilizers and organic manure. *Soil & Tillage Research*, *128*, 30–36. doi:10.1016/j.still.2012.10.001

Brimblecombe, P. (1996). Attitudes and responses towards air pollution in medieval England. *Journal of the Air Pollution Control Association*, *26*(10), 941–945. doi:10.1080/00022470.1976.10470341 PMID:789426

Brion, J. D., & Balahadia, F. F. (2017). Application of remote sensing and GIS for climate change and agriculture in Philippines. *IEEE 15th Student Conference on Research and Development (SCOReD)*, 229-233. 10.1109/SCORED.2017.8305388

Brooks, T. M., Mittermeier, R. A., da Fonseca, G. A., Gerlach, J., Hoffmann, M., Lamoreux, J. F., ... & Rodrigues, A. S. (2006). Global biodiversity conservation priorities. *Science*, *313*(5783), 58-61.

Brussard, P., Reed, J., & Tracy, R. (1998). Ecosystem management: What is it really? *Landscape and Urban Planning*, *40*(1-3), 9–20. doi:10.1016/S0169-2046(97)00094-7

Bull, J. W., Jobstvogt, N., Böhnke-Henrichs, A., Mascarenhas, A., Sitas, N., Baulcomb, C., Lambini, C. K., Rawlins, M., Baral, H., Zähringer, J., Carter-Silk, E., Balzan, M. V., Kenter, J. O., Häyhä, T., Petz, K., & Koss, R. (2016). Strengths, Weaknesses, Opportunities and Threats: A SWOT analysis of the ecosystem services framework. *Ecosystem Services*, *17*, 99–111. doi:10.1016/j.ecoser.2015.11.012

Burnett, A. P., Soreghan, M. J., Schols, C. A., & Brown, E. T. (2011). Tropical east African climate change and its relation to global climate: A record from Lake Tanganyika, tropical east Africa, over the past 90+kyr. *Palaeogeography, Palaeoclimatology, Palaeoecology*, *303*(1-4), 155–167. doi:10.1016/j.palaeo.2010.02.011

Cai, X., Wallington, K., Shafiee-Jood, M., & Marston, L. (2018). Understanding and managing the food-energy-water nexus – opportunities for water resources research. *Advances in Water Resources*, *11*, 259–273. doi:10.1016/j.advwatres.2017.11.014

Cao, M., Zhang, Q., & Shugart, H. H. (2001). Dynamic responses of African ecosystem carbon cycling to climate change. *Climate Research*, *17*, 183–193. doi:10.3354/cr017183

Carpenter, S., Kitchell, J., & Hodgson, J. (1985). Cascading Trophic Interactions and Lake Productivity. *Bioscience*, *35*(10), 634–639. doi:10.2307/1309989

Carter, S., & Gulati, M. (2014). *Climate change, the Food Energy Water Nexus and food security in South Africa. In Understanding the Food Energy Water Nexus*. WWF-SA.

Carter, T. R., & Saarikko, R. A. (1996). Estimating regional crop potential in Finland under a changing climate. *Agricultural and Forest Meteorology*, *79*(4), 301–313. doi:10.1016/0168-1923(95)02287-2

Chan, L. K. & Wu, M. L. (2002a). Quality function deployment: A comprehensive review +of its concepts and methods. *Quality Engineering, 15*(1), 23-35.

Chapin, F. S., Matson, P. A., & Vitousek, P. M. (2011). Landscape Heterogeneity and Ecosystem Dynamics. In *Principles of Terrestrial Ecosystem Ecology*. Springer. doi:10.1007/978-1-4419-9504-9_13

Charrouf, Z., & Guillaume, D. (2010). Should the amazigh diet (regular and moderate argan-oil consumption) have a beneficial impact on human health? *Critical Reviews in Food Science and Nutrition*, *50*(5), 473–477. doi:10.1080/10408390802544520 PMID:20373191

Chavan, P. B., Patil, D. K., & Dhondge, D. S. (2015). Design and development of manually operated reaper. *IOSR Journal of Mechanical and Civil Engineering*, *12*(3), 15–22.

Chaves, M. M., & Oliveira, M. M. (2004). Mechanisms underlying plant resilience to water deficits: Prospects for water-saving agriculture. *Journal of Experimental Botany*, *55*(407), 2365–2384. doi:10.1093/jxb/erh269 PMID:15475377

Chelleri, L., Minucci, G., Ruiz, A., & Karmaoui, A. (2014). Responses to Drought and Desertification in the Moroccan Drâa Valley Region: Resilience at the Expense of Sustainability? *International Journal of Climate Change: Impacts & Responses*, *5*(2), 17–33. doi:10.18848/1835-7156/CGP/v05i02/37199

Chen, B., Han, M. Y., Peng, K., Zhou, S. L., Shao, L., Wu, X. F., Wei, W. D., Liu, S. Y., Li, Z., Li, J. S., & Chen, G. Q. (2018). Global land-water nexus: Agricultural land and freshwater use embodied in worldwide supply chains. *The Science of the Total Environment*, *613*, 931–943. doi:10.1016/j.scitotenv.2017.09.138 PMID:28946381

Chen, F., Liu, Y., Liu, Q., & Li, X. (2014). International Journal of Remote Spatial downscaling of TRMM 3B43 precipitation considering spatial heterogeneity. *International Journal of Remote Sensing*, (August), 37–41. doi:10.1080/01431161.2014.902550

Chen, F., & Li, X. (2016). Evaluation of IMERG and TRMM 3B43 Monthly Precipitation Products over Mainland China. *Remote Sensing*, *8*(6), 1–18. doi:10.3390/rs8060472

Chen, S., Zhang, L., She, D., & Chen, J. (2019). Spatial Downscaling of Tropical Rainfall Measuring Mission (TRMM) Annual and Monthly Precipitation Data over the Middle and Lower Reaches of the Yangtze River Basin, China. *Water (Basel)*, *11*(3), 1–17. doi:10.3390/w11030568

Chieh-Tsung, C. (2012). A new electric brush cutter. *WSEAS Transactions on Systems and Control*, *3*(7), 2224–2856.

Chokkavarapu, N., & Ravibabu, V. (2019). Comparative study of GCMs, RCMs, downscaling and hydrological models : A review toward future climate change impact estimation. *SN Applied Sciences*, *1*(12), 1–15. doi:10.100742452-019-1764-x

Chu, C., Mandrak, N. E., & Minns, C. K. (2005). Potential impacts of climate change on the distributions of several common rare freshwater fishes in Canada. *Diversity & Distributions*, *11*(4), 299–310. doi:10.1111/j.1366-9516.2005.00153.x

Ciabatta, L., Brocca, L., Massari, C., Moramarco, T., Gabellani, S., Puca, S., & Wagner, W. (2016). International Journal of Applied Earth Observation and Geoinformation Rainfall-runoff modelling by using SM2RAIN-derived and state-of-the-art satellite rainfall products over Italy. *International Journal of Applied Earth Observation and Geoinformation*, *48*, 163–173. doi:10.1016/j.jag.2015.10.004

Clark, C. E., Horner, R. M., & Harto, C. B. (2013). Life Cycle Water Consumption for Shale Gas and Conventional Natural Gas. *Environmental Science & Technology*, *47*(20), 11829–11836. doi:10.1021/es4013855 PMID:24004382

Clarke, T. (2003). Delta blues. *Nature*, *422*(6929), 254–256. doi:10.1038/422254a PMID:12646891

Climent, Pelegri-Sebastia, Sogorb, Talens, & Chilo. (2017). Development of the MOOSY4 eNoseIoT for Sulphur-Based VOC Water Pollution Detection. *Journal of Sensors*, *17*(8). Advance online publication. doi:10.339017081917

Coast Information Team. (2004). *Ecosystem based management planning handbook.* Report to Coast Information Team. www.citbc.org/

Cohen, A. S., Gergurich, E. L., Kraemer, B. M., McGlue, M. M., McIntyre, P. B., Russell, J. M., & Swarzenski, P. W. (2016). Climate warming reduces fish production and benthic habitat in Lake Tanganyika, one of the most biodiverse freshwater ecosystems. *Proceedings of the National Academy of Sciences of the United States of America, 113*(34), 9563–9568. doi:10.1073/pnas.1603237113 PMID:27503877

Condés, S., & García-robredo, F. (2012). Forest Ecology and Management An empirical mixed model to quantify climate influence on the growth of Pinus halepensis Mill. stands in South-Eastern Spain. *Forest Ecology and Management Journal, 284*, 59–68. doi:10.1016/j.foreco.2012.07.030

Convention of Biological Diversity (CBC). (2004). *The ecosystem approach.* CBC.

Conway, D., Garderen, E., Deryng, D., Dorling, S., Krueger, T., Landman, W., Lankford, B., Lebek, K., Osborn, T., Ringler, C., Thurlow, J., Zhu, T., & Dalin, C. (2015). *Climate and southern Africa's water-energy-food nexus.* Available online:https://repository.up.ac.za/bitstream/handle/2263/52832/Conway_Climate_2015.pdf?sequence=1&isAllowed=y

Corman, J. R., McIntyre, P. B., Kujoba, B., Mbemba, W., Fink, D., Wheeler, C. W., Gans, C., Michel, E., & Flecker, A. S. (2010). Upwelling couples chemical and biological dynamics across the littoral and pelagic zones of Lake Tanganyika, East Africa. *Limnology and Oceanography, 55*(1), 214–234. doi:10.4319/lo.2010.55.1.0214

Cornelissen, T. (2011). Climate change and its effects on terrestrial insects and herbivory patterns. *Neotropical Entomology, 40*(2), 155–163. doi:10.1590/S1519-566X2011000200001 PMID:21584394

Cotrufo, M. F., Ineson, P., & Scott, A. (1998). Elevated CO2 reduces the nitrogen concentration of plant tissues. *Global Change Biology, 4*(1), 43–54. doi:10.1046/j.1365-2486.1998.00101.x

Coulter, G. W., & Spigel, R. H. (1991). Hydrodynamics, p. 49–75. In G. W. Coulter (Ed.), Lake Tanganyika and its life. Oxford Univ. Press.

Cramer, W., Bondeau, A., Woodward, F. I., Prentice, I. C., Betts, R. A., Brovkin, V., Cox, P. M., Fisher, V., Foley, J. A., Friend, A. D., Kucharik, C., Lomas, M. R., Ramankutty, N., Sitch, S., Smith, B., White, A., & Young-Molling, C. (2001). Global response of terrestrial ecosystem structure and function to CO_2 and climate change: Results from six dynamics global vegetation models. *Global Change Biology, 7*(4), 357–373. doi:10.1046/j.1365-2486.2001.00383.x

Crowley, T. J. (2000). Causes of climate change over the past 1000 years. *Science, 289*(5477), 270–277. doi:10.1126cience.289.5477.270 PMID:10894770

Culmsee, H. (2005). Vegetation and pastoral use in the Western High Atlas Mountains (Morocco) An assessment of sustainability from the geobotanical perspective. In M. Aït Hamza, & H. Popp (Eds.), Pour une nouvelle perception des fonctions des montagnes du Maroc (pp. 67-80). Rabat: Actes du 7e colloque maroco-allemand.

Curticle, C., Dunn, D., Roberts, J., Carr, S., & Halpin, P. (2012). Why ecosystem-based management may fail without changes to tool development and financing. *Bioscience, 62*(5), 508–515. doi:10.1525/bio.2012.62.5.13

D'Souza, J., & Rodrigues, B. F. (2013). Seasonal Diversity of ArbuscularMycorrhizal Fungi in Mangroves of Goa, India. *International Journal of Biodiversity, 2013*, 1–7. doi:10.1155/2013/196527

Dahan, R., Boughala, M., Mrabet, R., Laamari, A., Balaghi, R., & Laiouad, L. (2012). *A review of available knowledge on land degradation in Morocco OASIS—combating dryland degradation. International Center for Agricultural Research in the Dry Areas*. ICARDA.

Dale, V. H., Efroymson, R. A., & Kline, K. L. (2011). The land use climate change energy nexus. *Landscape Ecology, 26*(6), 755–773. doi:10.100710980-011-9606-2

Dandotiya, Jadon, & Sharma. (2018). Effects of meteorological parameters on gaseous air pollutant concentrations in urban area of Gwalior City, India. *Environmental Claims Journal, 1507508*. Advance online publication. doi:10.1080/10406026.2018

Dandotiya, B. (2019). *Health Effects of Air Pollution in Urban Environment*. Climate Change and Its Impact on Ecosystem Services and Biodiversity in Arid and Semi-Arid Zones., doi:10.4018/978-1-5225-7387-6.ch006

Dandotiya, B., Sharma, H. K., & Jadon, N. (2019, January). Role of Urban Vegetation in Particulate Pollution Control in Urban Areas of Gwalior City with Special Reference to SPM. *Adv. Biores, 10*(1), 97–103.

Dandotiya, B., Sharma, H. K., & Jadon, N. (2020). Ambient Air Quality and Meteorological Monitoring of Gaseous Pollutants in Urban Areas of Gwalior City India. *Environmental Claims Journal*, 1–16. Advance online publication. doi:10.1080/10406026.2020.1744854

Darand, M., Amanollahi, J., & Zandkarimi, S. (2017). Evaluation of the performance of TRMM Multi-satellite Precipitation Analysis (TMPA) estimation over Iran. *Atmospheric Research, 190*, 121–127. Advance online publication. doi:10.1016/j.atmosres.2017.02.011

David, P. (2017). Microalgae cultivation for carbon dioxide sequestration and protein production using a high-efficiency photobioreactor system. *Algal Research, 25*, 413–420. doi:10.1016/j.algal.2017.06.003

Davidson, E., & Janssens, I. (2006). Temperature sensitivity of soil carbon decomposition and feedbacks to climate change. *Nature, 440*(7081), 165–173. doi:10.1038/nature04514 PMID:16525463

Day, B. A., & Monroe, C. M. (1999). Environmental education and communication for sustainable world. In *Handbook for International Practitioners*. GreenCom.

De Haas, H. (2001). *Migration and agricultural transformations in the oases of Morocco and Tunisia*. KNAG.

De Jong, C., Makroum, K., & Leavesley, G. (2006). *Developing an oasis-based irrigation management tool for a large semi-arid mountainous catchment in Morocco*. Academic Press.

De Sassi, C., & Tylianakis, J. M. (2012). Climate Change Disproportionately Increases Herbivore over Plant or Parasitoid Biomass. *PLoS One*, *7*(7), e40557. doi:10.1371/journal.pone.0040557 PMID:22815763

De Silva, P. C. P., & De Silva, P. C. A. (2016). Ipanera: An Industry 4. 0 based Architecture for Distributed Soil-less Food Production Systems. *2016 Manufacturing & Industrial Engineering Symposium (MIES)*, 1–5. https://doi.org/10.1109/MIES.2016.7780266

De Waroux, Y. L. P., & Lambin, E. F. (2012). Monitoring degradation in arid and semi-arid forests and woodlands: The case of the argan woodlands (Morocco). *Applied Geography (Sevenoaks, England)*, *32*(2), 777–786. doi:10.1016/j.apgeog.2011.08.005

deMenocal, P. B. (2001). Cultural responses to climate change during the late Holocene. *Science*, *292*(5517), 667–673. doi:10.1126cience.1059827 PMID:11303088

Demeyer, R., & Turkelboom, F. (2014). The ecosystem services stakeholder matrix. In *OpenNESS manual: Stakeholder analysis for environmental decision-making at local level. EC FP7 Grant Agreement no. 308428*. INBO.

Deng, Q., Hui, D. F., Luo, Y. Q., Elser, J., Wang, Y. P., Loladze, I., Zhang, Q., & Dennis, S. (2014). Down-regulation of tissue N:P ratios in terrestrial plants by elevated CO_2. *Ecology*, *96*(12), 3354–3362. doi:10.1890/15-0217.1 PMID:26909440

Denton, F. (2014). *Does climate – smart agriculture hold the key to Africa banishing hunger by 2025? CCDA-LIVE (Wed. 8 Oct. 2014)*. ACPC/UNECA.

Department of Environmental Affairs (DEA). (2013). Long-Term Adaptation Scenarios Flagship Research Programme (LTAS) for South Africa. Pretoria, South Africa: DEA.

Descy, J. P., Hardy, M. A., Stenuite, S., Perlot, S., Leporco, B., Kimerei, I., Sekadende, B., Mwaitega, S. R., & Sinyenza, D. (2005). Phytoplankton pigments and community composition in Lake Tanganyika. *Freshwater Ecology*, *50*(4), 668–684. doi:10.1111/j.1365-2427.2005.01358.x

DFID, Department for International Development-UK. (2003). *Sustainable agriculture. Key sheet*. Retrieved on March 16 2018 from http://www.odi.org.uk/sites/odi.org.uk/files/odiassets/publications-opinion-files/3143.pdf

DFID. (2004). *Agricultural sustainability: Working / Supporting Paper*. Retrieved on May 10 2018 from http://dfid-agricultureconsultation.nri.org/summaries/wp12.pdf

Di Benedetto, N. A., Corbo, M. R., Campaniello, D., Cataldi, M. P., Bevilacqua, A., Sinigaglia, M., & Flagella, Z. (2017). *The role of plant growth promoting bacteria in improving nitrogen use efficiency for sustainable crop production: a focus on wheat*. Academic Press.

Diekkrüger, B., Busche, H., Klase, A., Klase, S., Rademacher, C., & Schulz, O. (2010). Impact of global change on hydrology and soil degradation–scenario analysis for the semi-arid Drâa catchment (South Morocco). *The global dimension of change in river basins: Threat, linkages and adaptation*, 5-13.

Dijkgraaf, K. H., Goddek, S., & Keesman, K. J. (2019). Modeling innovative aquaponics farming in Kenya. *Aquaculture International*, (May). Advance online publication. doi:10.100710499-019-00397-z

Dijkstra, F. A., Pendall, E., Morgan, J. A., Blumenthal, D. M., Carrillo, Y., LeCain, D. R., Follett, R. F., & Williams, D. G. (2012). Climate change alters stoichiometry of phosphorus and nitrogen in a semiarid grassland. *The New Phytologist*, *196*(3), 807–815. doi:10.1111/j.1469-8137.2012.04349.x PMID:23005343

Dill, L. M. (1983). Adaptive flexibility in the foraging behavior of fishes. *Canadian Journal of Fisheries and Aquatic Sciences*, *40*(4), 398–408. doi:10.1139/f83-058

Dintwe, K., & Okin, G. S. (2018). Soil organic carbon in savannas decreases with anthropogenic climate change. *Geoderma*, *309*, 7–16. doi:10.1016/j.geoderma.2017.08.030

Dintwe, K., Okin, G. S., D'Odorico, P., Hrast, T., Mladenov, N., Handorean, A., Bhattachan, A., & Caylor, K. K. (2014). Soil organic C and total N pools in the Kalahari: Potential impacts of climate change on C sequestration in savannas. *Plant and Soil*, *396*(1-2), 27–44. doi:10.100711104-014-2292-5

Dippenaar, M. (2018). The role of tax incentives in encouraging energy efficiency in the largest listed South African businesses. *Suid-Afrikaanse Tydskrif vir Ekonomiese en Bestuurswetenskappe*, *21*(1), a1723. doi:10.4102ajems.v21i1.1723

Dixon, G. R., & Biggs, M. P. (1996). The impact of environmental change on acclimation and dormancy in *Rhododendron. Aspects of Applied Biology*, *45*, 93–100.

Dobiesz, N. E., Hecky, R. E., Johnson, T. B., Sarvala, J., Dettmers, J. M., Lehtiniemi, M., Rudstam, L. G., Madenjian, C. P., & Witte, F. (2010). Metrics of ecosystem status for large aquatic systems – A global comparison. *Journal of Great Lakes Research*, *36*(1), 123–138. doi:10.1016/j.jglr.2009.11.003

Donato, D. C., Kauffman, J. B., Murdiyarso, D., Kurnianto, S., Stidham, M., & Kanninen, M. (2011). Mangroves among the most carbon-rich forests in the tropics. *Geoscience.*, *4*(5), 293–297. doi:10.1038/ngeo1123

Donaton, J., Durham, K., Cerrato, R., Schwerzmann, J., & Thorne, L. H. (2018). Long-term changes in loggerhead sea turtle diet indicate shifts in the benthic community associated with warming temperatures. *Estuarine, Coastal and Shelf Science*. Advance online publication. doi:10.1016/j.ecss

Donnelly, R., Neville, D., & Mous, P. (2003). *Report on a rapid ecological assessment of the Raja Ampat Islands, Papua, Eastern Indonesia*. The Nature Conservancy Southeast Asia Centre for Marine Protected Areas.

Donoghue, I., & Irvine, K. (2003). *Effects of sediment particle size composition on survivorship of benthic invertebrates from Lake Tanganyika, Africa*. ArchivfuerHydrobiologie.

Donohue, I., & Garcia Molinos, J. (2009). Impacts of increased sediment loads on the ecology of lakes. *Biological Reviews of the Cambridge Philosophical Society*, *84*(4), 517–531. doi:10.1111/j.1469-185X.2009.00081.x PMID:19485985

Donohue, T., Duck, R. W., & Irvine, K. (2003). Land use, sediment loads and dispersal pathways from two catchments at the southern end of Lake Tanganyika, Africa: Implications for lake management. *Environmental Geology*, *44*(4), 448–455. doi:10.100700254-003-0779-0

Doyle, L., Oliver, L., & Whitworth, C. (2020). Design of a Climate Smart Farming System in East Africa. *2018 IEEE Global Humanitarian Technology Conference (GHTC)*, 1–6.

Drexhage, J., & Murphy, D. (2010). *Sustainable development: from Brundtland to Rio 2012*. Background Paper for the High Level Panel on Global Sustainability, United Nations. Retrieved on April 10 2018 from https://www.un.org/wcm/webdav/site/climatechange/shared/gsp/docs/GSP1-6_Background%20on%20Sustainable%20Devt.pdf

Drinkwater, L. E., Wagoner, P., & Sarrantonlo, M. (2010). Legume- Based Cropping Systems have Reduced Carbon and Nitrogen Losses. *Nature*, *396*(6708), 262–265. doi:10.1038/24376

Driouech, F. (2010). *Distribution des précipitations hivernales sur le Maroc dans le cadre d'un changement climatique: descente d'échelle et incertitudes* (Thèse de doctorat). de l'Univers, de l'Environnement et de l'Espace, Université de Toulouse.

Driouech, F., Deque, M., & Mokssit, A. (2009). Numerical simulation of the probability distribution function of precipitation over Morocco. *Climate Dynamics*, *32*(7-8), 1055–1063. doi:10.100700382-008-0430-6

Duan, L., Fan, K., Li, W., & Liu, T. (2017). Spatial downscaling algorithm of TRMM precipitation based on multiple high-resolution satellite data for Inner Mongolia, China. *Theoretical and Applied Climatology*, *2001*, 15.

Duan, Z., & Bastiaanssen, W. G. M. (2013). Remote Sensing of Environment First results from Version 7 TRMM 3B43 precipitation product in combination with a new downscaling – calibration procedure. *Remote Sensing of Environment*, *131*, 1–13. doi:10.1016/j.rse.2012.12.002

Dudgeon, D., Arthington, A. H., Gessner, M. O., Kawabata, Z., Knowler, D., Lévêque, C., Naiman, R. J., Prieur-Richard, A.-H., Soto, D., Stiassny, M. L. J., & Sullivan, C. A. (2006). Freshwater biodiversity: Importance, threats, status and conservation challenges. *Biological Reviews of the Cambridge Philosophical Society*, *81*(2), 163–182. doi:10.1017/S1464793105006950 PMID:16336747

Easterling, D. R., Meehl, G. A., Parmesan, C., Changnon, S. A., Karl, T. R., & Mearns, L. O. (2000). Climate extremes: Observations, modeling, and impacts. *Science*, *289*(5487), 2068–2074. doi:10.1126cience.289.5487.2068 PMID:11000103

Eastside Forest Health Assessment Team (ESFHAT). (1993). Eastside Forest Health Assessment, Vol. 1, Executive Summary. U.S. Department of Agriculture, National Forest Service.

Eggert, A. (2012). Seaweed responses to temperature. In Seaweed Biology. Springer-Verlag. doi:10.1007/978-3-642-28451-9_3

Ekirch, A. A. Jr. (1993). *Man and Nature in America*. University of Nebraska Press.

El Benni, N., & Reviron, S. (2009). Geographical Indications: review of seven case-studies world wide. Swiss National Centre of Competence in Research.

El Goumi, Y., Fakiri, M., Lamsaouri, O., & Benchekroun, M. (2014). Salt stress effect on seed germination and some physiological traits in three Moroccan barley (Hordeum vulgare L.) cultivars. *Journal of Materials and Environmental Science*, *5*(2), 625–632.

El Khoumsi, W., Hammani, A., Kuper, M., & Bouaziz, A. (2014). Deterioration of groundwater in arid environments: what impact in oasis dynamics? Case study of Tafilalet, Morocco. World Academy of Science, Engineering and Technology, International Journal of Environmental, Chemical, Ecological, Geological and Geophysical Engineering, 8(11), 764-770.

El Khoumsi, W., Hammani, A., Kuper, M., & Bouaziz, A. (2017). La durabilité du système oasien face à la détérioration des ressources en eaux souterraines: Cas de la palmeraie de Tafilalet. *Revue Marocaine des Sciences Agronomiques et Vétérinaires*, *5*(1), 41–51.

El Mrabet, S., Ouahmane, L., El Mousadik, A., Msanda, F., & Abbas, Y. (2014). The Effectiveness of Arbuscular Mycorrhizal Inoculation and Bio-Compost Addition for Enhancing Reforestation with Argania spinosa in Morocco. *Open Journal of Forestry*, *4*(01), 14–23. doi:10.4236/ojf.2014.41003

Elboutahiri, N., Thami-Alami, I., & Udupa, S. M. (2010). Phenotypic and genetic diversity in Sinorhizobium meliloti and S. medicae from drought and salt affected regions of Morocco. *BMC Microbiology*, *10*(1), 15. doi:10.1186/1471-2180-10-15 PMID:20089174

Elhassnaoui, I., Moumen, Z., Serrari, I., Bouziane, A., Ouazar, D., & Hasnaoui, M. I. (2019). Generation of synthetic design storm hyetograph and hydrologic modeling under HEC HMS for Ziz watershed. *International Journal of Innovative Technology and Exploring Engineering*, *8*(10), 3308–3319. doi:10.35940/ijitee.J1214.0881019

Eloka-Eboka, A. C., & Inambao, F. L. (2017). Effects of CO_2 sequestration on lipid and biomass productivity in microalgal biomass production. *Applied Energy*, *195*, 1100–1111. doi:10.1016/j.apenergy.2017.03.071

Elwahidi, Bellefontaine, Ponette, & Defourny. (2014). Dynamics of change between the arganeraie on-purpose and social change: a socio-ecological balance opportunity? *Journal of Agriculture and Environment for International Development, 108*(2), 109-133 DOI: 10.12895/jaeid.20142.210

Endo, A., Tsurita, I., Burnett, K., & Orencio, P. M. (2017). A review of the current state of research on the water, energy, and food nexus. *Journal of Hydrology. Regional Studies*, *11*, 20–30. doi:10.1016/j.ejrh.2015.11.010

Enriquez, S., Duarte, C. M., & Sand-Jensen, K. (1993). Patterns in decomposition rates among photosynthetic organisms: the importance of detritus C:N:P content. *Oecologia*, *94*(4), 457–471. doi:10.1007/BF00566960 PMID:28313985

EPA. (2015). *Mississippi River/Gulf of Mexico Watershed Nutrient Task Force 2015 Report to Congress*. U.S. Environmental Protection Agency.

Era, N. (2016). *Age of aquaponics dawns on Namibia*. Available online: https://neweralive.na/posts/age-aquaponics-dawns-namibia

Esper, J., Frank, D., Büntgen, U., Verstege, A., Luterbacher, J., & Xoplaki, E. (2007). Long-term drought severity variations in Morocco. *Geophysical Research Letters*, *34*(17), L17702. doi:10.1029/2007GL030844

ESRI. (2017). *GIS Solution for Agriculture*. Retrieved from https://www.esri.com/library/brochures/pdfs/gis-sols-for-agriculture.pdf

Essery, C. I., & Wilcock, D. N. (2009). The variation in rainfall catch from standard UK Meteorological Office raingauges : A twelve year case study. *Hydrological Sciences Journal*, *6667*. Advance online publication. doi:10.1080/02626669109492482

European Commission. (2013). *Green Infrastructure (GI) Enhancing Europe's Natural Capital*. COM249. https://ec.europa.eu/environment/nature/ecosystems/docs/green_infrastructures/1_EN_ACT_part1_v5

European Union. (2000). *Caring for Our Future: Action for Europe's Environment* (3rd ed.). Luxembourg: EU.

Evans, M. R., & Moustakas, A. (2018). Plasticity in foraging behaviour as a possible response to climate change. *Ecological Informatics*, *47*, 61–66. doi:10.1016/j.ecoinf.2017.08.001

Eyotaru, O. (2013). East African Integration Hindered by Information Gaps – Minister. Daily Monitor Newspaper, p. 1.

Ezzine, H., Bouziane, A., Ouazar, D., & Hasnaoui, M. D. (2017). Downscaling of Open Coarse Precipitation Data through Spatial and Statistical Analysis, Integrating NDVI, NDWI, Elevation, and Distance from Sea. *Hindawi*, *2017*, 20. doi:10.1155/2017/8124962

Fang, H., Yu, G., Cheng, S., Mo, J., Yan, J., & Li, S. (2009). ^{13}C abundance, water-soluble and microbial biomass carbon as potential indicators of soil organic carbon dynamics in subtropical forests at different successional stages and subject to different nitrogen loads. *Plant and Soil*, *320*(1-2), 243–254. doi:10.100711104-009-9890-7

Fang, J., Du, J., Xu, W., Shi, P., Li, M., & Ming, X. (2013). Advances in Water Resources Spatial downscaling of TRMM precipitation data based on the orographical effect and meteorological conditions in a mountainous area. *Advances in Water Resources*, *61*, 42–50. doi:10.1016/j.advwatres.2013.08.011

FAO. (2014). *The Water-energy-food Nexus: a New Approach in Support of Food Security and Sustainable Agriculture*. Food Agric. Organ.

Farissi, M., Faghire, M., Bargaz, A., Bouizgaren, A., Makoudi, B., Sentenac, H., & Ghoulam, C. (2014). *Growth, nutrients concentrations, and enzymes involved in plants nutrition of alfalfa populations under saline conditions*. Academic Press.

Farissi, M., Bouizgaren, A., Faghire, M., Bargaz, A., & Ghoulam, C. (2011). Agro-physiological responses of Moroccan alfalfa (Medicago sativa L.) populations to salt stress during germination and early seedling stages. *Seed Science and Technology*, *39*(2), 389–401. doi:10.15258st.2011.39.2.11

Fatima, D. (2013). *Africa on the rise: can the opportunities from climate change spring the continent to transformative development?* An inaugural speech at the 3rd annual climate change and development from 21-23 October 2013 at United Nations conference centre, Addis Ababa, Ethiopia.

Fearnside, P. M. (2004). Are climate change impacts already affecting tropical forest biomass? *Global Environmental Change*, *14*(4), 299–302. doi:10.1016/j.gloenvcha.2004.02.001

Fedorov, A. V., & Philander, S. G. (2000). Is El Niño changing? *Science*, *288*(5473), 1997–2002. doi:10.1126cience.288.5473.1997 PMID:10856205

Feri, A., & Hong-Chao, Z. (2014). A fuzzy logic based aggregation method for LCIA. *Journal of Cleaner Production*, *67*, 159–172. doi:10.1016/j.jclepro.2013.12.010

Fertah, M., Belfkira, A., Taourirte, M., & Brouillette, F. (2017). Extraction and characterization of sodium alginate from Moroccan Laminaria digitata brown seaweed. *Arabian Journal of Chemistry*, *10*, S3707–S3714. doi:10.1016/j.arabjc.2014.05.003

Ficke, A. D., Myrick, C. A., & Hansen, L. J. (2007). Potential impacts of global climate change on freshwater fisheries. *Reviews in Fish Biology and Fisheries*, *17*(4), 581–613. doi:10.100711160-007-9059-5

Field, C. B., & Fung, I. Y. (1999). Enhanced: The not-so-big U.S. carbon sink. *Science*, *285*(5427), 544–545. doi:10.1126cience.285.5427.544

Fiore, A. M., Naik, V., & Leibensperger, E. M. (2015). Air Quality and Climate Connections. *Journal of the Air & Waste Management Association*, *65*(6), 645–685. doi:10.1080/10962247.2015.1040526 PMID:25976481

Forest Ecosystem Management Assessment Team (FEMAT). (1993). Forest Ecosystem Management: An Ecological, Economic, and Social Assessment. Report of the Ecosystem Management Assessment Team (A Federal Agency consortium), Portland, OR.

Fouad, M. O., Essahibi, A., Benhiba, L., & Qaddoury, A. (2014). Effectiveness of arbuscular mycorrhizal fungi in the protection of olive plants against oxidative stress induced by drought. *Spanish Journal of Agricultural Research*, *12*(3), 763–771. doi:10.5424jar/2014123-4815

Foulds, L. R. (1984). Linear Programming and Extensions 1.1. *Combinatorial Optimization for Undergraduates*, 67.

Fowler, H. J., & Ekstr, M. (2009). Multi-model ensemble estimates of climate change impacts on UK seasonal precipitation extremes. *International Journal of Climatology*, *416*(January), 385–416. doi:10.1002/joc.1827

Fox, H. R., Wilby, R. L., & Moore, H. M. (2001). The impact of river regulation and climate change on the barred estuary of the Oued Massa, southern Morocco. *Regulated Rivers: Research & Management: An International Journal Devoted to River Research and Management*, *17*(3), 235–250. doi:10.1002/rrr.613

Franke, A. C., Haverkort, A. J., & Steyn, J. M. (2013). Climate change and potato production in contrasting South African Agro-Ecosystems 2. Assessing risks and opportunities of adaptation strategies. *Potato Research*, *56*(1), 51–66. doi:10.100711540-013-9229-x

Fujisawa, M., New, M., & Johnston, P. (2013). Is Agricultural Sector Listening to Us? African Climate and Development Initiative, University of Cape Town.

Fung, I. (2000). Variable carbon sinks. *Science*, *290*(5495), 1313. doi:10.1126cience.290.5495.1313 PMID:17787235

Gandhewar, V. R., Moghe, S. M., Kapratwar, S., & Bobate, N. G. (2015). A review on human powered mini paddy harvester. *International Journal for Engineering Applications and Technology*.

Gandonou, C. B., Gnancadja, L. S., Abrini, J., & Skali-Senhaji, N. (2012). Salinity tolerance of some sugarcane (Saccharum sp.) cultivars in hydroponic medium. *International Sugar Journal*, *114*(1359), 190.

Gandonou, C., Abrini, J., Idaomar, M., & Skali-Senhaji, N. (2005). Effects of NaCl on growth and ion and proline accumulation in sugarcane (Saccharum sp.) callus culture. *Belgian Journal of Botany*, 173–180.

Gange, A. C., Ganage, E. G., Sparks, T. H., & Boddy, L. (2007). Rapid and recent changes in fungal fruiting patterns. *Science*, *316*(5821), 71.

Garg, I. K., & Sharma, V. K. (1991). Tractor-operated vertical-conveyer reaper windrower. *Indian Farming*, *40*(11), 32–36.

Garrett, D. S. (1957). *Biology of Root Invading Fungi*. Cambridge University Press.

Gasso, V., Oudshoorn, F. W., De Olde, E., & Sørensen, C. A. G. (2015). Generic sustainability assessment themes and the role of context: The case of Danish maize for German biogas. *Ecological Indicators*, *49*, 143–153. doi:10.1016/j.ecolind.2014.10.008

Gaviglio, A., Bertocchi, M., & Demartini, E. (2017). A Tool for the Sustainability Assessment of Farms: Selection, Adaptation and Use of Indicators for an Italian Case Study. *Resources*, *6*(4), 60. doi:10.3390/resources6040060

Geng, S. M., Yan, D. H., Zhang, T. X., Weng, B. S., Zhang, Z. B., & Qin, T. L. (2015). Effects of drought stress on agriculture soil. *Natural Hazards*, *75*(2), 1997–2011. doi:10.100711069-014-1409-8

Gephart, J. A., Troell, M., Henriksson, P. J. G., Beveridge, M. C. M., Verdegem, M., Metian, M., Mateos, L. D., & Deutsch, L. (2017). The 'seafood gap' in the food-water nexus literature—Issues surrounding freshwater use in seafood production chains. *Advances in Water Resources*, *110*, 505–514. doi:10.1016/j.advwatres.2017.03.025

Ghosh, P. K., & Mahanta, S. K. (2014). Carbon Sequestration in grassland systems. *Range Management and Agroforestry*, *35*, 173–181.

Gibelin, A. L., & Déqué, M. (2003). Anthropogenic climate change over the Mediterranean region simulated by a global variable resolution model. *Climate Dynamics*, *20*(4), 327–339. doi:10.100700382-002-0277-1

GIEC. (2007). Résumé à l'intention des décideurs. In Bilan 2007 des changements climatiques: Impacts, adaptation et vulnérabilité. Contribution du Groupe de travail II au quatrième Rapport d'évaluation. Cambridge University Press.

Gleadow, R. M., Evans, J. R., McCaffery, S., & Cavagnaro, T. R. (2009). Growth and nutritive value of cassava (ManihotesculentaCranz.) are reduced when grown in elevated CO2. *Plant Biology*, *11*, 76–82. doi:10.1111/j.1438-8677.2009.00238.x PMID:19778371

Goddek & Keesman. (2017). *Aquaponics - climate smart solutions to enhance food security in Namibia*. Wageningen University and Research.

Goddek, S., & Keesman, K. J. (2018). The necessity of desalination technology for designing and sizing multi-loop aquaponics systems. *Desalination*, *428*, 76–85. doi:10.1016/j.desal.2017.11.024

Goldenberg, S. B., Landsea, C. W., Mestas-Nunez, A. M., & Gray, W. M. (2001). The recent increase in Atlantic hurricane activity: Causes and implications. *Science*, *293*(5529), 474–479. doi:10.1126cience.1060040 PMID:11463911

Gong, L., Ran, Q., He, G., & Tiyip, T. (2015). A soil quality assessment under different land use types in Keriya river basin, Southern Xinjiang, China. *Soil & Tillage Research*, *146*, 223–229. doi:10.1016/j.still.2014.11.001

Gorain, P. C., Sengupta, S., Satpati, G. G., Paul, I., Tripathi, S., & Pal, R. (2018). Carbon sequestration in macroalgal mats of brackish-water habitats in Indian Sunderbans: Potential as renewable organic resource. *The Science of the Total Environment*, *626*, 689–702. doi:10.1016/j.scitotenv.2018.01.106 PMID:29898555

Government of Uganda. (1999). *A simplified version of the United Nations Convention on the elimination of all forms of discrimination against women*. Uganda Government.

Government of Uganda. (2015). *Climate change policy 2013. Ministry of water and environment.* Government of Uganda.

Grafton, R. Q., Libecap, G. D., Edwards, E. C., O'Brien, R. J., & Landry, C. (2012). Comparative assessment of water markets: Insights from the Murray–Darling Basin of Australia and the Western USA. *Water Policy*, *14*(2), 175–193. doi:10.2166/wp.2011.016

Graumlich, L. J., & Davis, M. B. (1993). Holocene variation in spatial scales of vegetation pattern in the upper Great Lakes. *Ecology*, *74*(3), 826–839. doi:10.2307/1940809

Gregg, W. W., & Casey, N. W. (2004). Global and regional evaluation of the SeaWiFS chlorophyll data set. *Remote Sensing of Environment*, *93*(4), 463–479. doi:10.1016/j.rse.2003.12.012

Gregory, J. M., & Oerlemans, J. (1998). Simulated future sea-level rise due to glacier melt based on regionally and seasonally resolved temperature changes. *Nature*, *391*(6666), 474–476. doi:10.1038/35119

Gren, I., & Zeleke, A. (2016). Policy design for forest carbon sequestration: A review of the literature. *Forest Policy and Economics*, *70*, 128–136. doi:10.1016/j.forpol.2016.06.008

Grieve, C. & Short, K. (2007). *Implementation of ecosystem-based management in marine capture fisheries. Case studies from World Wildlife Fund's marine ecoregions*. Academic Press.

Groenfeldt, D. (2010). The next nexus: Environmental ethics, water management and climate change. *Water Alternatives*, *3*, 575.

Gruijter, J., Walvoort, D. J., & Bragato, G. (2011). Application of fuzzy logic to boolean models for digital soil assessment. *Geoderma*, *166*(1), 15–33. doi:10.1016/j.geoderma.2011.06.003

Gulati, M., Jacobs, I., Jooste, A., Naidoo, D., & Fakir, S. (2014). The water-energy-food security nexus: Challenges and opportunities for food security in South Africa. *Aquatic Procedia*, *1*, 150–164. doi:10.1016/j.aqpro.2013.07.013

Gunkel, G. (2009). Hydropower - A Green Energy? Tropical Reservoirs and Greenhouse Gas Emissions. *CLEAN - Soil, Air. Water (Basel)*, *37*, 726–734.

Guofeng, Z., Dahe, Q., Yuanfeng, L., Fenli, C., & Pengfei, H. (2016). Accuracy of TRMM precipitation data in the southwest monsoon region of China. *Theoretical and Applied Climatology*, *10*. Advance online publication. doi:10.100700704-016-1791-0

Gupta, A. (2017). Water-energy-food (WEF) nexus and sustainable development. Geophysical Monograph Series, 221-241.

Gutjahr, O. (2013). Comparing precipitation bias correction methods for high-resolution regional climate simulations using COSMO-CLM Effects on extreme values and climate change signal. *Theoretical and Applied Climatology*, *2012*. Advance online publication. doi:10.100700704-013-0834-z

Haddouch, M. (1996). Situation actuelle et perspectives de développement du palmier dattier au Maroc. *Options Méditerranéennes*, *28*, 63–79.

Hadeel, A., Jabbar, M., & Chen, K. (2010). Application of remote sensing and GIS in the study of environmental sensitivity to desertification: A case study in Basrah Province, southern part of Iraq. Published at Springerlink.com. *Appl Geomat*, *2*(3), 101–112. doi:10.100712518-010-0024-y

Hannah, L., Midgley, G. F., & Millar, D. (2002). Climate change-integrated conservation strategies. *Global Ecology and Biogeography*, *11*(6), 485–495. doi:10.1046/j.1466-822X.2002.00306.x

Harris, Z. M., Milner, S., & Taylor, G. (2018). Biogenic Carbon-Capture and Sequestration. *Greenhouse Gas Balance of Bioenergy Systems*, 55-76.

Harrod, C., Ramirez, A., Valbo-Jørgensen, J., & Funge-Smith, S. (2018). How climate change impacts inland fisheries. In Impact of climate change in Fisheries and Aquaculture: Synthesis of current knowledge adaptation and mitigation options (pp. 375 –392). FAO Fisheries and Aquaculture Technical Paper No. 627.

Hatefi, S. M. (2018). Strategic planning of urban transportation system based on sustainable development dimensions using an integrated SWOT and fuzzy COPRAS approach. *Global J. Environ. Sci. Manage.*, *4*(1), 99–112. doi:10.22034/gjesm.2018.04.01.010

Hauck, J., Stein, C., Schiffer, E., & Vandewalle, M. (2015). Seeing the forest and the trees: Facilitating participatory network planning in environmental governance. *Global Environmental Change*, *35*, 400–410. doi:10.1016/j.gloenvcha.2015.09.022

He, P., Davy, D., Sciortino, J., Malcolm, C., Beveridge, M., Arnason, R., & Gudmundsson, A. (2018). Countering climate change: measures and tools to reduce energy use and greenhouse gas emission in fisheries and aquaculture. In Impact of climate change in Fisheries and Aquaculture: Synthesis of current knowledge adaptation and mitigation options (pp. 585-610). FAO Fisheries and Aquaculture Technical Paper No. 627.

Heidecke, C., & Heckelei, T. (2010). Impacts of changing water inflow distributions on irrigation and farm income along the Drâa River in Morocco. *Agricultural Economics*, *41*(2), 135–149. doi:10.1111/j.1574-0862.2009.00431.x

Hein, L., Koppen, K., Groot, R., & Ierland, E. (2006). Spatial scales, stakeholders and the valuation of ecosystem services. *Ecological Economics*, *57*(2), 209–228. doi:10.1016/j.ecolecon.2005.04.005

Hellig, T., Mertens, A., & Brandl, C. (2018). The interaction effect of working postures on muscle activity and subjective discomfort during static working postures and its correlation with OWAS. *International Journal of Industrial Ergonomics*, *68*, 25–33. doi:10.1016/j.ergon.2018.06.006

Hermann, S., Welsch, M., Segerstrom, R.E., Howells, M.I., Young, C., Alfstad, T., Rogner, H. & Steduto, P. (2012). Climate, land, energy and water (CLEW) interlinkages in Burkina Faso: an analysis of agricultural intensification and bioenergy production. *Natural Resources Forum, 36,* 245-262.

Hien, H. T., Marchand, C., Aimé, J., Nhon, D. H., Hong, P. N., Tung, N. X., & Cuc, N. T. (2018). Belowground carbon sequestration in a mature planted mangroves (Northern Viet Nam). *Forest Ecology and Management.*

Hirich, A., Choukr-Allah, R., & Jacobsen, S. E. (2014). Deficit irrigation and organic compost improve growth and yield of quinoa and pea. *Journal Agronomy & Crop Science, 200*(5), 390–398. doi:10.1111/jac.12073

Hirich, A., Choukr-Allah, R., & Jacobsen, S. E. (2014). The combined effect of deficit irrigation by treated wastewater and organic amendment on quinoa (Chenopodium quinoa Willd.) productivity. *Desalination and Water Treatment, 52*(10-12), 2208–2213. doi:10.1080/19443994.2013.777944

Hochachka, P. W., & Somero, G. N. (2002). *Biochemical adaptation: mechanism and process in physiological evolution.* Oxford University Press.

Hodell, D. A., Brenner, M., Curtis, J. H., & Guilderson, T. (2001). Solar forcing of drought frequency in the Maya lowlands. *Science, 292*(5520), 1367–1370. doi:10.1126cience.1057759 PMID:11359010

Hoffmann, J. E., & Dall, E. P. (2018). Integrating desalination with concentrating solar thermal power: A Namibian case study. *Renewable Energy, 115*, 423–432. doi:10.1016/j.renene.2017.08.060

Homewood, K., Kristjanson, P., & Trench, P. C. (Eds.). (2009). Staying Maasai? Livelihoods, Conservation and Development in East Africa's Rangeland. Springer. doi:10.1007/978-0-387-87492-0

Hop, H., & Graham, M. (1995). Respiration of juvenile Arctic cod (*Boreogadussaida*): Effects of acclimation, temperature and food intake. *Polar Biology, 15*(5), 359–367. doi:10.1007/BF00238487

Hori, M., Gashagaza, M. M., Nshombo, M., & Kawanabe, H. (1993). Littoral fish communities in Lake Tanganyika: Irreplaceable diversity supported by intricate interactions among species. *Conservation Biology, 1993*(7), 657–666. doi:10.1046/j.1523-1739.1993.07030657.x

Horton, R., Yohe, G., Easterling, W., Kates, R., Ruth, M., Sussman, E., Whelchel, A., Wolfe, D., & Lipschultz, F. (2014). Ch. 16: Northeast. Climate Change Impacts in the United States: The Third National Climate Assessment. U.S. Global Change Research Program.

Hossain, M. A., Hoque, M. A., Wohab, M. A., Miah, M. A. M., and Hassan, M. S. (2015). Technical and economic performance of combined harvester in farmers' field. *Bangladesh J. Agril. Res., 40*(2), 291-304.

Hossain, M. D. (2003). *Fabrication of self-propelled reaper by locally available materials* (MS thesis). Department of Farm Power and Machinery, BAU, Mymensingh, Bangladesh.

Houghton, R. A., Hackler, J. L., & Lawrence, K. T. (1999). The U.S. carbon budget: Contributions from land-use change. *Science, 285*(5427), 574–578. doi:10.1126cience.285.5427.574 PMID:10417385

Housh, M., Yaeger, M. A., Cai, X., McIsaac, G. F., Khanna, M., Sivapalan, M., Ouyang, Y., Al-Qadi, I., & Jain, A. K. (2015). Managing Multiple Mandates: A System of Systems Model to Analyze Strategies for Producing Cellulosic Ethanol and Reducing Riverine Nitrate Loads in the Upper Mississippi River Basin. *Environmental Science & Technology*, *49*(19), 11932–11940. doi:10.1021/acs.est.5b02712 PMID:26348783

Houshyar, E., Smith, P., Mahmoodi-Eshkaftaki, M., & Azadi, H. (2017). Sustainability of wheat production in Southwest Iran: A fuzzy-GIS based evaluation by ANFIS. *Cogent Food & Agriculture*, *3*(1), 1327682. doi:10.1080/23311932.2017.1327682

Howden, M., Soussana, J. F., Tubiello, F. N., Chhetri, N., Dunlop, M., & Meinke, H. (2007). Adaptation strategies for climate change. *Proceedings of the National Academy of Sciences of the United States of America*, *104*(50), 19691–19698. doi:10.1073/pnas.0701890104 PMID:18077402

Hungate Bruce, A. (2009). Assessing the effect of elevated carbon dioxide on soil carbon: A comparison of four meta-analyses. *Global Change Biology*, *15*(8), 2020–2034. doi:10.1111/j.1365-2486.2009.01866.x

Hunt, D. (1995). *Farm power and machinery management, Cost determination* (9th ed.). Iowa State University Press.

Hunt, J. C. R., Aktas, Y. D., Mahalov, A., Moustaoui, M., Salamanca, F., & Georgescu, M. (2018). Climate change and growing megacities: Hazards and vulnerability. *Proceedings of the Institution of Civil Engineers. Engineering Sustainability*, *171*(6), 314–326. doi:10.1680/jensu.16.00068

Huq, S. (2001). Climate change and Bangladesh. *Science*, *294*(5547), 1617. doi:10.1126cience.294.5547.1617 PMID:11721020

Ilangumaran, G., & Smith, D. L. (2017). Plant growth promoting rhizobacteria in amelioration of salinity stress: A systems biology perspective. *Frontiers in Plant Science*, *8*, 1768. doi:10.3389/fpls.2017.01768 PMID:29109733

Immerzeel, W. W., Rutten, M. M., & Droogers, P. (2009). Remote Sensing of Environment Spatial downscaling of TRMM precipitation using vegetative response on the Iberian Peninsula. *Remote Sensing of Environment*, *113*(2), 362–370. doi:10.1016/j.rse.2008.10.004

Ioannis, R. (2018a). Carbon sequestration via enhanced weathering of peridotites and basalts in seawater. *Applied Geochemistry*, *91*, 197–207. doi:10.1016/j.apgeochem.2017.11.001

IPCC. (1997). *The Regional Impacts of Climate Change*. IPCC.

IPCC. (2001a). *Climate change 2001: the scientific basis*. Contribution of Working Group I to the Third Assessment Report of the IPCC. Cambridge University Press. Available from http://www.ipcc.ch

IPCC. (2002). *Climate change and biodiversity.* IPCC Technical Paper V. https://www.ipcc.ch/pub/tpbiodiv.pdf

IPCC. (2007). *Climate Change 2007: The Physical Science Basis: Contribution of Working Group I to the Fourth Assessment Report of the Intergovernmental Panel on Climate Change*. Cambridge University Press.

IPCC. (2007). Summary for Policymakers. In *The physical Science Basis. Contribution of Working Group I to the Fourth Assessment Report of the Intergovernmental Panel on Climate Change*. Cambridge University Press.

IPCC. (2013). Summary for policy makers. In *Climate Change 2013: The Physical Science Basis. Contribution of Working Group I to the Fifth Assessment Report of the Intergovernmental Panel on Climate Change* (pp. 1–32). Cambridge University Press.

IPCC. (2013). Summary for Policymakers. In Climate Change 2013: The Physical Science Basis. Contribution of Working Group I to the Fifth Assessment Report of the Intergovernmental Panel on Climate Change. Cambridge University Press.

IPCC. (2013). *Working Group I Contribution to the IPCC Fifth Assessment Report Climate Change 2013:The Physical Science Basis Summary for Policymakers*. Cambridge University Press.

IPCC. (2014). *IPCC Synthesis of Scientific-Technical Information relevant to interpreting Article 2 of the UN Framework Convention on Climate Change*. IPCC.

IPCC. (2018). Summary for Policymakers. In *Global Warming of 1.5°C. An IPCC Special Report on the impacts of global warming of 1.5°C above pre-industrial levels and related global greenhouse gas emission pathways, in the context of strengthening the global response to the threat of climate change, sustainable development, and efforts to eradicate poverty*. World Meteorological Organization.

Irfan, M., Hassan, M., & Hassan, N. (2018). Unravelling the Fuzzy Effect of Economic, Social and Environmental Sustainability on the Corporate Reputation of Public-Sector Organizations: A Case Study of Pakistan. *Sustainability*, *10*(3), 769. doi:10.3390u10030769

Irvem, A., & Ozbuldu, M. (2019). Evaluation of Satellite and Reanalysis Precipitation Products Using GIS for All Basins in Turkey. *Hindawi*, *12*, 1–11. Advance online publication. doi:10.1155/2019/4820136

Jackson, R. B., Canadell, J. G., Le Quéré, C., Andrew, R. M., Korsbakken, J. I., Peters, G. P., & Nakicenovic, N. (2016). Reaching peak emissions. *Nature Climate Change*, *6*(1), 7–9. doi:10.1038/nclimate2892

Jackson-Smith, D. (2010). *Toward Sustainable Agricultural Systems in the 21st Century*. National Academies Press.

Jarvie, H. P., Sharpley, A. N., Flaten, D., Kleinman, P. J., Jenkins, A., & Simmons, T. (2015). The Pivotal Role of Phosphorus in a Resilient Water–Energy–Food Security Nexus. *Journal of Environmental Quality*, *44*(4), 1049–1062. doi:10.2134/jeq2015.01.0030 PMID:26437086

Jedli, H., Jbara, A., Hedfi, H., Bouzgarrou, S., & Slimi, K. (2017). Carbon dioxide adsorption isotherm study on various cap rocks in a batch reactor for CO_2 sequestration processes. *Applied Clay Science, 136*, 199–207. doi:10.1016/j.clay.2016.11.022

Jefferies, R. L., & Maron, J. L. (1997). The embarrassment of riches: Atmospheric deposition of nitrogen and community and ecosystem processes. *Trends in Ecology & Evolution, 12*(2), 74–78. doi:10.1016/S0169-5347(96)20125-9 PMID:21237983

Jehanzeb, M., & Cheema, M. (2012). International Journal of Remote Local calibration of remotely sensed rainfall from the TRMM satellite for different periods and spatial scales in the Indus Basin. *International Journal of Remote Sensing*, (November), 37–41. doi:10.1080/01431161.2011.617397

Jenkins, V. (2018). Sustainable management of natural resources: Lessons from Wales. *Journal of Environmental Law, 30*(3), 399–423. doi:10.1093/jel/eqy012

Jensen, M. N. (2003). Consensus on ecological impacts remains elusive. *Science, 299*(5603), 38. doi:10.1126cience.299.5603.38 PMID:12511629

Jia, S., Zhu, W., Aifeng, L., & Yan, T. (2011). Remote Sensing of Environment A statistical spatial downscaling algorithm of TRMM precipitation based on NDVI and DEM in the Qaidam Basin of China. *Remote Sensing of Environment, 115*(12), 3069–3079. doi:10.1016/j.rse.2011.06.009

Jilali, A., & Zarhloule, Y. (2015). Structural control of the hydrogeology in Figuig Oasis in the eastern High Atlas of Morocco. *Proceedings of the Geologists' Association, 126*(2), 232-243. 10.1016/j.pgeola.2014.12.003

Jing, W., Zhang, P., Jiang, H., & Zhao, X. (2017). Reconstructing Satellite-Based Monthly Precipitation over Northeast China Using Machine Learning Algorithms. *Remote Sensing, 9*(8), 1–17. doi:10.3390/rs9080781

Johnston, P., Hachigonta, S., Sibanda, L. M., & Thomas, T. S. (2012). Southern African Agriculture and Climate Change: A Comprehensive Analysis -South Africa. International Food Policy Research Institute (IFPRI).

Jónasdóttir, S. H. (2019). Fatty Acid Profiles and Production in Marine Phytoplankton. *Marine Drugs, 17*(3), 151. doi:10.3390/md17030151 PMID:30836652

Joyce, A., Goddek, S., Kotzen, B., & Wuertz, S. (2019). Aquaponics: Closing the Cycle on Limited Water, Land and Nutrient Resources. In S. Goddek, A. Joyce, B. Kotzen, & G. Burnell (Eds.), *Aquaponics Food Production Systems*. Springer. doi:10.1007/978-3-030-15943-6_2

Junge, R., Bulc, T. G., & Anseeuw, D. (2019). Aquaponics as an Educational Tool. In Aquaponics as an Educational Tool. https://doi.org/ doi:10.1007/978-3-030-15943-6

Kabbadj, A., Makoudi, B., Mouradi, M., Pauly, N., Frendo, P., & Ghoulam, C. (2017). Physiological and biochemical responses involved in water deficit tolerance of nitrogen-fixing Vicia faba. *PLoS One, 12*(12), e0190284. doi:10.1371/journal.pone.0190284 PMID:29281721

Kabiri, L. (2004, June). Impact des changements climatiques et anthropiques sur les ressources en Eau dans les Oasis de Sud Marocain: Cas de Ferkla (Tinjdad, Errachidia, Maroc). In *Colloque International, "Quaternaire et Changements Globaux: Bilan et Perspectives" en Hommage au Professeu Hugues FAURE, Orléans, France* (*Vol. 3*). Academic Press.

Kabiri, L. (2005). Impact des changements climatiques et anthropiques sur les ressources en eau dans l'Oasis de Ferkla (Tinjdad, Goulmima, Errachidia, Maroc). Academic Press.

Karaska, M. A., Huguenin, R. L., Beacham, J. L., Wang, M., Jensen, J. R., & Kaufmann, R. S. (2004). AVIRIS Measurements of Chlorophyll, Suspended Minerals, Dissolved Organic Carbon, and Turbidity in the Neuse River, North Carolina. *Photogrammetric Engineering and Remote Sensing*, *70*(1), 125–133. doi:10.14358/PERS.70.1.125

Karhu, O., Harkonen, R., Sorvali, P., & Vepsalainen, P. (1981). Observing working postures in industry: Examples of OWAS application. *Applied Ergonomics, 12*(1), 13-17. doi:10.1016/0003-6870(81)90088-0

Karhu, O., Kansi, P., & Kuorinka, I. (1977). Correcting working postures in industry: A practical method for analysis. *Applied Ergonomics*, *8*(4), 199–201. doi:10.1016/0003-6870(77)90164-8 PMID:15677243

Karmaoui, A. (2019). Drought and desertification in Moroccan Pre-Sahara, Draa valleys: Exploring from the perspective of young people. *Geoenvironmental Disasters*, *6*(1), 2. doi:10.118640677-019-0118-8

Karmaoui, A., Ifaadassan, I., Messouli, M., & Khebiza, M. Y. (2015). Characterization of common environmental indicators of the Moroccan Oasean Biome, Pilot study in the reserve biosphere of Oases in Southern Morocco. *Advances in Research*, *5*(5), 1–15. doi:10.9734/AIR/2015/19369

Karmaoui, A., Ifaadassan, I., Messouli, M., & Khebiza, M. Y. (2015). Sustainability of the Moroccan oasean system (case study: Middle Draa Valley). *Global Journal of Technology and Optimization*, *6*(01), 170. doi:10.4172/2229-8711.1000170

Karmaoui, A., Messouli, M., Khebiza, Y. M., & Ifaadassan, I. (2014). Environmental vulnerability to climate change and anthropogenic impacts in dryland,(pilot study: Middle Draa Valley, South Morocco). *Journal of Earth Science & Climatic Change*, (S11), 1.

Karmaoui, A., & Moumane, A. (2016). Changes in the environmental vulnerability of Oasean system (desert oasis), Pilot study in Middle Draa Valley, Morocco. *Expert Opinion on Environmental Biology*, *5*(3). Advance online publication. doi:10.4172/2325-9655.1000135

Karoly, D. J., Braganza, K., Stott, P. A., Arblaster, J. M., Meehl, G. A., Broccoli, A. J., & Dixon, K. W. (2003). Detection of a human influence on North American climate. *Science*, *302*(5648), 1200–1203. doi:10.1126cience.1089159 PMID:14615536

Kates, R. W., Parris, T. M., & Leiserowitz, A. A. (2005). What is sustainable development? Goals, indicators, values, and practice. *Environment, 47*(3), 8-21.

Keesman, K. J., Körner, O., Wagner, K., Urban, J., Karimanzira, D., Rauschenbach, T., & Goddek, S. (2019). *Aquaponics Systems Modelling*. doi:10.1007/978-3-030-15943-6_11

Keith, P., Eric, L., Jeffery, K., Ernie, M., & Amy, S. (2019). *Soil C Sequestration as a Biological Negative Emission Strategy*. Frontiers in Climate. doi:10.3389/fclim.2019.00008

Kerr, R. A. (1999). Big El Niños ride the back of slower climate change. *Science*, *283*(5405), 1108–1109. doi:10.1126cience.283.5405.1108

Kerr, R. A. (2001). It's official: Humans are behind most of global warming. *Science*, *291*(5504), 566. doi:10.1126cience.291.5504.566A PMID:11229388

Keskinen, M., Guillaume, J. H. A., Kattelus, M., Porkka, M., Rasanen, T. A., & Varis, O. (2016). The water-energy-food nexus and the transboundary context: Insights from large Asian rivers. *Water-Sui*, *8*(5), w8050193. doi:10.3390/w8050193

Khadraji, A., Mouradi, M., & Ghoulam, C. (2017). Growth and Mineral Nutrition of the Chickpea (Cicer arietinum L.)-Rhizobia Symbiosis under Water Deficit. *Brazilian Archives of Biology and Technology*, 60.

Khdary, N. H., Ghanem, M. A., Abdesalam, M. E., & Al-Garadah, M. M. (2016). Sequestration of CO_2 using Cu nanoparticles supported on spherical and rod-shape mesoporous silica. *Journal of Saudi Chemical Society*, *22*(3), 343–351. doi:10.1016/j.jscs.2016.05.004

Kidd, C. (2001). Satellite Rainfall Climatology : A Review. *International Journal of Climatology*, *1066*(9), 1041–1066. doi:10.1002/joc.635

Kidd, C., & Levizzani, V. (2011). Status of satellite precipitation retrievals. *Hydrology and Earth System Sciences*, *15*(4), 1109–1116. doi:10.5194/hess-15-1109-2011

Kim, L. (2014). *Effect of increased temperature and decreased food quality on metabolism and growth of an algivorous cichlid, Tropheusduboisi and effect of food habit on the field metabolism of African Cichlids* (Dissertation). Retrieved from https://etd.ohiolink.edu/

Kimaro, E., Mor, S., Jenny-Ann, L., & Toribio, M. (2018). Climate change perception and impacts on cattle production in pastoral communities of northern Tanzania. *Pastoralism Research. Policy & Practice*, *8*, 19.

Kim, J. P., Jung, I. W., Park, K. W., Yoon, S. K., & Lee, D. (2016). Hydrological Utility and Uncertainty of Multi-Satellite Precipitation Products in the Mountainous Region of South Korea. *Remote Sensing*, *19*(7), 608. Advance online publication. doi:10.3390/rs8070608

Kimmerer, R. W. (1989). Environmental determinants of spatial pattern in the vegetation of Pb-Zn mines in southwestern Wisconsin. *American Midlaiid Nattrralist*, *121*(1), 134–143. doi:10.2307/2425664

Kinicki, R., Eddy, B., Shockey, I., Boecker, S., Ghahremani, S., Kuck, M., & O'Dell, S. (2017). Organic Farming in Morocco: An Investigation of Agricultural Extension Centers. Academic Press.

Knippertz, P., Christoph, M., & Speth, P. (2003). Long-term precipitation variability in Morocco and the link to the large-scale circulation in recent and future climates. *Meteorology and Atmospheric Physics*, *83*(1-2), 67–88. doi:10.100700703-002-0561-y

Knorr, M., Frey, S. D., & Curtis, P. S. (2005). Nitrogen addition and litter decomposition: A meta-analysis. *Ecology*, *8*(12), 3252–3257. doi:10.1890/05-0150

Koenker, R., Bassett, G., & Jan, N. (1978). Regression Quantiles. *Econometrica*, *46*(1), 33–50. doi:10.2307/1913643

Kofi, A. D. (2014). CCDA- Live. New initiative will help finance efforts to harness information to fight climate change. *Daily news*, *10*, 1.

Kothamasi, D., Kothamasi, S., Bhattacharyya, A., & Babu, C. R. (2006). Arbuscular mycorrhizae and phosphate solubilising bacteria of the rhizosphere of the mangrove ecosystem of Great Nicobar island, India. *Biology and Fertility of Soils*, *42*(4), 358–361. doi:10.100700374-005-0035-8

Kozarevića, S., & Puška, A. (2018). Use of fuzzy logic for measuring practices and performances of supply chain. *Operations Research Perspectives*, *5*, 150–160. doi:10.1016/j.orp.2018.07.001

Koźmińska, A., Wiszniewska, A., Hanus-Fajerska, E., Boscaiu, M., Al Hassan, M., Halecki, W., & Vicente, O. (2019). Identification of Salt and Drought Biochemical Stress Markers in Several Silene vulgaris Populations. *Sustainability*, *11*(3), 800. doi:10.3390u11030800

Kraemer, B. M., Hook, S., Huttula, T., Kotilainen, P., O'Reilly, C. M., Peltonen, A., Plisnier, P.-D., Sarvala, J., Tamatamah, R., Vadeboncoeur, Y., Wehrli, B., & McIntyre, P. B. (2015). Century-Long Warming Trends in the Upper Water Column of Lake Tanganyika. *PLoS One*, *10*(7), e0134537. doi:10.1371/journal.pone.0132490 PMID:26225567

Kristensen, E., Bouillon, S., Dittmar, T., & Marchand, C. (2008). Organic carbon dynamics in mangrove ecosystems: A review. *Aquatic Botany*, *89*(2), 201–219. doi:10.1016/j.aquabot.2007.12.005

Kro, S., & Law, R. J. (2005). Sensing the sea. *Trends in Biotechnology*, *23*(5). Advance online publication. doi:10.1016/j.tibtech.2005.03.004 PMID:15866003

Kuhn, A., Heidecke, C., Roth, A., Goldbach, H., Burkhardt, J., Linstädter, A., Kemmerling, B. & Gaiser, T. (2010). *II-5.1 Importance of resource management for livelihood security under Climate Change in Southern Morocco*. Academic Press.

Kumar, D., Al Hassan, M., Naranjo, M. A., Agrawal, V., Boscaiu, M., & Vicente, O. (2017). Effects of salinity and drought on growth, ionic relations, compatible solutes and activation of antioxidant systems in oleander (Nerium oleander L.). *PLoS One*, *12*(9), e0185017. doi:10.1371/journal.pone.0185017 PMID:28922384

Kumar, M., Sundaram, S., Gnansounou, E., Larroche, C., & Thakur, I. S. (2017). Carbon dioxide capture, storage and production of biofuel and biomaterials by bacteria: A review. *Bioresource Technology*. PMID:28951132

Kunkel, K. E., Easterling, D. R., Kristovich, D. R., Gleason, B., Stoecker, L., & Smith, R. (2012). Meteorological causes of the secular variations in observed extreme precipitation events for the conterminous United States. *Journal of Hydrometeorology*, *13*(3), 1131–1141. doi:10.1175/JHM-D-11-0108.1

Kurokawa, J., Ohara, T., Morikawa, T., Hanayama, S., Greet, J. M., Fukui, T., Kawashima, K., & Akimoto, H. (2013). Emissions of air pollutants and greenhouse gases over Asian regions during 2000–2008: Regional Emission inventory in Asia (REAS) version 2. *Atmospheric Chemistry and Physics*, *13*(21), 11019–11058. doi:10.5194/acp-13-11019-2013

Kwon, Y. S., Li, F., Chung, N., Bae, M. J., Hwang, S. J., Byoen, M. S., Park, Y. S., & Park, Y.-S. (2012). Response of fish communities to various environmental variables across multiple spatial scales. *International Journal of Environmental Research and Public Health*, *9*(10), 3629–3653. doi:10.3390/ijerph9103629 PMID:23202766

Kyukwang, K., & Myung, H. (2015). Sensor Node for Remote Monitoring of Waterborne Disease-Causing Bacteria. *Journal of Sensors*, *15*(3), 10569–10579. doi:10.3390150510569 PMID:25951340

Lal, R., Lorenz, K., Hüttl, R. F., & Schneider, B. U. (2012). Carbon Sequestration in Temperate Forests, Recarbonization of the Biosphere. Academic Press.

Lal, R. (2001). Potential of desertification control to sequester carbon and mitigate the greenhouse effect. *Climatic Change*, *51*(1), 35–72. doi:10.1023/A:1017529816140

Lal, R. (2003). Carbon sequestration in dryland ecosystems. *Environmental Management*, *33*, 528–544. PMID:15453406

Lamaoui, M., Jemo, M., Datla, R., & Bekkaoui, F. (2018). Heat and drought stresses in crops and approaches for their mitigation. *Frontiers in Chemistry*, *6*, 26. doi:10.3389/fchem.2018.00026 PMID:29520357

Lamastra, L., Balderacchi, M., Di Guardo, A., Monchiero, M., & Trevisa, M. (2016). A novel fuzzy expert system to assess the sustainability of the viticulture at the wine-estate scale. *The Science of the Total Environment*, *572*, 724–733. doi:10.1016/j.scitotenv.2016.07.043 PMID:27601053

Lamqadem, A. A., Saber, H., & Rahimi, A. (2017). Spatiotemporal changes of vegetation in the Middle Draa Valley Oasis: A study case of M'hamid El Ghizlane Oasis (Morocco). *European Scientific Journal*, *13*(24), 115–132. doi:10.19044/esj.2017.v13n24p115

Lamqadem, A., Saber, H., & Pradhan, B. (2018). Quantitative assessment of desertification in an arid Oasis using remote sensing data and spectral index techniques. *Remote Sensing*, *10*(12), 1862. doi:10.3390/rs10121862

Lance, B. C. E. (2014). Mallows ' Cp Statistic I. *Wiley StatsRef: Statistics Reference Online*, 1–2.

Lechenet, M., Bretagnolle, V., Bockstaller, C., Boissinot, F., Petit, M.-S., Petit, S., & Munier-Jolain, N. M. (2014). Reconciling Pesticide Reduction with Economic and Environmental Sustainability in Arable Farming. *PLoS One*, *9*(6), e97922. doi:10.1371/journal.pone.0097922 PMID:24887494

Leech, S., Wiensczyk, A., & Turner, J. (2009). Ecosystem management: A practitioner's guide. *BC Journal of Ecosystems and Management*, *10*(2), 1–12.

Lee, H. W., Ban, S., Ikeda, T., & Matsuishi, T. (2003). Effect of temperature on development, growth and reproduction in the marine copepod Pseudocalanusnewmani at satiating food condition. *Journal of Plankton Research*, *25*(3), 261–271. doi:10.1093/plankt/25.3.261

Lehodey, P., Alheit, J., Barange, M., Baumgartner, T., Beaugrand, G., Drinkwater, K., & Roy, C. (2006). Climate variability, fish, and fisheries. *Journal of Climate*, *19*(20), 5009–5030. doi:10.1175/JCLI3898.1

Lemoine, N. (2015). *The Effects of Climate Warming on Plant-Herbivore Interactions. Electronic Dissertation.* Florida International University. Doi:10.25148/etd.FIDC000056

Lemoine, N. P., Burkepile, D. E., & Parker, J. D. (2014). Variable effects of temperature on insect herbivory. *PeerJ*, *2*, e376. doi:10.7717/peerj.376 PMID:24860701

Li, Q. (2010). *A new Likert Scale Based on Fuzzy Sets Theory* (Ph.D. dissertation). University of Connecticut.

Lichtenbelt, W. D. (1992). *Digestion in an Ectothermic Herbivore, the Green Iguana (Iguanaiguana): Effect of Food Composition and Body Temperature*. Academic Press.

Li, Q. (2013). A novel Likert scale based on fuzzy sets theory. *Expert Systems with Applications*, *40*(5), 1609–1618. doi:10.1016/j.eswa.2012.09.015

Liu, J., Zhang, W., & Nie, N. (2018). Spatial Downscaling of TRMM Precipitation Data Using an Optimal Subset Regression Model with NDVI and Terrain Factors in the Yarlung Zangbo River Basin, China. [Research]. *Hindawi*, *13*, 1–13. Advance online publication. doi:10.1155/2018/3491960

Liu, P., Di, L., Du, Q., & Wang, L. (2018). Remote Sensing Big Data : Theory, Methods and Applications. *Remote Sensing*, *4*(5), 711. Advance online publication. doi:10.3390/rs10050711

Liu, S., Jiang, Z., Zhang, J., Wu, Y., Huang, X., & Macreadie, P. (2017). Sediment microbes mediate the impact of nutrient loading on blue carbon sequestration by mixed sea grass meadows. *The Science of the Total Environment*, *599-600*, 1479–1484. doi:10.1016/j.scitotenv.2017.05.129 PMID:28531956

Li, X., Feng, K., Siu, Y. L., & Hubacek, K. (2012). Energy-water nexus of wind power in China: The balancing act between CO_2 emissions and water consumption. *Energy Policy*, *45*, 440–448. doi:10.1016/j.enpol.2012.02.054

Lobell, D. B., & Asner, G. P. (2003). Climate and management contributions to recent trends in U.S. agricultural yields. *Science*, *299*(5609), 1032. doi:10.1126cience.1077838 PMID:12586935

Lopes, C. (2013). *Can climate change spring Africa's transformative development? A Statement by the UN Undersecretary-General and ECA Executive Secretary on 21 October 2013,on third annual conference on climate change and development in Africa from 21-23 October 2013 at UN conference centre*. Addis Ababa: UNECA.

Lutterschmidt, W. I., & Hutchison, V. H. (1997). The critical thermal maximum: Data to support the onset of spasms as the definitive end point. *Canadian Journal of Zoology*, *75*(10), 1553–1560. doi:10.1139/z97-782

Lu, Y., Wang, R., Zhang, Y., Su, H., Wang, P., Jenkins, A., Ferrier, R. C., Bailey, M., & Squire, G. (2015). Ecosystem health towards sustainability. *Ecosystem Health and Sustainability*, *1*(1), 2. doi:10.1890/EHS14-0013.1

Lybbert, T. J., Aboudrare, A., Chaloud, D., Magnan, N., & Nash, M. (2011). Booming markets for Moroccan argan oil appear to benefit some rural households while threatening the endemic argan forest. *Proceedings of the National Academy of Sciences of the United States of America*, *108*(34), 13963–13968. doi:10.1073/pnas.1106382108 PMID:21873185

M'Hirit, Benchekroum, El Yousfi, & Bendaanoun. (1998). The argan tree: a forest species with multiple uses. Mardaga.

Maes, J., Zulian, G., Thijssen, M., Castell, C., Baro, F., Ferreira, A., & Melo, J. (2016). Mapping and Assessment of Ecosystems and Their Services, Urban Ecosystems: 4th Report- Final. Luxembourg: Office for Official Publications of the European Union.

Maggioni, V., & Meyers, P. C. M. D. R. (2016). A Review of Merged High Resolution Satellite Precipitation Product Accuracy During the Tropical Rainfall Measuring Mission (TRMM) - Era. *Journal of Hydrometeorology*. Advance online publication. doi:10.1175/JHM-D-15-0190.1

Mahmood, S., Hemdi, A. R., Saman, M. Z. M., & Noordin, M. Y. (2014). Fuzzy Logic Approach for Assessing Sustainability: Methodology Development for Hollow Fiber Membrane Module. *Advanced Materials Research*, *845*, 579–583. doi:10.4028/www.scientific.net/AMR.845.579

Maidment, D. R. (1993). Handbook of Hydrology. McGraw-Hill Professional.

Mainguet, M., Dumay, F., Kabiri, L., & Rémini, B. (2011). Prodromes of desertification in the Oasis of Tafilalet (Morocco) and specific local solutions. In *Coping with Global Environmental Change, Disasters and Security* (pp. 901–919). Springer. doi:10.1007/978-3-642-17776-7_53

Mallows, C. L. (1973). Some Comments on Cp. *Technometrics,* 37–41. doi:10.1080/00401706.1973.10489103

Mancino, & Ferrara. (2020). Landsat 7 (ETM +) Derived Vegetation Indices in a Mediterranean Environment. *Remote Sensing*, *8*, 20. doi:10.3390/rs12020291

Mann, M. E., & Jones, P. D. (2003). Global surface temperatures over the past two millennia. *Geophysical Research Letters*, *30*(15), 5–4. doi:10.1029/2003GL017814

Manuel, M., Lightfoot, D., & Fattahi, M. (2018). The sustainability of ancient water control techniques in Iran: An overview. *Water History*, *10*(1), 13–30. doi:10.100712685-017-0200-7

Maraun, Wetterhall, Ireson, Chandler, Kendon, Widmann, Brienen, Rust, Sauter, Themeßl, Venema, Chun, Goodess, Jones, Onof, & Vrac. (2010). Precipitation downscaling under climate change: Recent developments to bridge the gap between dynamical models and the end user. *Rev. Geophys*, *48*(2009), 1–34. doi:10.1029/2009RG000314

Marchand, F., Debruyne, L., Triste, L., Gerrard, C., Padel, S., & Lauwers, L. (2014). Key characteristics for tool choice in indicator based sustainability assessment at farm level. *Ecology and Society, 19*(3), 46.

Mariotte, P., Robroek, B. J., Jassey, V. E., & Buttler, A. (2015). Subordinate plants mitigate drought effects on soil ecosystem processes by stimulating fungi. *Functional Ecology*, *29*(12), 1578–1586. doi:10.1111/1365-2435.12467

Markantonis, V., Reynaud, A., Karabulut, A., El-Hajj, R., Altinbilek, D., Awad, I., Bruggeman, A., Constantianos, V., Mysiak, J., Lamaddalena, N., Salah, M., Monteiro, H., Pistocchi, A., & ... . (2019). Can the implementation of the water-energy-food nexus support economic growth in the Mediterranean region? The current status and the way forward. *Frontiers in Environmental Science*, *7*, 1–11. doi:10.3389/fenvs.2019.00084

Marland, G., & Boden, T., A., (1997). *Trends: A Compendium of Data on Global Change*. GHG Concentrations.

Masipa, T. S. (2017). The impact of climate change on food security in South Africa: Current realities and challenges ahead. *Jamba: Journal of Disaster Risk Studies*, *9*(1), a411. doi:10.4102/jamba.v9i1.411 PMID:29955344

Mason, R. (2011). Ecoregional planning. *Journal of Planning Literature*, *26*(4), 405–419. doi:10.1177/0885412211411091

Mbomba, N. B. (1983). Comparative morphology of the feeding apparatus in cichlidian algal feeders of Lake Tanganyika. *African Study Monographs*, *3*, 1–23.

McDowell, N. (2002). Developing countries to gain from carbon trading fund. [London]. *Nature*, *420*(6911), 4. doi:10.1038/420004a PMID:12422171

Mcgrail, P., Nune, S., Motkuri, R., Glezakou, V., Koech, P., Adint, T., Fifield, L., Fernandaz, C., Liu, J., (2017). *System and process for production of magnesium metal and magnesium hydride from magnesium-containing salts and brines*. Academic Press.

McIntyre, P. B., Michel, E., France, K., Rivers, A., Hakizimana, P., & Cohen, A. S. (2005). Individual- and Assemblage-Level Effects of Anthropogenic Sedimentation on Snails in Lake Tanganyika. *Conservation Biology*, *19*(1), 171–181. doi:10.1111/j.1523-1739.2005.00456.x

McLeod, K., & Leslie, H. (2009). Why ecosystem-based management? In K. L. McLeod & H. M. Leslie (Eds.), *Ecosystem-Based Management for the Oceans* (pp. 3–12). Island Press.

McNab, B. K. (2002). *The Physiological Ecology of Vertebrates: a View from Energetics*. Cornell University Press.

Meatco, N. (2019). *Hydroponics Fodder Production: a glimmer of hope for farming*. Available online: https://www.meatco.com.na/news/559/Hydroponics-Fodder-Production-a-glimmer-of-hope-for-farming/

Meddich, A., Jaiti, F., Bourzik, W., El Asli, A., & Hafidi, M. (2015). Use of mycorrhizal fungi as a strategy for improving the drought tolerance in date palm (Phoenix dactylifera). *Scientia Horticulturae*, *192*, 468–474. doi:10.1016/j.scienta.2015.06.024

Mehta, A. V., Yang, S., Group, M. S., & Springs, C. (2008). Advances in Geosciences Precipitation climatology over Mediterranean Basin from ten years of TRMM measurements. *Advances in Geosciences*, *17*, 87–91. doi:10.5194/adgeo-17-87-2008

Meisner, C. A., Petter, H., Badruddin, M., Razzaque, M. A., Giri, G. S., & Scott, J. (1997). Mechanical revolution among small landholders of South Asia: The growing use of Chinese hand tractors. *Proc. of the Joint Intl. Conf. on Agril. Eng. & Tech. Exhibition Vol.'97, 3*, 781-787.

Mekonnen, M. M., Gerbens-Leenes, P. W., & Hoekstra, A. Y. (2015). The consumptive water footprint of electricity and heat: A global assessment. *Environmental Science. Water Research & Technology*, *1*(3), 285–297. doi:10.1039/C5EW00026B

Mekonnen, M. M., & Hoekstra, A. Y. (2016). Four billion people facing severe water scarcity. *Science Advances*, *2*(2), e1500323. doi:10.1126ciadv.1500323 PMID:26933676

Milewski, A., Elkadiri, R., & Durham, M. (2015). Assessment and Comparison of TMPA Satellite Precipitation Products in Varying Climatic and Topographic Regimes in Morocco. *Remote Sensing*, *21*(5), 5697–5717. Advance online publication. doi:10.3390/rs70505697

Mills, A. J., Fey, M. V., Gröngröft, A., Petersen, A., & Medinski, T. V. (2006). Unravelling the effects of soil properties on water infiltration : Segmented quantile regression on a large data set from arid south-west Africa. *Australian Journal OfSoil Research*, *44*(8), 783–797. doi:10.1071/SR05180

Mills, A., Fey, M., Donaldson, J., Todd, S., & Theron, L. (2009). *Soil infiltrability as a driver of plant cover and species richness in the semi-arid Karoo*. doi:10.100711104-009-9904-5

Ministry of Environment and Tourism. (2013). *National Climate Change Strategy & Action Plan 2013 – 2020*. Available online Namibia Food insecurity https://reliefweb.int/report/namibia/namibia-food-insecurity-emergency-plan-action-final-report-operation-n-mdrna009

Minson, D. J., & McLeod, M. N. (1970). The digestibility of temperate and tropical grasses. *Proc. 11th Int. Grassl. Congr.*, 719-722.

Mirsaeidi, A., Moradi, H., & Zekri, M. (2016). Fuzzy-based Prioritization of Health, Safety, and Environmental Risks: The Case of a Large Gas Refinery. *Iranian Journal of Health, Safety & Environment, 4*(1), 641-653.

Mishra, R., Singh, Y. P., Mishra, Y. D., Singh, S., & Singh, H. (2013). Dissemination of improved sickles for female agriculture workers for crop harvesting. *A Journal of Multidisciplinary Advance Research, 2*(1), 118-123.

Mitra, A., Sengupta, K., & Banerjee, K. (2011). Standing biomass and carbon storage of above-ground structures in dominant mangrove trees in the Sundarbans. *Forest Ecology and Management, 261*(7), 1325–1335. doi:10.1016/j.foreco.2011.01.012

Mohamed, B., Mounia, K., Aziz, A., Ahmed, H., Rachid, B., & Lotfi, A. (2018). Sewage sludge used as organic manure in Moroccan sunflower culture: Effects on certain soil properties, growth and yield components. *The Science of the Total Environment, 627*, 681–688. doi:10.1016/j.scitotenv.2018.01.258 PMID:29426193

Mohtar, H., Assi, T., & Daher, B. (2015). *Bridging the Water and Food Gap: The Role of the Water-Energy-Food Nexus*. Working Paper. United Nations University Institute for Integrated Management of Material Fluxes and of Resources (UNU-FLORES).

Mokhtari, N., Mrabet, R., Lebailly, P., & Bock, L. (2014). Spatialisation des bioclimats, de l ' aridité et des étages de végétation du Maroc. *Rev. Mar. Sci. Agron. Vét*, 50–66.

Mölsä, H., Reynolds, J. E., Coenen, E. J., & Lindqvist, O. V. (1999). Fisheries research towards resource management on Lake Tanganyika. *Hydrobiologia, 407*, 1–24. doi:10.1023/A:1003712708969

Mondal, P., Jain, M., Defries, R. S., Galford, G. L., & Small, C. (2014). Sensitivity of crop cover to climate variability : Insights from two Indian agro-ecoregions. *Journal of Environmental Management*. Advance online publication. doi:10.1016/j.jenvman.2014.02.026 PMID:24680541

Morocco. (2014). Drought Management in Morocco. *Workshop on Developing and Implementing Mitigation and Preparedness Of Water Scarcity and Drought (WS&D) Management Plans Zaragoza*, 38.

Moumen, Z. (2019). *Test the Capacity of SWAT Model to Simulate the Hydrological Behavior of Watershed in Semi-Arid Climate Hydrologic Modeling Using SWAT*. doi:10.4018/978-1-5225-9771-1.ch008

Moumen, Z., El, N., El, A., Tvaronavičienė, M., & Lahrach, A. (2019). Water security and sustainable development To cite this version : HAL Id : hal-02342701. *Insights into Regional Development, 1*(4), 301–317. doi:10.9770/ird.2019.1.4(2)

Mpandeli, S., Naidoo, D., Mabhaudhi, T., Nhemachena, C., Nhamo, L., Liphadzi, S., Hlahla, S., & Modi, A. (2018). Climate change adaptation through the water-energy-food nexus in Southern Africa. *International Journal of Environmental Research and Public Health, 15*(10), 2306. doi:10.3390/ijerph15102306 PMID:30347771

Muchapondwa, E., Biggs, H., Driver, A., Matose, F., Mungatana, E., & Scheepers, K. (2012). Providing economic incentives for biodiversity conservation in an emrging bioregional context. *Journal of Sustainable Development, 5*(11), 118–129. doi:10.5539/jsd.v5n11p118

Mueller, P., & Diamond, J. (2001). Metabolic rate and environmental productivity: Well-provisioned animals evolved to run and idle fast. *Proceedings of the National Academy of Sciences of the United States of America*, *98*(22), 12550–12554. doi:10.1073/pnas.221456698 PMID:11606744

Mu, J., & Khan, S. (2009). The effect of climate change on the water and food nexus in China. *Food Security*, *1*(4), 413–430. doi:10.100712571-009-0042-y

Mulla, D. J. (2013). Twenty five years of remote sensing in precision agriculture: Key advances and remaining knowledge gaps. *Biosystems Engineering*, *114*(4), 358–371. doi:10.1016/j.biosystemseng.2012.08.009

Muller, M. (2015). The 'nexus' as a step back towards a more coherent water resource management paradigm. *Water Alternatives*, *8*(1), 675–694.

Muñoz-Rojas, M., Román, J. R., Roncero-Ramos, B., Erickson, T. E., Merritt, D. J., Aguila-Carricondo, P., & Cantón, Y. (2018). Cyanobacteria inoculation enhances carbon sequestration in soil substrates used in dryland restoration. *The Science of the Total Environment*, *636*, 1149–1154. doi:10.1016/j.scitotenv.2018.04.265 PMID:29913577

Munubi, R. (2015). *Algal quality controls the distribution, behavior and growth of algivorous cichlids in Lake Tanganyika* (Electronic Thesis or Dissertation). Retrieved from https://etd.ohiolink.edu/

Munubi, R. N., McIntyre, P. B., & Vadeboncoeur, Y. (2018). Do grazers respond to or control food quality? Cross-scale analysis of algivorous fish in littoral Lake Tanganyika. *Oecologia*, *188*(3), 889–900. doi:10.100700442-018-4240-1 PMID:30099606

Mupambwa, Hausiku, Dakarai, Nciizah, & Dube. (2019). The unique Namib desert-coastal region and its opportunities for climatesmart agriculture: A review. *Cogent Food & Agriculture, 5*(1)., https://doi.org/ doi:10.1080/23311932.2019.1645258

Murmu, S., & Biswas, S. (2015). Application of fuzzy logic and neural network in crop classification: A review. *Aquatic Procedia*, *4*, 1203–1210. doi:10.1016/j.aqpro.2015.02.153

Mzabri, I., Legsayer, M., Kouddane, N., Boukroute, A., & Berrichi, A. (2017). *Salt stress effects on some morphological, physiological and biochemical parameters of saffron plant (Crocus sativus L.) in Eastern Morocco*. Academic Press.

Nag, P. K., Goswami, A., Ashtekar, S. P., & Pradhan, C. K. (1988). Ergonomics in sickle operation. *Applied Ergonomics, 19*(3), 233-239.

Nagayo, M., Vega, & Rodrigo. (2017). An Automated Solar-Powered Aquaponics System towards Agricultural Sustainability in the Sultanate of Oman. *IEEE International Conf. on Smart Grid and Smart Cities*, 1–8

Nair, P. K. R., Nair, V. D., Kumar, B. M., & Haile, S. G. (2009). Soil carbon sequestration in tropical agroforestry systems: A feasibility appraisal. *Environmental Science & Policy, 12*(8), 1099–1111. doi:10.1016/j.envsci.2009.01.010

Najja, S., Zad, M., Zulkafli, Z., & Muharram, F. M. (2018). Satellite Rainfall (TRMM 3B42-V7) Performance Assessment and Adjustment over Pahang River. *Remote Sensing*, 1–24. doi:10.3390/rs10030388

Nalawade, S. M., Turbatmath, P. A., & Gajakos, A. V. (2009). Design and development of tractor operated Jowar reaper windrower. *New Agriculturist, 20*(12), 75-81.

Namibian. (2019). *Windhoek declares water crisis.* https://www.namibian.com.na/188436/archive-read/Windhoek-declares-water-crisis

NamibianSun. (2017). *Aquatics the answer.* Available online, https://www.namibiansun.com/news/aquaponics-the-answer

National Environment Management Agency. (2002). *State of the environment report for Uganda 2002.* Ministry of Water, Lands and Environment. Kampala: Government of Uganda.

National Environment Management Authority (NEMA). (2003). *Guidelines for District State of Environment Report (DSOER).* Preparation in Uganda [Final Report]. Kampala: NEMA.

National Environment Management Authority. (2000/2001). *State of environment report 2000/2001.* Government of Uganda. Kampala: NEMA.

Naumann, Barbosa, Carrao, & Singleton. (2012). Monitoring Drought Conditions and Their Uncertainties in Africa Using TRMM Data. *American Meteorological Society*, 1867–1874. doi:10.1175/JAMC-D-12-0113.1

Navarro, E., Iglesias, J. I. P., & Ortega, M. M. (1992). Natural sediment as a food source for the cockle Cerastodermaedule (L.): Effect of variable particle concentration on feeding, digestion and the scope for growth. *Journal of Experimental Marine Biology and Ecology*, *156*(1), 69–87. doi:10.1016/0022-0981(92)90017-5

Ndebele-Murisa, M. R., Musil, C. F., & Raitt, L. (2010). A review of phytoplankton dynamics in tropical African lakes. *South African Journal of Science*, *106*(1-2), 13–18. doi:10.4102ajs.v106i1/2.64

Nedea, P. S. (2012). The Evolving Concept of Sustainable Agriculture. *Anale. Seria Stiinte Economice. Timisoara*, *18*, 64–69.

Neogi, S., Roy, K. S., Panda, B. B., Nayak, A. K., Rao, K. S., Manna, M. C., & Bhattacharyya, P. (2014). Soil respiration, labile carbon pools and enzyme activities as affected by tillage practices in a tropical rice-maize-cowpea cropping system. *Environmental Monitoring and Assessment*, *186*(7), 4223–4236. doi:10.100710661-014-3693-x PMID:24609455

Neset, T. S., Wiréhn, L., Opach, T., Glaas, E., & Linnér, B. O. (2019). Evaluation of indicators for agricultural vulnerability to climate change: The case of Swedish agriculture. *Ecological Indicators*, *105*, 571–580. doi:10.1016/j.ecolind.2018.05.042

Nezar, K., Khdary, H., Ghanem, M. A., Abdesalam, M. E., & Al-Garadah, M. A. (2016). Sequestration of CO_2 using Cu nanoparticles supported on spherical and rod-shape mesoporous silica. *Journal of Saudi Chemical Society, 16*, 1319-6103.

Nhamo, L., Ndlela, B., Nhemachena, C., Mabhaudhi, T., Mpandeli, S., & Matchaya, G. (2018). The water-energy-food nexus: Climate risks and opportunities in Southern Africa. *Water (Basel), 10*(5), 567. doi:10.3390/w10050567

Nichol, J. E., & Abbas, S. (2015). Science of the Total Environment Integration of remote sensing datasets for local scale assessment and prediction of drought. *The Science of the Total Environment, 505*, 503–507. doi:10.1016/j.scitotenv.2014.09.099 PMID:25461052

Nistor, M. M., Cheval, S., Gualtieri, A. F., Dumitrescu, A., Boţan, V. E., Berni, A., Hognogi, G., Irimuş, I. A., & Porumb-Ghiurco, C. G. (2017). Crop evapotranspiration assessment under climate change in the Pannonian basin during 1991-2050. *Meteorological Applications, 24*(1), 84–91. doi:10.1002/met.1607

Nkotagu, H. H. (2008). Lake Tanganyika ecosystem management strategies. *Aquatic Ecosystem Health & Management, 11*(1), 36–41. doi:10.1080/14634980801891373

NOAA National Centers for Environmental Information. (2019). *State of the Climate: Global Climate Report for May 2019*. www.ncdc.noaa.gov/sotc/global/201905

Noby, M., Hasan, M. K., Ali, M. R., Saha, C. K., Alam, M. M., & Hossain, M. M. (2018). Performance evaluation of modified BAU self-propelled reaper for paddy. *J Bangladesh Agril Univ, 16*(2), 171–177. doi:10.3329/jbau.v16i2.37956

Nouaim, R. (2005). *The argan tree in Morocco: between myths and re'alite's*. L'Harmattan.

O' Marra, F. P. (2012). The role of grasslands in food security and climate change. *Annals of Botany, 110*(6), 1263–1270. doi:10.1093/aob/mcs209 PMID:23002270

O'Connor, M. I., Gilbert, B., & Brown, C. J. (2011). Theoretical predictions for how temperature affects the dynamics of interacting herbivores and plants. *American Naturalist, 178*(5), 626–638. doi:10.1086/662171 PMID:22030732

O'Reilly, C. M., Alin, S. R., Plisnier, P. D., Cohen, A. S., & McKee, B. A. (2003). Climate change decreases aquatic ecosystem productivity of Lake Tanganyika, Africa. *Nature, 424*(6950), 766–768. doi:10.1038/nature01833 PMID:12917682

Ogungbe, A. S., Alabi, C. O. A., & Ometan, O. O. (2015). Comparison of Heat Retention in Fresh Water and Salt Water Samples. *European Journal of Academic Essays, 2*(1), 10–14.

Okaka, W. (2002). An environmental education program: Uganda Polytechnic Kyambogo. Applied Environmental Education and Communication, 1(1), 45-52.

Omer, A. M. (2008). Energy, environment and sustainable development. *Renewable & Sustainable Energy Reviews, 12*(9), 2265–2300. doi:10.1016/j.rser.2007.05.001

ONEE. (2015). *Environmental and social impact study of solar photovoltaic project of Tafilalt, Morocco: site Zagora*. Author.

Ormann, C. A. F. D., Alabrese, J. U. M. C., Rroita, G. U. G. U., Atechou, E. L. M., & Ahn, V. O. B. (2018). Model averaging in ecology : A review of Bayesian, information-theoretic, and tactical approaches for predictive inference. *Ecological Monographs*, *0*(0), 1–20. doi:10.1002/ecm.1309

Osano, P. (2011). *Life at the Crossroads: How Climate Change Threatens the Existence of the Maasai*. Centre for International Governance Innovation. Available at www.africaportal.org/articles/2011/09/06/life-crossroads-how-climate-change-threatensexistence-maasai

Osano, P., Said, M., Moiko, S., Ole Kaelo, D., Schomers, S., Birner, R., & Ogutu, J. (2013). Pastoralism and ecosystem-based adaptation in Kenyan Maasailand. *International Journal of Climate Change Strategies and Management*, *5*(2), 198–214. doi:10.1108/17568691311327596

Ouraich, I., & Tyner, W. E. (2014). Moroccan agriculture, climate change, and the Moroccan Green Plan: A CGE analysis. *African Journal of Agricultural and Resource Economics, 13*(311-2019-681), 307-330.

Oyiga, B. C., Sharma, R. C., Shen, J., Baum, M., Ogbonnaya, F. C., Léon, J., & Ballvora, A. (2016). Identification and characterization of salt tolerance of wheat germplasm using a multivariable screening approach. *Journal Agronomy & Crop Science*, *202*(6), 472–485. doi:10.1111/jac.12178

Oyugi, D. O., Cucherousset, J., Baker, D. J., & Britton, J. R. (2012). Effects of temperature on the foraging and growth rate of juvenile common carp, Cyprinuscarpio. *Journal of Thermal Biology*, *37*(1), 89–94. doi:10.1016/j.jtherbio.2011.11.005

Palm, H. W., Knaus, U., Appelbaum, S., Strauch, S. M., & Kotzen, B. (2019). Coupled Aquaponics Systems. In S. Goddek, A. Joyce, B. Kotzen, & G. Burnell (Eds.), *Aquaponics Food Production Systems*. Springer. doi:10.1007/978-3-030-15943-6_7

Palomares, M., & Heymans, J. (2006). Historical Ecology of the Raja Ampat Archipelago, Papua Province, Indonesia. Vancouver, Canada: Fisheries Centre Research Reports 14(7), Fisheries Centre, University of British Columbia.

Pandey, V. C., Sahu, N., Behera, S. K., & Singh, N. (2016). Carbon sequestration in fly ash dumps: Comparative assessment of three plant association. *Ecological Engineering*, *95*, 198–205. doi:10.1016/j.ecoleng.2016.06.010

Pappa, Kapsis, Mente, & Berillis. (2017). Aquaponics Software in Greece. *Journal of Fisheries*, *11*(3), 1–4.

Pardoe, J., Conway, D., Namaganda, E., Vincent, K., Dougill, A., & Kashaigili, J. (2018). Climate change and the water-energy-food nexus: Insights from policy and practice in Tanzania. *Climate Policy*, *18*(7), 863–877. doi:10.1080/14693062.2017.1386082

Pardo, G., del Prado, A., Martínez-Mena, M., Bustamante, M. A., Martín, J. A. R., Álvaro-Fuentes, J., & Moral, R. (2017). Orchard and horticulture systems in Spanish Mediterranean coastal areas: Is there a real possibility to contribute to C sequestration? *Agriculture, Ecosystems & Environment*, *238*, 153–167. doi:10.1016/j.agee.2016.09.034

Paredes-trejo, F., Barbosa, H., & Santos, C. A. C. (2019). Evaluation of the Performance of SM2RAIN-Derived Rainfall Products over Brazil. *Remote Sensing*, *11*(9), 1–28. doi:10.3390/rs11091113

Park, N. (2013). Spatial Downscaling of TRMM Precipitation Using Geostatistics and Fine Scale Environmental Variables. *Hindawi*, *2013*, 1–9. doi:10.1155/2013/237126

Pasha, A. K., Ramdhani, M. A., & Mulyana, E. (2018). System Design of Controlling and Monitoring on Aquaponic Based on Internet of Things. *2018 4th International Conference on Wireless and Telematics (ICWT)*, 1–5.

Patidar, S. K., & Mishra, S. (2017). *Carbon Sequestration by Microalgae: A Green Approach for Climate Change Mitigation*. Earth Systems and Environmental Sciences.

Patterson, J. H., Newbigin, E. D., Tester, M., Bacic, A., & Roessner, U. (2009). Metabolic responses to salt stress of barley (Hordeum vulgare L.) cultivars, Sahara and Clipper, which differ in salinity tolerance. *Journal of Experimental Botany*, *60*(14), 4089–4103. doi:10.1093/jxb/erp243 PMID:19666960

Paudel, M., Sah, S. K., McDonald, A., & Chaudhary, N. K. (2014). Soil Organic Carbon Sequestration in Rice-Wheat System under Conservation and Conventional Agriculture in Western Chitwan, Nepal. *World Journal of Agricultural Research.*, *2*(6A), 1–5. doi:10.12691/wjar-2-6A-1

Peck, L. (1974). Accuracy of Precipitation Measurements for Hydrologic Modelin. *Water Resources Research*, *1*(4), 857–863.

Pegna, F. G., Bartolini, P., El Rhaffari, L., Fahim, S., Bonaiuti, E., Le, Q. B., & Zucca, C. (2019). Sustaining Moroccan oasis agricultural system through small mechanization inputs. *JUNCO| Journal of UNiversities and international development COoperation, 1*(1).

Pérez-Piqueres, A., Edel-Hermann, V., Alabouvette, C., & Steinberg, C. (2006). Response of soil microbial communities to compost amendments. *Soil Biology & Biochemistry*, *38*(3), 460–470. doi:10.1016/j.soilbio.2005.05.025

Person-Le, R. J., Buchet, V., Vincent, B., Le, D. H., & Quemener, L. (2006). Effects of temperature on the growth of Pollack (*Pollachiuspollachius*) juveniles. *Aquaculture (Amsterdam, Netherlands)*, *251*(2-4), 340–345. doi:10.1016/j.aquaculture.2005.06.029

Pirot, J. Y., Meynell, P., & Elder, D. (2000). *Ecosystem management: Lessons from around the world – A guide for development and conservation practitioners*. IUCN. data.iucn.org/dbtw-wpd/edocs/2000-051.pdf

PNUD: Programme des Nations Unies pour le Développement. (2011a). *Adaptation au changement climatique pour les oasis résilientes. Etude « évaluation du changement climatique futur au niveau des zones oasienne marocaine».* Author.

PNUD: Programme des Nations Unies pour le Développement. (2011b). *Evaluation prospective des vulnérabilités et risques climatiques aux horizons 2030 et 2050. Projet d'évaluation de la vulnérabilité et des impacts du changement climatique dans les oasis du Maroc et structuration de stratégies territoriales d'adaptation.* Author.

Porcel, R., Aroca, R., & Ruiz-Lozano, J. M. (2012). Salinity stress alleviation using arbuscular mycorrhizal fungi. A review. *Agronomy for Sustainable Development, 32*(1), 181–200. doi:10.100713593-011-0029-x

Poussart, J.-N. (2004). Verification of Soil Carbon Sequestration: Sample Requirement. *Environmental Management, 33*, 416–425. doi:10.100700267-003-9149-7

Power, I. M., Dipple, G. M., & Francis, P. S. (2017). Assessing the carbon sequestration potential of magnesium oxychloride cement building materials. *Cement and Concrete Composites, 78*, 97–107. doi:10.1016/j.cemconcomp.2017.01.003

Prado, R. & Da Silva, G. (2017). *Ecological Response to Global Change: Changes in C:N:P Stoichiometry in Environmental Adaptations of Plants.* doi:10.5772/intechopen.69246

Prentice, I.C., Farquhar, G.D., Fasham, M.J.R., Goulden, M.L., Heimann, M., Jaramillo, V.J., Kheshgi, H.S., Le Que're´, C., Scholes, R.J., & Wallace, D.W.R. (2001). The carbon cycle and atmospheric carbon dioxide. In Climate Change 2001: The Scientific Basis. Contribution of Working Group I to the Third Assessment Report of the Intergovernmental Panel on Climate Change. Cambridge University Press.

Proksch, G., Ianchenko, A., & Kotzen, B. (2019). *Aquaponics in the Built Environment.* Aquaponics in the Built Environment. doi:10.1007/978-3-030-15943-6_21

Pupkewitz. (2019). *PupkewitzMegaBuild Garden Talk: Introducing Aquaponics.* Available on line, https://www.megabuildonline.com/articles/media-centre/news/46

Quinn, M. S. (2002). Ecosystem-based management. In D. Thompson (Ed.), *Tools for environmental management* (pp. 370–382). New Society Publishers.

Radomski, P., & Goeman, T. J. (2001). Consequences of human lakeshore development on emergent and floating-leaf vegetation abundance. *North American Journal of Fisheries Management, 21*(1), 46–61. doi:10.1577/1548-8675(2001)021<0046:COHLDO>2.0.CO;2

Rahman, S., Rashid, M. A., & Sanajpati, J. N. (1980). Design and development of hand-operated grain harvester. *AMA, 11*(1), 79-80.

Ramaraj, R., Tsai, D. D. W., & Chen, P. H. (2014a). An exploration of the relationships between microalgae biomass growth and related environmental variables. *Journal of Photochemistry and Photobiology. B, Biology, 135*, 44–47. doi:10.1016/j.jphotobiol.2014.04.001 PMID:24792572

Ramaraj, R., Tsai, D. D. W., & Chen, P. H. (2014b). Freshwater microalgae niche of air carbon dioxide mitigation. *Ecological Engineering*, *68*, 47–52. doi:10.1016/j.ecoleng.2014.03.058

Ramaraj, R., Tsai, D. D. W., & Chen, P. H. (2015). Carbon dioxide fixation of freshwater microalgae growth on natural water medium. *Ecological Engineering*, *75*, 86–92. doi:10.1016/j.ecoleng.2014.11.033

Rasche, L. (2014). The importance of climate variability changes for future levels of tree-based forest ecosystem services. *The International Journal of Biodiversity Science, Ecosystem Services & Management*, *10*(3), 187–197. doi:10.1080/21513732.2014.939719

Rashid, H., Scholes, R., & Ash, N. (2005). Ecosystems and human well-being: Current State and Trends: Vol. 1. *Findings of the Condition and Trends Working Group of the Millennium Ecosystem Assessment Washington*. Island Press.

Rath, K. M., & Rousk, J. (2015). Salt effects on the soil microbial decomposer community and their role in organic carbon cycling: A review. *Soil Biology & Biochemistry*, *81*, 108–123. doi:10.1016/j.soilbio.2014.11.001

Ravy-Jonsen, L. (2009). Ecosystem management a management view. Department of Environmental and Business Economics. IME Working Paper 86/09.

Ray, D. K., Gerber, J. S., MacDonald, G. K., & West, P. C. (2015). Climate variation explains a third of global crop yield variability. *Nature Communications*, *6*(1), 5989. doi:10.1038/ncomms6989 PMID:25609225

Ray, R., & Jana, T. K. (2017). Carbon sequestration by mangrove forest: One approach for managing carbon dioxide emission from coal-based power plant. *Atmospheric Environment*, *171*, 149–154. doi:10.1016/j.atmosenv.2017.10.019

Reliefweb. (2019). *President declares state of emergency over drought*. https://reliefweb.int/report/namibia/president-declares-state-emergency-over-drought

Rigopoulos, I., Delimitis, A., Ioannou, I., Efstathiou, A. M., & Kyratsi, T. (2018b). Effect of ball milling on the carbon sequestration efficiency of serpentinized peridotites. *Minerals Engineering*, *120*, 66–74. doi:10.1016/j.mineng.2018.02.011

Rochdane, S., Bounoua, L., Zhang, P., Imhoff, M., Messouli, M., & Yacoubi-Khebiza, M. (2014). Combining satellite data and models to assess vulnerability to climate change and its impact on food security in Morocco. *Sustainability*, *6*(4), 1729–1746. doi:10.3390u6041729

Rodda, J. C., & Dixon, H. (2011). Rainfall measurement revisited. *Royal Meteorogical Society*, *1842*, 131–136. doi:10.1002/wea.1884

Rodda, J. C., & Smith, S. W. (1986). Technical note the significance of the systematic error in rainfall measurement for assessing wet deposition. *Atmospheric Environment*, *20*(5), 1059–1064. doi:10.1016/0004-6981(86)90293-3

Rodrígueza, E., Pechea, R., Garbisub, C., Gorostizac, I., Epeldeb, L., Artetxed, U., Irizare, A., Sotoe, M., & Mabecerrild, J. (2016). Dynamic quality index for agricultural soils based on fuzzy logic. *Ecological Indicators*, *60*, 678–692. doi:10.1016/j.ecolind.2015.08.016

Rogers, P., Silva, R., & Bhatia, R. (2002). Water is an economic good: How to use prices to promote equity, efficiency, and sustainability. *Water Policy*, *4*(1), 1–17. doi:10.1016/S1366-7017(02)00004-1

Roman-Liu, D. (2014). Comparison of concepts in easy-to-use methods for MSD risk assessment. *Applied Ergonomics*, *45*(3), 420–427. doi:10.1016/j.apergo.2013.05.010 PMID:23849897

Rondeaux, G., Steven, M., & Baret, F. (1996). Optimization of Soil-Adjusted Vegetation Indices. *Remote Sensing and Environment, 107*(August), 95–107.

Rosenbauer, R. J., Thomas, B., Bischoff, J. L., & Palandri, J. (2012). Carbon sequestration via reaction with basaltic rocks: Geochemical modeling and experimental results. *Geochimica et Cosmochimica Acta*, *89*, 116–133. doi:10.1016/j.gca.2012.04.042

Rosenberg, A., & McLeod, K. (2005). Implementing ecosystem-based approaches to management for the conservation of ecosystem services. *Marine Ecology Progress Series*, *300*, 241–296. doi:10.3354/meps300270

Rosenfeld, J. S., Van Leeuwen, T. E., Richards, J. G., & Allen, D. (2015). Relationship between growth and standard metabolic rate: Measurement artefacts and implications for habitat use and life history adaptation in salmonids. *Journal of Animal Ecology*, *84*(1), 4–20. doi:10.1111/1365-2656.12260 PMID:24930825

Rosenzweig, C (2007). Assessment of Observed Changes and Responses in Natural and Managed Systems. In *Climate Change 2007: Impacts, Adaptation and Vulnerability. Contribution of Working Group II to the Fourth Assessment Report of the Intergovernmental Panel on Climate Change* (p. 98). Cambridge University Press.

Rowe, J. S. (1992). The ecosystem approach to forestland management. *Forestry Chronicle*, *68*(2), 222–224. doi:10.5558/tfc68222-2

Rowell, D. P. (2005). A scenario of European climate change for the late twenty-first century: Seasonal means and interannual variability. *Climate Dynamics*, *25*(7-8), 837–849. doi:10.100700382-005-0068-6

Royaume du Maroc. (2010). *Seconde communication nationale à la Convention-Cadre des Nations Unies sur les Changements Climatiques*. Ministère de l'Environnement.

Roy, K. S., Neogi, S., Nayak, A. K., & Bhattacharyya, P. (2014). Effect of nitrogen fertilization on methane and carbon dioxide production potential in relation to labile carbon pools in tropical flooded rice soils in eastern India. *Archives of Agronomy and Soil Science*, *60*(10), 1329–1344. doi:10.1080/03650340.2013.869673

Rufli, H. (2001). Seasonal abundance of zooplankton and the planktivorous fish, Stolothrissa, in Tanzanian waters of Lake Tanganyika. *The great lakes of the world (GLOW): food-web, health & integrity*, 97.

Salam, A., Pandey, P., Shrestha, S., & Anal, A. K. (2017). The need for the nexus approach. In P. Abdul Salam, S. Shrestha, V. Prasad Pandey, & A. K. Anal (Eds.), *Water-Energy-Food Nexus: Principles and Practices*. John Wiley and Sons, Inc. doi:10.1002/9781119243175.ch1

Sami, M., Shiekhdavoodi, M. J., Almassi, M., & Marzban, A. (2013). Assessing the sustainability of agricultural production systems using fuzzy logic. *Journal of Central European Agriculture*, *14*(3), 318–330. doi:10.5513/JCEA01/14.3.1323

Sanders, K. T., & Webber, M. E. (2012). Evaluating the energy consumed for water use in the United States. *Environmental Research Letters*, *7*(3), 34034. doi:10.1088/1748-9326/7/3/034034

Sardans, J., Rivas-Ubach, A., & Peñuelas, J. (2012). The C: N: P stoichiometry of organisms and ecosystems in a changing world: a review and perspectives. *Perspectives in Plant Ecology, Evolution and Systematics*, *14*(1), 33–47. doi:10.1016/j.ppees.2011.08.002

SastryMusti. (2020). *Industry 4.0-Based Enterprise Information System for Demand-Side Management and Energy Efficiency*. Novel Approaches to Information Systems Design. doi:10.4018/978-1-7998-2975-1.ch007

Sathaye, J., Najam, A., Cocklin, C., Heller, T., Lecocq, F., Llanes-Regueiro, J., Pan, J., … Winkler. (2007). Sustainable development and mitigation. In *Climate change 2007: Mitigation. Contribution of working group iii to the fourth assessment report of the intergovernmental panel on climate change*. Cambridge University Press. Retrieved on April 29 2018 from https://www.ipcc.ch/pdf/assessmentreport/ar4/wg3/ar4-wg3-chapter12.pdf

Sati, V. P. (2015). Climate change and socio-ecological transformation in high mountains: An empirical study of Garhwal Himalaya. *Change and Adaptation in Socio-Ecological Systems*, *2*(1). Advance online publication. doi:10.1515/cass 2015 0005

Savadova, K., Mazur-Marzec, H., Karosienė, J., Kasperovičienė, J., Vitonytė, I., Toruńska-Sitarz, A., & Koreivienė, J. (2018). Effect of increased temperature on native and alien nuisance cyanobacteria from temperate lakes: An experimental approach. *Toxins*, *10*(11), 445. doi:10.3390/toxins10110445 PMID:30380769

Sayre, R. (2010). Microalgae: The Potential for Carbon Capture. *Bioscience*, *60*(9), 722–727. doi:10.1525/bio.2010.60.9.9

Scattini, N., & Maj, S. P. (2017). *Aquaponics – A Process Control Approach*. https://doi.org/doi:10.5539/mas.v11n11p43

Schilling, J., Freier, K. P., Hertig, E., & Scheffran, J. (2012). Climate change, vulnerability and adaptation in North Africa with focus on Morocco. *Agriculture, Ecosystems & Environment*, *156*, 12–26. doi:10.1016/j.agee.2012.04.021

Schilling, J., Hertig, E., Tramblay, Y., & Scheffran, J. (2020). Climate change vulnerability, water resources and social implications in North Africa. *Regional Environmental Change*, *20*(1), 1–12. doi:10.100710113-020-01597-7

Schindler, D. E., & Eby, L. A. (1997). Stoichiometry of fishes and their prey: Implications for nutrient recycling. *Ecology*, *78*(6), 1816–1831. doi:10.1890/0012-9658(1997)078[1816:SOFATP]2.0.CO;2

Schneider, U., Becker, A., Finger, P., Meyer-christoffer, A., Ziese, M., & Rudolf, B. (2014). GPCC ' s new land surface precipitation climatology based on quality-controlled in situ data and its role in quantifying the global water cycle. *Theoretical and Applied Climatology*, *115*(1-2), 15–40. doi:10.100700704-013-0860-x

Schnier, S., Cai, X., & Cao, Y. (2016). Importance of natural and anthropogenic environmental factors to fish communities of the Fox River in Illinois. *Environmental Management*, *57*(2), 389–411. doi:10.100700267-015-0611-0 PMID:26404430

Schoo, K. L., Aberle, N., Malzahn, A. M., & Boersma, M. (2012). Food Quality Affects Secondary Consumers Even at Low Quantities: An Experimental Test with Larval European Lobster. *PLoS One*, *7*(3), e33550. doi:10.1371/journal.pone.0033550 PMID:22442696

Schreck, C. B. (2010). Stress and fish reproduction: The roles of allostasis and hormesis. *General and Comparative Endocrinology*, *165*(3), 549–556. doi:10.1016/j.ygcen.2009.07.004 PMID:19596332

Schwaller, M. R., & Morris, K. R. (2011). A Ground Validation Network for the Global Precipitation Measurement Mission. *Journal of Atmospheric and Oceanic Technology*, *28*(3), 301–320. doi:10.1175/2010JTECHA1403.1

Schweiger, O., Settele, J., Kudrna, O., Klotz, S., & Kühn, I. (2008). Climate change can cause spatial mismatch of trophically interacting species. *Ecology*, *89*(12), 3472–3479. doi:10.1890/07-1748.1 PMID:19137952

Scott, C. A., & Sugg, Z. P. (2015). Global energy development and climate-induced water scarcity-Physical limits, sectoral constraints, and policy imperatives. *Energies*, *8*(8), 8211–8225. doi:10.3390/en8088211

Scotti, R., Bonanomi, G., Scelza, R., Zoina, A., & Rao, M. A. (2015). Organic amendments as sustainable tool to recovery fertility in intensive agricultural systems. *Journal of Soil Science and Plant Nutrition*, *15*(2), 333–352. doi:10.4067/S0718-95162015005000031

Seeliger, L., Clercq, W., Hoffmann, W., Cullis, J., Horn, A., & Witt, M. (2018). Applying the water-energy-food nexus to farm profitability in the Middle Breede catchment, South Africa. *South African Journal of Science*, *114*(11/12), 1–10. doi:10.17159ajs.2018/5062

Segond, M., Wheater, H. S., & Onof, C. (2007). The significance of spatial rainfall representation for flood runoff estimation : A numerical evaluation based on the Lee catchment, UK. *Journal of Hydrology (Amsterdam)*, *347*(1-2), 116–131. doi:10.1016/j.jhydrol.2007.09.040

Sellers, P. J., Randall, D. A., Collatz, G. J., Berry, J. A., Field, C. B., Dazlich, D. A., Zhang, C., Collelo, G. D., & Bounoua, L. A. (1996). Revised Land Surface Parameterization (SiB2) for Atmospheric GCMS. Part I: Model Formulation. *American Meteorological Society*. Advance online publication. doi:10.1175/1520-0442(1996)009<0676:ARLSPF>2.0.CO;2

Serraj, R., & Sinclair, T. R. (2002). Osmolyte accumulation: Can it really help increase crop yield under drought conditions? *Plant, Cell & Environment*, *25*(2), 333–341. doi:10.1046/j.1365-3040.2002.00754.x PMID:11841674

Settele, J., Scholes, R., Betts, R., Bunn, S., Leadley, P., Nepstad, D., Overpeck, J. T., & Taboada, M. A. (2014). Terrestrial and Inland Water Systems. In Climate Change (2014) Impacts, Adaptation and Vulnerability. Part A: Global and Sectoral Aspects. Contribution of Working Group II to the Fourth Assessment Report of the Intergovernmental Panel on Climate Change. Cambridge University Press.

Shah, T., Scott, C., Kishore, A. & Sharma, A. (2007). Energy-irrigation nexus in South Asia: improving groundwater conservation and power sector viability. *BMJ Clinical Resources, 2*, 992.

Shah, T. (2009). *Taming the Anarchy: Groundwater Governance in South Asia*. Routledge.

Shalini, P. S., Wankhade, A. M., & Syed, M. N. (2015). Design and fabrication of crop reaper. *International Journal for Engineering Applications and Technology*.

Sharma, H. K., Dandotiya, B., & Jadon, N. (2017). Exposure of Air Pollution and Its Health Effects in Traffic Police Persons of Gwalior City, India. *Environmental Claims Journal*, *29*(4), 305–315. doi:10.1080/10406026.2017.1390357

Sheng, Y., & Zhu, L. (2018). Biochar alters microbial community and carbon sequestration potential across different soil pH. *The Science of the Total Environment*, *622-623*, 1391–1399. doi:10.1016/j.scitotenv.2017.11.337 PMID:29890604

Sherchan, Miles, Ikner, Yu, Snyder, & Ian. (2018). Pepper Near Real-Time Detection of E. coli in Reclaimed Water. *Journal of Sensors*, *18*(8). Advance online publication. doi:10.339018072303

Shoji, J., Toshito, S., Mizuno, K., Kamimura, Y., Hori, M., & Hirakawa, K. (2011). Possible effects of global warming on fish recruitment: Shifts in spawning season and latitudinal distribution can alter growth of fish early life stages through changes in daylength. *ICES Journal of Marine Science*, *68*(6), 1165–1169. doi:10.1093/icesjms/fsr059

Shugart, H. H. (1998). *Terrestrial ecosystems in changing environments*. Cambridge University Press.

Shukla, S. K., Shee, S., Maity, S. K., Solomon, S., Awasthy, S. K., Gaur, A., Pathak, A. D., & Jaiswal, V. P. (2017). Soil Carbon Sequestration and Crop Yields in Rice-Wheat and Sugarcane-Ratoon-Wheat Cropping Systems through Crop Residue Management and Inoculation of Trichodermaviride in Subtropical India. *Sugar Tech*, *19*(4), 347–358. doi:10.100712355-016-0470-x

Shuo, L., Li, Y., Li, X., Tian, X., Zhao, A., Wang, S., Wang, S., & Shi, J. (2016). Effect of straw management on carbon sequestration and grain production in a maize–wheat cropping system in Anthrosol of the Guanzhong Plain. *Soil & Tillage Research*, *157*, 43–51. doi:10.1016/j.still.2015.11.002

Silva, E., & Lopes, S. (2017). Validation of Satellite Rainfall Products over a Mountainous Watershed in a Humid Subtropical Climate Region of Brazil. *Remote Sensing*, 1–18. doi:10.3390/rs9121240

Simonneaux, V., Cheggour, A., Deschamps, C., Mouillot, F., Cerdan, O., & Le Bissonnais, Y. (2015). Land use and climate change effects on soil erosion in a semi-arid mountainous watershed (High Atlas, Morocco). *Journal of Arid Environments*, *122*, 64–75. doi:10.1016/j.jaridenv.2015.06.002

Simpson, G., & Jewitt, G. (2019). The development of the water-energy-food nexus as a framework for achieving resource security: A review. *Frontiers in Environmental Science*, *7*(8), 1–9. doi:10.3389/fenvs.2019.00008

Simpson, T. W., Sharpley, A. N., Howarth, R. W., Paerl, H. W., & Mankin, K. R. (2008). The new gold rush: Fueling ethanol production while protecting water quality. *Journal of Environmental Quality*, *37*(2), 318–324. doi:10.2134/jeq2007.0599 PMID:18268293

Singh, R., Khakha, J., & Singh, G. (2009). *Development and testing of manually operated vertical conveyor reaper* (Unpublished, B-Tech. thesis). Indira Gandhi Agricultural University, Raipur.

Singh, S. P. (2012). Physiological workload of farm women while evaluating sickles for paddy harvesting. *Agric Eng. Int: CIGR Journal, 14*(1), 1540.

Singh, S., Vatsa, D. K., & Verma, M. K. (2007). Feasibility and Performance Evaluation of Power Tiller Operated Reaper in Hills of Himachal Pradesh. *Agricultural Engineering Today, 31*(2), 6-10.

Singh, T. P., & Singh, B. (1995). Design and development of animal-drawn, engine-operated reaper. *AMA, 26*(2), 29-34.

Singh, B. R., & Lal, R. (2005). The potential of soil carbon sequestration through improved management practices in Norway. *Environment, Development and Sustainability*, *7*(1), 161–184. doi:10.100710668-003-6372-6

SIWM. (2001). *Integrated management of water resources project in the Souss Massa report Achievement of Pilot Projects*. USAID Government of Morocco.

Smajgl, A., Ward, J., & Pluschke, L. (2016). The water-food-energy nexus-realising a new paradigm. *Journal of Hydrology (Amsterdam)*, *533*, 533–540. doi:10.1016/j.jhydrol.2015.12.033

Smith, C. M., David, M. B., Mitchell, C., Masters, M. D., Anderson-Teixeira, K. J., Bernacchi, C. J., & DeLucia, E. H. (2013). Reduced Nitrogen losses after conversion of row crop agriculture to perennial biofuel crops. *Journal of Environmental Quality*, *42*(1), 219–228. doi:10.2134/jeq2012.0210 PMID:23673757

Smith, G. (2018). Step away from stepwise. *Journal of Big Data*, 1–12. doi:10.118640537-018-0143-6

Smith, P., Martino, D., Cai, Z., Gwary, D., Janzen, H., Kumar, P., McCarl, B., Ogle, S., O'Mara, F., Rice, C., Scholes, B., & Sirotenko, O. (2007). Agriculture. In *Climate Change 2007: Mitigation. Contribution of Working Group III to the Fourth Assessment Report of the Intergovernmental Panel on Climate Change*. Cambridge University Press.

Smith, P., & Smith, P. (2004). Monitoring and verification of soil C changes under Article 3.4 of the Kyoto Protocol. *Soil Use and Management, 20*(2), 264–270. doi:10.1079/SUM2004239

Smith, R., & Maltby, E. (2003). *Using the ecosystem approach to implement the convention on biological diversity: Key issues and case studies*. International Union for Conservation of Nature and Natural Resources. doi:10.2305/IUCN.CH.2003.CEM.2.en

Smith, S. E., & Read, D. J. (2008). *Mycorrhizal symbiosis* (3rd ed.). Academic.

Snoeks, J. (2000). How well known is the ichthyodiversity of the large East African lakes? *Advances in Ecological Research, 31*, 17–38. doi:10.1016/S0065-2504(00)31005-4

Society of American Foresters. (1993). *Sustaining Long Term Forest Health and Productivity*. Society of American Foresters.

Song Yang, E. A. S. (2008). Convective – Stratiform Precipitation Variability at Seasonal Scale from 8 Yr of TRMM Observations : Implications for Multiple Modes of Diurnal Variability. *American Meteorological Society*. doi:10.1175/2008JCLI2096.1

Song, Y., Cervarich, M., Jain, A. K., Kheshgi, H. S., Landuyt, W., & Cai, X. (2016). The interplay between bioenergy grass production and water resources in the United States of America. *Environmental Science & Technology, 50*(6), 3010–3019. doi:10.1021/acs.est.5b05239 PMID:26866460

Spring, S., & Peck, E. L. (1980). Robert E. Norton Memorial Lecture. *American Meteorological Society, 61*(8), 894–902.

Stenuite, S., Pirlot, S., Hardy, M. A., Sarmento, H., Tarbe, A. L., Leporcq, B., & Descy, J. P. (2007). Phytoplankton production and growth rate in Lake Tanganyika: Evidence of adecline in primary productivity in recent decades. *Freshwater Biology, 52*(11), 2226–2239. doi:10.1111/j.1365-2427.2007.01829.x

Stott, P. A., Tett, S. F. B., Jones, G. S., Allen, M. R., Mitchell, J. F. B., & Jenkins, G. J. (2000). External control of 20th century temperature by natural and anthropogenic forcings. *Science, 290*(5499), 2133–2137. doi:10.1126cience.290.5499.2133 PMID:11118145

Stringer, C. E., Trettin, C. C., Zarnoch, S. J., & Tang, W. (2015). Carbon stocks of mangroves within the Zambezi River Delta, Mozambique. *Forest Ecology and Management, 354*, 139–148. doi:10.1016/j.foreco.2015.06.027

Sturmbauer, C., Mark, W., &Dallinger, R. (1992). Ecophysiology of aufwuchs-eating cichlids in Lake Tanganyika: niche separation by trophic specialization. *Environ Biol Fishes, 35*, 283-290.

Suen, J. P., Eheart, J. W., Herricks, E. E., & Chang, F. J. (2009). Evaluating the potential impact of reservoir operation on fish communities. *Journal of Water Resources Planning and Management, 135*(6), 475–483. doi:10.1061/(ASCE)0733-9496(2009)135:6(475)

Suhl, J., Dannehl, D., Kloas, W., Baganz, D., Jobs, S., Scheibe, G., & Schmidt, U. (2016). Advanced aquaponics: Evaluation of intensive tomato production inaquaponics vs. conventional hydroponics. *Agricultural Water Management, 178*, 335–344. doi:10.1016/j.agwat.2016.10.013

Sumaila, U. R., Cheung, W. W. L., Lam, V. W. Y., Pauly, D., & Herrick, S. (2011). Climate change impacts on the biophysics and economics of world fisheries. *Nature Climate Change, 1*(9), 449–456. doi:10.1038/nclimate1301

Sunday, J., Bates, A., & Dulvy, N. (2011). Global analysis of thermal tolerance and latitude in ectotherms. *Proceedings. Biological Sciences, 278*(1713), 1823–1830. doi:10.1098/rspb.2010.1295 PMID:21106582

Takeuchi, Y., Ochi, H., Kohda, M., Sinyinza, D., & Hori, M. (2010). 20-year census of a rocky littoral fish community in Lake Tanganyika. *Ecology Freshwater Fish, 19*(2), 239–248. doi:10.1111/j.1600-0633.2010.00408.x

Taleb, S., Bellaka, M., & Achour, A. (2007). Effect of the fence on biodiversity and the dynamics of argan tree. *Annals of Forest Research in Morocco, 38*, 171-185.

Tallis, H., Levin, P., Ruckelshaus, M., Lester, S., McLeod, K., Fluharty, D., & Halpern, B. (2009). The many faces of ecosystem-based management: Making the process work today in real places. *Marine Policy, 34*(2), 340–348. doi:10.1016/j.marpol.2009.08.003

Tang, G., Ma, Y., Long, D., Zhong, L., & Hong, Y. (2016). Evaluation of GPM Day-1 IMERG and TMPA Version-7 legacy products over Mainland China at multiple spatiotemporal scales. *Journal of Hydrology, 533*, 152–167. doi:10.1016/j.jhydrol.2015.12.008

Tang, L., Tian, Y., Yan, F., & Habib, E. (2015). An improved procedure for the validation of satellite-based precipitation estimates. *Atmospheric Research, 163*, 61–73. Advance online publication. doi:10.1016/j.atmosres.2014.12.016

Thiemig, V., Rojas, R., Zambrano-bigiarini, M., & De Roo, A. (2013). Hydrological evaluation of satellite-based rainfall estimates over the Volta and Baro-Akobo Basin. *Journal of Hydrology (Amsterdam), 499*, 324–338. doi:10.1016/j.jhydrol.2013.07.012

Thomas, J. W. (1996). Forest Service perspective on ecosystem management. *Ecological Applications, 6*(3), 703–705. doi:10.2307/2269465

Thompson, M., Gamage, D., Hirotsu, N., Martin, A., & Seneweera, S. (2017). Effects of elevated carbon dioxide on photosynthesis and carbon partitioning: A perspective on root sugar sensing and hormonal crosstalk. *Frontiers in Physiology, 8*, 578. doi:10.3389/fphys.2017.00578 PMID:28848452

Tian, Peters-Lidard, & Choudhury. (2007). Multitemporal Analysis of TRMM-Based Satellite Precipitation Products for Land. *Journal of Hydrometeorology*, 1165–1183. doi:10.1175/2007JHM859.1

Timmermann, A., Oberhuber, J., Bacher, A., Esch, E., Latif, M., & Roeckner, E. (1999). Increased El Niño frequency in a climate model forced by future greenhouse warming. [London]. *Nature*, *398*(6729), 694–697. doi:10.1038/19505

Travel News Namibia. (2016). *Aquaponics; the way forward for Namibian agriculture?* Available online, http://www.travelnewsnamibia.com/news/aquaponics-way-forward-namibian-agriculture/

Tripathi, S., Kumari, S., Chakraborty, A., Gupta, A., Chakrabarti, K., & Bandyapadhyay, B. K. (2006). Microbial biomass and its activities in salt-affected coastal soils. *Biology and Fertility of Soils*, *42*(3), 273–277. doi:10.100700374-005-0037-6

Trivedi, P., Anderson, I. C., & Singh, B. K. (2013). Microbial modulators of soil carbon storage: Integrating genomic and metabolic knowledge for global prediction. *Trends in Microbiology*, *21*(12), 12. doi:10.1016/j.tim.2013.09.005 PMID:24139848

Tsai, D. D., Chen, P. H., Chou, C. M., Hsu, C., & Ramaraj, R. (2015). Carbon sequestration by alga ecosystems. *Ecological Engineering*, *84*, 386–389. doi:10.1016/j.ecoleng.2015.09.024

Tsai, D. D., Chen, P. H., & Ramaraj, R. (2017). The potential of carbon dioxide capture and sequestration with algae. *Ecological Engineering*, *98*, 17–28. doi:10.1016/j.ecoleng.2016.10.049

Tsanis, D. D. A. I. K. (2016). Comparison of multiple linear regression and artificial neural network models for downscaling TRMM precipitation products using MODIS data. *Environmental Earth Sciences*. Advance online publication. doi:10.100712665-016-5883-z

Tsujimoto, M., Kajikawa, Y., Tomita, J., & Matsumoto, Y. (2017). A review of the ecosystem concept— Towards coherent ecosystem design. *Technological Forecasting and Social Change*, *136*, 49–58. doi:10.1016/j.techfore.2017.06.032

Tuo, Y., Duan, Z., Disse, M., & Chiogna, G. (2016). Science of the Total Environment Evaluation of precipitation input for SWAT modeling in Alpine catchment : A case study in the Adige river basin (Italy). *The Science of the Total Environment*, *573*, 66–82. doi:10.1016/j.scitotenv.2016.08.034 PMID:27552731

Turpie, J., Winkler, H., Spalding-Fecher, R., & Midgley, G. (2002). *Economic Impacts of Climate Change in South Africa: A Preliminary Analysis of Unmitigated Damage Costs*. Southern Waters Ecological Research & Consulting & Energy & Development Research Centre, University of Cape Town.

UN. (2015). *Sustainable Development Goals*. Retrieved from: http://www.un.org/sustainabledevelopment/sustainable-development-goals/ 10.18356/90048b6a-en

UNDP. (2005). *Uganda human development report 2005. Linking environment to development: A deliberate choice*. UNDP.

UNECA / ACPC. (2014). *Fourth Conference On Climate Change and Development in Africa Summary and Recommendations.* Marrakech, Morocco: UNECA / ACPC.

United Nations Framework Convention on Climate Change (UNFCCC). (2011). *Fact sheet: climate change science-the status of climate change science today.* https://unfccc.int/files/press/backgrounders/application/pdf/press_factsh_science.pdf

Upendar, G., Singh, S., Chakrabarty, J., Chandra Ghanta, K., Dutta, S., & Dutta, A. (2018). Sequestration of carbon dioxide and production of biomolecules using cyanobacteria. *Journal of Environmental Management, 218*, 234–244. doi:10.1016/j.jenvman.2018.04.031 PMID:29680755

Van der Putten, W. H., de Ruiter, P. C., Bezemer, T. M., Harvey, J. A., Wassen, M., & Wolters, V. (2004). Trophic interactions in a changing world. *Basic and Applied Ecology, 5*(6), 487–494. doi:10.1016/j.baae.2004.09.003

Van der Waals, Kruger, Franke, Havenkort, & Steyn. (2013). Climate change and potato production in contrasting South African agro-ecosystems 3. Effects on relative development rates of selected pathogens and pests. *Journal of the European Association for Potato Research.*

Vass, M. M., (2017). Renewable energies cannot compete with forest carbon sequestration to cost-efficiently meet the EU carbon target for 2050. *Renewable Energy.*

Vaughan, D. G., Marshall, G. J., Connolley, W. M., King, J. C., & Mulvaney, R. (2001). Devil in the detail. *Science, 293*(5536), 1777–1779. doi:10.1126cience.1065116 PMID:11546858

Vengosh, A., Jackson, R. B., Warner, N., Darrah, T. H., & Kondash, A. (2014). A critical review of the risks to water resources from unconventional shale gas development and hydraulic fracturing in the United States. *Environmental Science & Technology, 48*(15), 8334–8348. doi:10.1021/es405118y PMID:24606408

Verburg, P., Hecky, R. E., & Kling, H. (2003). Ecological consequences of a century of warming in Lake Tanganyika. *Science, 301*(5632), 505–507. doi:10.1126cience.1084846 PMID:12829786

Vernandhes, W., Salahuddin, N. S., Kowanda, A., & Sari, S. P. (2017). Smart Aquaponic with Monitoring and Control System Based On IoT. *Second International Conference on Informatics and Computing (ICIC)*, Jayapura, Indonesia. 10.1109/IAC.2017.8280590

Vinodh S. & Chintha S. K. (2011). Application of fuzzy QFD for enabling leanness in a manufacturing organization. *International Journal of Production Research, 49*(6), 1627–1644.

Vitt, L. J., & Caldwell, J. P. (2014). *Herpetology: an introductory biology of amphibians and reptiles.* Academic Press.

Vurukonda, S. S. K. P., Vardharajula, S., Shrivastava, M., & SkZ, A. (2016). Enhancement of drought stress tolerance in crops by plant growth promoting rhizobacteria. *Microbiological Research, 184*, 13–24. doi:10.1016/j.micres.2015.12.003 PMID:26856449

Wafula, D. (2012). *Where do policy-makers get information? Science Communication Training* Workshop *for Bio-Innovate Projects.* BIPCEA.

Wagner, C. E., McIntyre, P. B., Buels, K. S., Gilbert, D. M., & Michel, E. (2009). Diet predicts intestine length in Lake Tanganyika's cichlid fishes. *Functional Ecology*, *23*(6), 1122–1131. doi:10.1111/j.1365-2435.2009.01589.x

Wagner, C., & Adrian, R. (2009). Cyanobacteria dominance: Quantifying the effects of climate change. *Limnology and Oceanography*, *54*(6part2), 2460–2468. doi:10.4319/lo.2009.54.6_part_2.2460

Walczyńska, A., & Sobczyk, Ł. (2017). The underestimated role of temperature-oxygen relationship in large-scale studies on size-to-temperature response. *Ecology and Evolution*, *7*(18), 7434–7441. doi:10.1002/ece3.3263 PMID:28944028

Waldrop, M. P., & Firestone, M. K. (2006). Response of microbial community composition and function to soil climate change. *Microbial Ecology*, *52*(4), 716–724. doi:10.100700248-006-9103-3 PMID:17061172

Walker, B., Steffen, W., Canadell, J., & Ingram, J. (1999). *The terrestrial biosphere and global change: implications for natural and managed ecosystems. Synthesis volume*. Cambridge University Press.

Walsh, M. J., Van Doren, L. G., Shete, N., Prakash, A., & Salim, U. (2018). Financial trade-offs of energy and food uses of algal biomass under stochastic conditions. *Applied Energy*, *210*, 591–603. doi:10.1016/j.apenergy.2017.08.060

Wang, J. (2013). Temporal responses of NDVI to precipitation and temperature in the central Great Plains, USA. *International Journal of Remote Sensing*, (March), 37–41. doi:10.1080/01431160210154812

Wang, J., Xiujun Wang, X., Xu, M., Feng, G., Zhang, W., Yang, X., & Huang, S. (2015). Contributions of wheat and maize residues to soil organic carbon under long-term rotation in north China. *Scientific Reports*, *5*(1), 1140. doi:10.1038rep11409 PMID:26100739

Wang, K., Xu, T., Wang, F., & Tian, H. (2016). Experimental study of CO_2–brine–rock interaction during CO_2 sequestration in deep coal seams. *International Journal of Coal Geology*, *154-155*, 265–274. doi:10.1016/j.coal.2016.01.010

Wang, Q., Li, Y., & Alva, A. (2010). Cropping Systems to Improve Carbon Sequestration for Mitigation of Climate Change. *Journal of Environmental Protection*, *1*(03), 207–215. doi:10.4236/jep.2010.13025

Wang, Q., Lu, H., Chen, J., Hong, H., Liu, J., Li, J., & Yan, C. (2018). Spatial distribution of glomalin-related soil protein and its relationship with sediment carbon sequestration across a mangrove forest. *The Science of the Total Environment*, *613-614*, 548–556. doi:10.1016/j.scitotenv.2017.09.140 PMID:28926809

Wang, W., Lu, H., Yang, D., Sothea, K., Jiao, Y., Gao, B., Peng, X., & Pang, Z. (2016). Modelling Hydrologic Processes in the Mekong River Basin Using a Distributed Model Driven by Satellite Precipitation and Rain Gauge Observations. *PLoS One*, *11*(3), 1–19. doi:10.1371/journal.pone.0152229 PMID:27010692

Warren, R., VanDerWal, J., Price, J., Walbergen, J. A., Atkinson, I., Ramirez-Villegas, J., Osborn, T. J., Jarvis, A., Shoo, L. P., Williams, S. E., & Lowe, J. (2013). Quantifying the benefit of early climate change mitigation in avoiding biodiversity loss. *Nature Climate Change*, *3*(7), 678–682. doi:10.1038/nclimate1887

Water Research Commission. (2018). *Assessing the state of the water-energy-food (WEF) nexus in South Africa*. Available Online: http://www.wrc.org.za/wp-content/uploads/mdocs/KV%20 365-18.pdf

Watson, A. (2013). Reduce, reuse, recycle: green technologies and practices at work. *Beyond the Numbers, 2*, 19-31.

Webber, M. E. (2015). A puzzle for the planet. *Scientific American, 312*(2), 62–67. doi:10.1038scientificamerican0215-62

Weijerman, M., Lindeboom, H., & Zuur, A. F. (2005). Regime shifts in marine ecosystems of the North Sea and Wadden Sea. *Marine Ecology Progress Series, 298*, 21–39. doi:10.3354/meps298021

Weitz, N., Strambo, C., Kemp-Benedict, E., & Nilsson, M. (2017). Closing the governance gaps in the water-energy-food nexus: Insights from integrative governance. *Global Environmental Change*, *45*, 165–173. doi:10.1016/j.gloenvcha.2017.06.006

Wetzel, R. G. (2001). *Limnology: Lake andRiver Ecosystems* (3rd ed.). AcademicPress.

Wheater, H. S. (2017). Modelling hydrological processes in arid and semi-arid areas : an introduction to the workshop. Cambridge University Press. doi:10.1017/CBO9780511535734.002

Williams, B. (2010). Adaptive management of natural resources-framework and issues. *Journal of Environmental Management*, *92*(5), 1346–1353. doi:10.1016/j.jenvman.2010.10.041 PMID:21075505

Williams, B., & Brown, E. (2014). Adaptive management: From the talk to real action. *Environmental Management*, *53*(2), 465–479. doi:10.100700267-013-0205-7 PMID:24271618

Wilson, O. (2007). *The Role of Media Communications in Developing Tourism Policy and Cross-Cultural Communication for Peace, Security for Sustainable* Tourism Industry in Africa. A Paper at the 4th International Institute of Peace through Tourism African Conference on Peace through Tourism at Educators' Forum, Kampala.

Wilson, O. (2010). Developing Regional Communications Campaigns Strategy For Environment and Natural Resources Management Policy Awareness for The East African Community. *Research Journal of Environmental and Earth Sciences*, *2*(2), 106–111.

Wilson, O. (2013). *The dawn of community radio in Bangladesh*. Lambert Academic publishers.

Woo, D. K., & Kumar, P. (2016). Mean age distribution of inorganic soil-nitrogen. *Water Resources Research*, *52*(7), 5516–5536. doi:10.1002/2015WR017799

World Food Programme / Namibia. (n.d.). https://www1.wfp.org/countries/namibia

World Future Council. (2016). *Growing food in Windhoek, Urban agriculture policy and practice.* Available online: https://www.worldfuturecouncil.org/wp-content/uploads/2017/03/Food-Handbook-final-web.pdf

Wright, I. J., & Westoby, M. (2003). Nutrient concentration, resorption and lifespan: Leaf traits of Australian sclerophyll species. *Functional Ecology*, *17*(1), 10–19. doi:10.1046/j.1365-2435.2003.00694.x

Xiaomin Yang, X., Songa, Z., Liu, H., Zwieten, L. V., Song, A., Lie, Z., Hao, A., Zhanga, X., & Wang, H. (2018). Phytolith accumulation in broadleaf and conifer forests of northern China: Implications for phytolith carbon sequestration. *Geoderma*, *312*, 36–44. doi:10.1016/j.geoderma.2017.10.005

Xu, D., & Guo, X. (2013). A Study of Soil Line Simulation from Landsat Images in. *Remote Sensing*, *1*(9), 4533–4550. doi:10.3390/rs5094533

Xue, C., Song, W., Qin, L., Dong, Q., & Wen, X. (2015). A spatiotemporal mining framework for abnormal association patterns in marine environments with a time series of remote sensing images. *International Journal of Applied Earth Observation and Geoinformation*, *38*, 105–114. doi:10.1016/j.jag.2014.12.009

Xu, Z., Yin, X., Sun, T., Cai, Y., Ding, Y., Yang, W., & Yang, Z. (2017). Labyrinths in large reservoirs: An invisible barrier to fish migration and the solution through reservoir operation. *Water Resources Research*, *53*(1), 817–831. doi:10.1002/2016WR019485

Yaffee, S. (1999). Three faces of ecosystem management. *Conservation Biology*, *13*(4), 713–725. doi:10.1046/j.1523-1739.1999.98127.x

Yamaoka, K. (1983). Morphology and feeding behavior of five species of genus Petrochromis (Teleostei: Cichlidae). *PhysiolEcolJpn*, *19*(1), 57–75.

Yamaoka, K. (1987). Comparative osteology of the jaw of algal-feeding cichlids (Pisces, Teleostei) from Lake Tanganyika. *Kochi Daigaku Kaiyo Seibutsu Kyoiku Kenkyu Senta Kenkyu Hokoku*, *9*, 87–137.

Yamaoka, K., Hori, M., & Kuratani, S. (1986). Ecomorphology of feeding in ‘goby-like’ cichlid fishes in Lake Tanganyika. *Physiol. Ecol. Jap.*, *23*, 17–29.

Yana, G., Xinga, Y., Wang, J., Lic, Z., Wang, L., Wanga, Q., Xua, L., Zhanga, Z., Zhange, J., Donga, X., Shana, W., Guoa, L., & Han, S. (2018). Sequestration of atmospheric CO_2 in boreal forest carbon pools in northeastern China: Effects of nitrogen deposition. *Agricultural and Forest Meteorology*, *248*, 70–81. doi:10.1016/j.agrformet.2017.09.015

Yan, Y. (2018). Integrate carbon dynamic models in analyzing carbon sequestration impact of forest biomass harvest. *The Science of the Total Environment*, *615*, 581–587. doi:10.1016/j.scitotenv.2017.09.326 PMID:28988094

Yermolaev, O., Lisetskii, F., Marinina, O., & Buryak, A. (2015). Basin and eco-regional approach to optimise the use of water and land resources. *Biosciences Biotechnology Research Asia*, *12*(2), 145–158. doi:10.13005/bbra/2185

Yoo, C., Jun, S. Y., Lee, J. Y., Ahn, C. Y., & Oh, H. M. (2010). Selection of microalgae for lipid production under high levels carbon dioxide. *Bioresource Technology*, *101*(1), 71–74. doi:10.1016/j.biortech.2009.03.030 PMID:19362826

Yousfi, E. L. M. (1988). Degradation of forest cover in southern Morocco: Case of Admine arganeraie. Rabat: 3rd cycle Memory IAV Hassan II.

Yuan, F., Zhang, L., Wah, K., Win, W., Ren, L., & Zhao, C. (2017). Assessment of GPM and TRMM Multi-Satellite Precipitation Products in Streamflow Simulations in a Data-Sparse Mountainous Watershed in Myanmar. *Remote Sensing*, *9*(3), 1–23. doi:10.3390/rs9030302

Yuan, X., Mu, R., Zuo, J., & Wang, Q. (2015). Economic Development, Energy Consumption, and Air Pollution: A Critical Assessment in China. *Human and Ecological Risk Assessment*, *21*(3), 781–798. doi:10.1080/10807039.2014.932204

Zahra, S. I., Abbas, F., Ishaq, W., Ibrahim, M., Hammad, H. M., Akram, B., & Salik, M. R. (2016). Carbon Sequestration potential of soils under maize production in Irrigated Agriculture of the Punjab province of Pakistan. *The Journal of Animal & Plant Sciences.*, *26*, 706–715.

Zeng, R., Cai, X., Ringler, C., & Zhu, T. (2017). Hydropower versus irrigation—An analysis of global patterns. *Environmental Research Letters*, *12*(3), 34006. doi:10.1088/1748-9326/aa5f3f

Zhang, C., Chen, X., Li, Y., Ding, W., & Fu, G. (2018). Water-energy-food nexus: Concepts, questions and methodologies. *Journal of Cleaner Production*, *195*, 625–639. doi:10.1016/j.jclepro.2018.05.194

Zhang, T., Li, B., Yuan, Y., Gao, X., Sun, Q., Xu, L., & Jiang, Y. (2018). Remote Sensing of Environment Spatial downscaling of TRMM precipitation data considering the impacts of macro-geographical factors and local elevation in the Three-River Headwaters Region. *Remote Sensing of Environment*, *215*(May), 109–127. doi:10.1016/j.rse.2018.06.004

Zhang, Y., Li, Y., Ji, X., Luo, X., & Li, X. (2018). Fine-Resolution Precipitation Mapping in a Mountainous Watershed : Geostatistical Downscaling of TRMM Products Based on Environmental Variables. *Remote Sensing*, *10*(1), 1–27. doi:10.3390/rs10010119

Zhao, B., Su, Y., Zhang, Y., & Cui, G. (2015). Carbon dioxide fixation and biomass production from combustion flue gas using energy microalgae. *Energy*, *89*, 347–357. doi:10.1016/j.energy.2015.05.123

Zhao, C., Chen, J., Du, P., & Yuan, H. (2018). Characteristics of climate change and extreme weather from 1951 to 2011 in China. *International Journal of Environmental Research and Public Health, 15*(11), 2540. doi:10.3390/ijerph15112540 PMID:30428540

Zhao, H., Yang, B., Yang, S., & Huang, Y. (2017). NU SC Institute of Environment and Sustainable Development in Agriculture, Chinese Academy of. *Atmospheric Research*. Advance online publication. doi:10.1016/j.atmosres.2017.11.005

Zheng, X., & Zhu, J. (2015). A methodological approach for spatial downscaling of TRMM precipitation data in North China. *International Journal of Remote Sensing, 36*(January), 37–41. doi:10.1080/01431161.2014.995275

Zhou, Z., Guo, B., Su, Y., Chen, Z., & Wang, J. (2020). Multidimensional evaluation of the TRMM 3B43V7 satellite-based precipitation product in mainland China from 1998 – 2016. *PeerJ, 8*, 1–24. doi:10.7717/peerj.8615 PMID:32117637

Zhu, B., Chen, G., Cao, X., & Wei, D. (2017). Molecular characterization of CO_2 sequestration and assimilation in microalgae and its biotechnological applications. *Bioresource Technology, 244*, 1207–1215. doi:10.1016/j.biortech.2017.05.199 PMID:28606753

Zhu, T., & Ringler, C. (2012). Climate change impacts on water availability and use in the Limpopo river basin. *Water (Basel), 4*(1), 64–86. doi:10.3390/w4010063

Ziervogel, G., New, M., Garderen, E., Midgley, G., Taylor, A., Hamann, R., Stuart-Hill, S., Myers, J., & Warburton, M. (2014). Climate change impacts and adaptation in South Africa. Royal Geographical Society, 605-620.

Zucca, C. (2016). *Agropastoral System Diagnosis: Description of Agricultural Production Systems and Restoration Interventions with a Stress on their Weaknesses and Constraints*. Academic Press.

Related References

To continue our tradition of advancing academic research, we have compiled a list of recommended IGI Global readings. These references will provide additional information and guidance to further enrich your knowledge and assist you with your own research and future publications.

Abresch, J. (2003). Geographic Information Systems Research and Data Centers. In A. Hanson & B. Levin (Eds.), *Building a Virtual Library* (pp. 52–64). Hershey, PA: IGI Global. doi:10.4018/978-1-59140-106-3.ch004

Ajiboye, O. E., & Yusuff, O. S. (2017). Foreign Land Acquisition: Food Security and Food Chains – The Nigerian Experience. In I. Management Association (Ed.), Natural Resources Management: Concepts, Methodologies, Tools, and Applications (pp. 1524-1545). Hershey, PA: IGI Global. doi:10.4018/978-1-5225-0803-8.ch072

Alapiki, H. E., & Amadi, L. A. (2018). Sustainable Food Consumption in the Neoliberal Order: Challenges and Policy Implications. In A. Obayelu (Ed.), *Food Systems Sustainability and Environmental Policies in Modern Economies* (pp. 90–123). Hershey, PA: IGI Global. doi:10.4018/978-1-5225-3631-4.ch005

Altaş, A. (2018). Geographical Information System Applications Utilized in Museums in Turkey Within the Scope of the Cultural Heritage Tourism: A Case Study of Mobile Application of Müze Asist. In S. Chaudhuri & N. Ray (Eds.), *GIS Applications in the Tourism and Hospitality Industry* (pp. 42–60). Hershey, PA: IGI Global. doi:10.4018/978-1-5225-5088-4.ch002

Andreea, I. R. (2018). Beyond Macroeconomics of Food and Nutrition Security. *International Journal of Sustainable Economies Management*, *7*(1), 13–22. doi:10.4018/IJSEM.2018010102

Anwar, J. (2017). Reproductive and Mental Health during Natural Disaster: Implications and Issues for Women in Developing Nations – A Case Example. In I. Management Association (Ed.), Gaming and Technology Addiction: Breakthroughs in Research and Practice (pp. 446-472). Hershey, PA: IGI Global. doi:10.4018/978-1-5225-0778-9.ch021

Awadh, H., Aksissou, M., Benhardouze, W., Darasi, F., & Snaiki, J. (2018). Socioeconomic Status of Artisanal Fishers in the West Part of Moroccan Mediterranean. *International Journal of Social Ecology and Sustainable Development*, *9*(1), 40–52. doi:10.4018/IJSESD.2018010104

Aye, G. C., & Haruna, R. F. (2018). Effect of Climate Change on Crop Productivity and Prices in Benue State, Nigeria: Implications for Food Security. In V. Erokhin (Ed.), *Establishing Food Security and Alternatives to International Trade in Emerging Economies* (pp. 244–268). Hershey, PA: IGI Global. doi:10.4018/978-1-5225-2733-6.ch012

Azizan, S. A., & Suki, N. M. (2017). Consumers' Intentions to Purchase Organic Food Products. In T. Esakki (Ed.), *Green Marketing and Environmental Responsibility in Modern Corporations* (pp. 68–82). Hershey, PA: IGI Global. doi:10.4018/978-1-5225-2331-4.ch005

Barakabitze, A. A., Fue, K. G., Kitindi, E. J., & Sanga, C. A. (2017). Developing a Framework for Next Generation Integrated Agro Food-Advisory Systems in Developing Countries. In I. Management Association (Ed.), Agri-Food Supply Chain Management: Breakthroughs in Research and Practice (pp. 47-67). Hershey, PA: IGI Global. doi:10.4018/978-1-5225-1629-3.ch004

Beachcroft-Shaw, H., & Ellis, D. (2018). Using Successful Cases to Promote Environmental Sustainability: A Social Marketing Approach. In I. Management Association (Ed.), Sustainable Development: Concepts, Methodologies, Tools, and Applications (pp. 936-953). Hershey, PA: IGI Global. doi:10.4018/978-1-5225-3817-2.ch042

Behnassi, M., Kahime, K., Boussaa, S., Boumezzough, A., & Messouli, M. (2017). Infectious Diseases and Climate Vulnerability in Morocco: Governance and Adaptation Options. In I. Management Association (Ed.), Public Health and Welfare: Concepts, Methodologies, Tools, and Applications (pp. 91-109). Hershey, PA: IGI Global. doi:10.4018/978-1-5225-1674-3.ch005

Bekele, F., & Bekele, I. (2017). Social and Environmental Impacts on Agricultural Development. In W. Ganpat, R. Dyer, & W. Isaac (Eds.), *Agricultural Development and Food Security in Developing Nations* (pp. 21–56). Hershey, PA: IGI Global. doi:10.4018/978-1-5225-0942-4.ch002

Benaouda, A., & García-Peñalvo, F. J. (2018). Towards an Intelligent System for the Territorial Planning: Agricultural Case. In F. García-Peñalvo (Ed.), *Global Implications of Emerging Technology Trends* (pp. 158–178). Hershey, PA: IGI Global. doi:10.4018/978-1-5225-4944-4.ch010

Bhaskar, A., Rao, G. B., & Vencatesan, J. (2017). Characterization and Management Concerns of Water Resources around Pallikaranai Marsh, South Chennai. In P. Rao & Y. Patil (Eds.), *Reconsidering the Impact of Climate Change on Global Water Supply, Use, and Management* (pp. 102–121). Hershey, PA: IGI Global. doi:10.4018/978-1-5225-1046-8.ch007

Bianchini, E., & Sicurella, S. (2010). GIS: A New Tool for Criminology and Victimology's Studies. In R. Sette (Ed.), *Cases on Technologies for Teaching Criminology and Victimology: Methodologies and Practices* (pp. 87–110). Hershey, PA: IGI Global. doi:10.4018/978-1-60566-872-7.ch006

Bogataj, D., & Drobne, D. (2017). Control of Perishable Goods in Cold Logistic Chains by Bionanosensors. In S. Joo (Ed.), *Applying Nanotechnology for Environmental Sustainability* (pp. 376–402). Hershey, PA: IGI Global. doi:10.4018/978-1-5225-0585-3.ch016

Bogataj, D., & Drobne, D. (2017). Control of Perishable Goods in Cold Logistic Chains by Bionanosensors. In I. Management Association (Ed.), Materials Science and Engineering: Concepts, Methodologies, Tools, and Applications (pp. 471-497). Hershey, PA: IGI Global. doi:10.4018/978-1-5225-1798-6.ch019

Bogueva, D., & Marinova, D. (2018). What Is More Important: Perception of Masculinity or Personal Health and the Environment? In D. Bogueva, D. Marinova, & T. Raphaely (Eds.), *Handbook of Research on Social Marketing and Its Influence on Animal Origin Food Product Consumption* (pp. 148–162). Hershey, PA: IGI Global. doi:10.4018/978-1-5225-4757-0.ch010

Bouzid, M. (2017). Waterborne Diseases and Climate Change: Impact and Implications. In M. Bouzid (Ed.), *Examining the Role of Environmental Change on Emerging Infectious Diseases and Pandemics* (pp. 89–108). Hershey, PA: IGI Global. doi:10.4018/978-1-5225-0553-2.ch004

Bowles, D. C. (2017). Climate Change-Associated Conflict and Infectious Disease. In M. Bouzid (Ed.), *Examining the Role of Environmental Change on Emerging Infectious Diseases and Pandemics* (pp. 68–88). Hershey, PA: IGI Global. doi:10.4018/978-1-5225-0553-2.ch003

Buck, J. J., & Lowry, R. K. (2017). Oceanographic Data Management: Quills and Free Text to the Digital Age and "Big Data". In P. Diviacco, A. Leadbetter, & H. Glaves (Eds.), *Oceanographic and Marine Cross-Domain Data Management for Sustainable Development* (pp. 1–22). Hershey, PA: IGI Global. doi:10.4018/978-1-5225-0700-0.ch001

Buse, C. G. (2017). Are Climate Change Adaptation Policies a Game Changer?: A Case Study of Perspectives from Public Health Officials in Ontario, Canada. In M. Bouzid (Ed.), *Examining the Role of Environmental Change on Emerging Infectious Diseases and Pandemics* (pp. 230–257). Hershey, PA: IGI Global. doi:10.4018/978-1-5225-0553-2.ch010

Calderon, F. A., Giolo, E. G., Frau, C. D., Rengel, M. G., Rodriguez, H., Tornello, M., ... Gallucci, R. (2018). Seismic Microzonation and Site Effects Detection Through Microtremors Measures: A Review. In N. Ceryan (Ed.), *Handbook of Research on Trends and Digital Advances in Engineering Geology* (pp. 326–349). Hershey, PA: IGI Global. doi:10.4018/978-1-5225-2709-1.ch009

Carfì, D., Donato, A., & Panuccio, D. (2018). A Game Theory Coopetitive Perspective for Sustainability of Global Feeding: Agreements Among Vegan and Non-Vegan Food Firms. In I. Management Association (Ed.), Game Theory: Breakthroughs in Research and Practice (pp. 71-104). Hershey, PA: IGI Global. doi:10.4018/978-1-5225-2594-3.ch004

Carr, D. A., & Carr, T. R. (2007). Geographic Information System Applications in the Public Sector. In G. Garson (Ed.), *Modern Public Information Technology Systems: Issues and Challenges* (pp. 293–311). Hershey, PA: IGI Global. doi:10.4018/978-1-59904-051-6.ch015

Castagnolo, V. (2018). Analyzing, Classifying, Safeguarding: Drawing for the Borgo Murattiano Neighbourhood of Bari. In G. Carlone, N. Martinelli, & F. Rotondo (Eds.), *Designing Grid Cities for Optimized Urban Development and Planning* (pp. 93–108). Hershey, PA: IGI Global. doi:10.4018/978-1-5225-3613-0.ch006

Chekima, B. (2018). The Dilemma of Purchase Intention: A Conceptual Framework for Understanding Actual Consumption of Organic Food. *International Journal of Sustainable Economies Management*, *7*(2), 1–13. doi:10.4018/IJSEM.2018040101

Chen, Y. (2017). Sustainable Supply Chains and International Soft Landings: A Case of Wetland Entrepreneurship. In B. Christiansen & F. Kasarcı (Eds.), *Corporate Espionage, Geopolitics, and Diplomacy Issues in International Business* (pp. 232–247). Hershey, PA: IGI Global. doi:10.4018/978-1-5225-1031-4.ch013

Çıtak, L., Akel, V., & Ersoy, E. (2018). Investors' Reactions to the Announcement of New Constituents of BIST Sustainability Index: An Analysis by Event Study and Mean-Median Tests. In M. Risso & S. Testarmata (Eds.), *Value Sharing for Sustainable and Inclusive Development* (pp. 270–289). Hershey, PA: IGI Global. doi:10.4018/978-1-5225-3147-0.ch012

Crossland, M. D. (2009). Geographic Information Systems as Decision Tools. In M. Khosrow-Pour, D.B.A. (Ed.), Encyclopedia of Information Science and Technology, Second Edition (pp. 1630-1633). Hershey, PA: IGI Global. doi:10.4018/978-1-60566-026-4.ch257

Crossland, M. D., Herschel, R. T., Perkins, W. C., & Scudder, J. N. (2002). Geographic Information Systems: How Cognitive Style Impacts Decision-Making Effectiveness. In M. Mahmood (Ed.), *Advanced Topics in End User Computing* (Vol. 1, pp. 285–301). Hershey, PA: IGI Global. doi:10.4018/978-1-930708-42-6.ch017

D'Aleo, V., D'Aleo, F., & Bonanno, R. (2018). New Food Industries Toward a New Level of Sustainable Supply: Success Stories, Business Models, and Strategies. In V. Erokhin (Ed.), *Establishing Food Security and Alternatives to International Trade in Emerging Economies* (pp. 74–97). Hershey, PA: IGI Global. doi:10.4018/978-1-5225-2733-6.ch004

Dagevos, H., & Reinders, M. J. (2018). Flexitarianism and Social Marketing: Reflections on Eating Meat in Moderation. In D. Bogueva, D. Marinova, & T. Raphaely (Eds.), *Handbook of Research on Social Marketing and Its Influence on Animal Origin Food Product Consumption* (pp. 105–120). Hershey, PA: IGI Global. doi:10.4018/978-1-5225-4757-0.ch007

Deenapanray, P. N., & Ramma, I. (2017). Adaptations to Climate Change and Climate Variability in the Agriculture Sector in Mauritius: Lessons from a Technical Needs Assessment. In I. Management Association (Ed.), Natural Resources Management: Concepts, Methodologies, Tools, and Applications (pp. 655-680). Hershey, PA: IGI Global. doi:10.4018/978-1-5225-0803-8.ch030

Deenapanray, P. N., & Ramma, I. (2017). Adaptations to Climate Change and Climate Variability in the Agriculture Sector in Mauritius: Lessons from a Technical Needs Assessment. In I. Management Association (Ed.), Natural Resources Management: Concepts, Methodologies, Tools, and Applications (pp. 655-680). Hershey, PA: IGI Global. doi:10.4018/978-1-5225-0803-8.ch030

Deshpande, S., Basu, S. K., Li, X., & Chen, X. (2017). Smart, Innovative and Intelligent Technologies Used in Drug Designing. In I. Management Association (Ed.), Pharmaceutical Sciences: Breakthroughs in Research and Practice (pp. 1175-1191). Hershey, PA: IGI Global. doi:10.4018/978-1-5225-1762-7.ch045

Dlamini, P. N. (2017). Use of Information and Communication Technologies Tools to Capture, Store, and Disseminate Indigenous Knowledge: A Literature Review. In P. Ngulube (Ed.), *Handbook of Research on Theoretical Perspectives on Indigenous Knowledge Systems in Developing Countries* (pp. 225–247). Hershey, PA: IGI Global. doi:10.4018/978-1-5225-0833-5.ch010

Dolejsova, M., & Kera, D. (2017). The Fermentation GutHub Project and the Internet of Microbes. In S. Konomi & G. Roussos (Eds.), *Enriching Urban Spaces with Ambient Computing, the Internet of Things, and Smart City Design* (pp. 25–46). Hershey, PA: IGI Global. doi:10.4018/978-1-5225-0827-4.ch002

Dolunay, O. (2018). A Paradigm Shift: Empowering Farmers to Eliminate the Waste in the Form of Fresh Water and Energy Through the Implementation of 4R+T. In I. Management Association (Ed.), Sustainable Development: Concepts, Methodologies, Tools, and Applications (pp. 882-892). Hershey, PA: IGI Global. doi:10.4018/978-1-5225-3817-2.ch039

Dube, P., Heijman, W. J., Ihle, R., & Ochieng, J. (2018). The Potential of Traditional Leafy Vegetables for Improving Food Security in Africa. In V. Erokhin (Ed.), *Establishing Food Security and Alternatives to International Trade in Emerging Economies* (pp. 220–243). Hershey, PA: IGI Global. doi:10.4018/978-1-5225-2733-6.ch011

Duruji, M. M., & Urenma, D. F. (2017). The Environmentalism and Politics of Climate Change: A Study of the Process of Global Convergence through UNFCCC Conferences. In I. Management Association (Ed.), *Natural Resources Management: Concepts, Methodologies, Tools, and Applications* (pp. 77-108). Hershey, PA: IGI Global. 10.4018/978-1-5225-0803-8.ch004

Dutta, U. (2017). Agro-Geoinformatics, Potato Cultivation, and Climate Change. In S. Londhe (Ed.), *Sustainable Potato Production and the Impact of Climate Change* (pp. 247–271). Hershey, PA: IGI Global. doi:10.4018/978-1-5225-1715-3.ch012

Edirisinghe, R., Stranieri, A., & Wickramasinghe, N. (2017). A Taxonomy for mHealth. In N. Wickramasinghe (Ed.), *Handbook of Research on Healthcare Administration and Management* (pp. 596–615). Hershey, PA: IGI Global. doi:10.4018/978-1-5225-0920-2.ch036

Ekpeni, N. M., & Ayeni, A. O. (2018). Global Natural Hazard and Disaster Vulnerability Management. In A. Eneanya (Ed.), *Handbook of Research on Environmental Policies for Emergency Management and Public Safety* (pp. 83–104). Hershey, PA: IGI Global. doi:10.4018/978-1-5225-3194-4.ch005

Ene, C., Voica, M. C., & Panait, M. (2017). Green Investments and Food Security: Opportunities and Future Directions in the Context of Sustainable Development. In M. Mieila (Ed.), *Measuring Sustainable Development and Green Investments in Contemporary Economies* (pp. 163–200). Hershey, PA: IGI Global. doi:10.4018/978-1-5225-2081-8.ch007

Escamilla, I., Ruíz, M. T., Ibarra, M. M., Soto, V. L., Quintero, R., & Guzmán, G. (2018). Geocoding Tweets Based on Semantic Web and Ontologies. In M. Lytras, N. Aljohani, E. Damiani, & K. Chui (Eds.), *Innovations, Developments, and Applications of Semantic Web and Information Systems* (pp. 372–392). Hershey, PA: IGI Global. doi:10.4018/978-1-5225-5042-6.ch014

Escribano, A. J. (2018). Marketing Strategies for Trendy Animal Products: Sustainability as a Core. In F. Quoquab, R. Thurasamy, & J. Mohammad (Eds.), *Driving Green Consumerism Through Strategic Sustainability Marketing* (pp. 169–203). Hershey, PA: IGI Global. doi:10.4018/978-1-5225-2912-5.ch010

Eudoxie, G., & Roopnarine, R. (2017). Climate Change Adaptation and Disaster Risk Management in the Caribbean. In W. Ganpat & W. Isaac (Eds.), *Environmental Sustainability and Climate Change Adaptation Strategies* (pp. 97–125). Hershey, PA: IGI Global. doi:10.4018/978-1-5225-1607-1.ch004

Farmer, L. S. (2017). Data Analytics for Strategic Management: Getting the Right Data. In V. Wang (Ed.), *Encyclopedia of Strategic Leadership and Management* (pp. 810–822). Hershey, PA: IGI Global. doi:10.4018/978-1-5225-1049-9.ch056

Fattal, L. R. (2017). Catastrophe: An Uncanny Catalyst for Creativity. In R. Shin (Ed.), *Convergence of Contemporary Art, Visual Culture, and Global Civic Engagement* (pp. 244–262). Hershey, PA: IGI Global. doi:10.4018/978-1-5225-1665-1.ch014

Forti, I. (2017). A Cross Reading of Landscape through Digital Landscape Models: The Case of Southern Garda. In A. Ippolito (Ed.), *Handbook of Research on Emerging Technologies for Architectural and Archaeological Heritage* (pp. 532–561). Hershey, PA: IGI Global. doi:10.4018/978-1-5225-0675-1.ch018

Gharbi, A., De Runz, C., & Akdag, H. (2017). Urban Development Modelling: A Survey. In S. Faiz & K. Mahmoudi (Eds.), *Handbook of Research on Geographic Information Systems Applications and Advancements* (pp. 96–124). Hershey, PA: IGI Global. doi:10.4018/978-1-5225-0937-0.ch004

Ghosh, I., & Ghoshal, I. (2018). Implications of Trade Liberalization for Food Security Under the ASEAN-India Strategic Partnership: A Gravity Model Approach. In V. Erokhin (Ed.), *Establishing Food Security and Alternatives to International Trade in Emerging Economies* (pp. 98–118). Hershey, PA: IGI Global. doi:10.4018/978-1-5225-2733-6.ch005

Glaves, H. M. (2017). Developing a Common Global Framework for Marine Data Management. In P. Diviacco, A. Leadbetter, & H. Glaves (Eds.), *Oceanographic and Marine Cross-Domain Data Management for Sustainable Development* (pp. 47–68). Hershey, PA: IGI Global. doi:10.4018/978-1-5225-0700-0.ch003

Godulla, A., & Wolf, C. (2018). Future of Food: Transmedia Strategies of National Geographic. In R. Gambarato & G. Alzamora (Eds.), *Exploring Transmedia Journalism in the Digital Age* (pp. 162–182). Hershey, PA: IGI Global. doi:10.4018/978-1-5225-3781-6.ch010

Gomes, P. P. (2018). Food and Environment: A Review on the Sustainability of Six Different Dietary Patterns. In A. Obayelu (Ed.), *Food Systems Sustainability and Environmental Policies in Modern Economies* (pp. 15–31). Hershey, PA: IGI Global. doi:10.4018/978-1-5225-3631-4.ch002

Gonzalez-Feliu, J. (2018). Sustainability Evaluation of Green Urban Logistics Systems: Literature Overview and Proposed Framework. In A. Paul, D. Bhattacharyya, & S. Anand (Eds.), *Green Initiatives for Business Sustainability and Value Creation* (pp. 103–134). Hershey, PA: IGI Global. doi:10.4018/978-1-5225-2662-9.ch005

Goodland, R. (2017). A Fresh Look at Livestock Greenhouse Gas Emissions and Mitigation. In I. Management Association (Ed.), Natural Resources Management: Concepts, Methodologies, Tools, and Applications (pp. 124-139). Hershey, PA: IGI Global. doi:10.4018/978-1-5225-0803-8.ch006

Goundar, S., & Appana, S. (2018). Mainstreaming Development Policies for Climate Change in Fiji: A Policy Gap Analysis and the Role of ICTs. In I. Management Association (Ed.), Sustainable Development: Concepts, Methodologies, Tools, and Applications (pp. 402-432). Hershey, PA: IGI Global. doi:10.4018/978-1-5225-3817-2.ch020

Granell-Canut, C., & Aguilar-Moreno, E. (2018). Geospatial Influence in Science Mapping. In M. Khosrow-Pour, D.B.A. (Ed.), Encyclopedia of Information Science and Technology, Fourth Edition (pp. 3473-3483). Hershey, PA: IGI Global. doi:10.4018/978-1-5225-2255-3.ch302

Grigelis, A., Blažauskas, N., Gelumbauskaitė, L. Ž., Gulbinskas, S., Suzdalev, S., & Ferrarin, C. (2017). Marine Environment Data Management Related to the Human Activity in the South-Eastern Baltic Sea (The Lithuanian Sector). In P. Diviacco, A. Leadbetter, & H. Glaves (Eds.), *Oceanographic and Marine Cross-Domain Data Management for Sustainable Development* (pp. 282–302). Hershey, PA: IGI Global. doi:10.4018/978-1-5225-0700-0.ch012

Guma, I. P., Rwashana, A. S., & Oyo, B. (2018). Food Security Policy Analysis Using System Dynamics: The Case of Uganda. *International Journal of Information Technologies and Systems Approach, 11*(1), 72–90. doi:10.4018/IJITSA.2018010104

Guma, I. P., Rwashana, A. S., & Oyo, B. (2018). Food Security Indicators for Subsistence Farmers Sustainability: A System Dynamics Approach. *International Journal of System Dynamics Applications*, *7*(1), 45–64. doi:10.4018/IJSDA.2018010103

Gupta, P., & Goyal, S. (2017). Wildlife Habitat Evaluation. In A. Santra & S. Mitra (Eds.), *Remote Sensing Techniques and GIS Applications in Earth and Environmental Studies* (pp. 258–264). Hershey, PA: IGI Global. doi:10.4018/978-1-5225-1814-3.ch013

Hanson, T., & Hildebrand, E. (2018). GPS Travel Diaries in Rural Transportation Research: A Focus on Older Drivers. In I. Management Association (Ed.), Intelligent Transportation and Planning: Breakthroughs in Research and Practice (pp. 609-625). Hershey, PA: IGI Global. doi:10.4018/978-1-5225-5210-9.ch027

Hartman, M. B. (2017). Research-Based Climate Change Public Education Programs. In I. Management Association (Ed.), Natural Resources Management: Concepts, Methodologies, Tools, and Applications (pp. 992-1003). Hershey, PA: IGI Global. doi:10.4018/978-1-5225-0803-8.ch046

Hashim, N. (2017). Zanzibari Seaweed: Global Climate Change and the Promise of Adaptation. In I. Management Association (Ed.), Natural Resources Management: Concepts, Methodologies, Tools, and Applications (pp. 365-391). Hershey, PA: IGI Global. doi:10.4018/978-1-5225-0803-8.ch019

Herrera, J. E., Argüello, L. V., Gonzalez-Feliu, J., & Jaimes, W. A. (2017). Decision Support System Design Requirements, Information Management, and Urban Logistics Efficiency: Case Study of Bogotá, Colombia. In G. Jamil, A. Soares, & C. Pessoa (Eds.), *Handbook of Research on Information Management for Effective Logistics and Supply Chains* (pp. 223–238). Hershey, PA: IGI Global. doi:10.4018/978-1-5225-0973-8.ch012

Huizinga, T., Ayanso, A., Smoor, M., & Wronski, T. (2017). Exploring Insurance and Natural Disaster Tweets Using Text Analytics. *International Journal of Business Analytics*, *4*(1), 1–17. doi:10.4018/IJBAN.2017010101

Hung, K., Kalantari, M., & Rajabifard, A. (2017). An Integrated Method for Assessing the Text Content Quality of Volunteered Geographic Information in Disaster Management. *International Journal of Information Systems for Crisis Response and Management*, *9*(2), 1–17. doi:10.4018/IJISCRAM.2017040101

Husnain, A., & Avdic, A. (2018). Identifying the Contemporary Status of E-Service Sustainability Research. In I. Management Association (Ed.), Sustainable Development: Concepts, Methodologies, Tools, and Applications (pp. 467-485). Hershey, PA: IGI Global. doi:10.4018/978-1-5225-3817-2.ch022

Iarossi, M. P., & Ferro, L. (2017). "The Past is Never Dead. It's Not Even Past": Virtual Archaeological Promenade. In A. Ippolito & M. Cigola (Eds.), *Handbook of Research on Emerging Technologies for Digital Preservation and Information Modeling* (pp. 228–255). Hershey, PA: IGI Global. doi:10.4018/978-1-5225-0680-5.ch010

Ignjatijević, S., & Cvijanović, D. (2018). Analysis of Serbian Honey Production and Exports. In *Exploring the Global Competitiveness of Agri-Food Sectors and Serbia's Dominant Presence: Emerging Research and Opportunities* (pp. 109–139). Hershey, PA: IGI Global. doi:10.4018/978-1-5225-2762-6.ch005

Jana, S. K., & Karmakar, A. K. (2017). Globalization, Governance, and Food Security: The Case of BRICS. In I. Management Association (Ed.), Natural Resources Management: Concepts, Methodologies, Tools, and Applications (pp. 692-712). Hershey, PA: IGI Global. doi:10.4018/978-1-5225-0803-8.ch032

Jana, S. K., & Karmakar, A. K. (2017). Food Security in Asia: Is There Convergence? In I. Management Association (Ed.), Natural Resources Management: Concepts, Methodologies, Tools, and Applications (pp. 109-123). Hershey, PA: IGI Global. doi:10.4018/978-1-5225-0803-8.ch005

John, J., & Kumar, S. (2018). A Locational Decision Making Framework for Shipbreaking Under Multiple Criteria. In I. Management Association (Ed.), Operations and Service Management: Concepts, Methodologies, Tools, and Applications (pp. 504-527). Hershey, PA: IGI Global. doi:10.4018/978-1-5225-3909-4.ch024

John, J., & Srivastava, R. K. (2018). Decision Insights for Shipbreaking using Environmental Impact Assessment: Review and Perspectives. *International Journal of Strategic Decision Sciences*, *9*(1), 45–62. doi:10.4018/IJSDS.2018010104

Joshi, Y., & Rahman, Z. (2018). Determinants of Sustainable Consumption Behaviour: Review and Conceptual Framework. In A. Paul, D. Bhattacharyya, & S. Anand (Eds.), *Green Initiatives for Business Sustainability and Value Creation* (pp. 239–262). Hershey, PA: IGI Global. doi:10.4018/978-1-5225-2662-9.ch011

Juma, D. W., Reuben, M., Wang, H., & Li, F. (2018). Adaptive Coevolution: Realigning the Water Governance Regime to the Changing Climate. In I. Management Association (Ed.), Hydrology and Water Resource Management: Breakthroughs in Research and Practice (pp. 346-357). Hershey, PA: IGI Global. doi:10.4018/978-1-5225-3427-3.ch014

K., S., & Tripathy, B. K. (2018). Neighborhood Rough-Sets-Based Spatial Data Analytics. In M. Khosrow-Pour, D.B.A. (Ed.), *Encyclopedia of Information Science and Technology, Fourth Edition* (pp. 1835-1844). Hershey, PA: IGI Global. doi:10.4018/978-1-5225-2255-3.ch160

Kabir, F. (2018). Towards a More Gender-Inclusive Climate Change Policy. In N. Mahtab, T. Haque, I. Khan, M. Islam, & I. Wahid (Eds.), *Handbook of Research on Women's Issues and Rights in the Developing World* (pp. 354–369). Hershey, PA: IGI Global. doi:10.4018/978-1-5225-3018-3.ch022

Kabir, F. (2018). Towards a More Gender-Inclusive Climate Change Policy. In I. Management Association (Ed.), Climate Change and Environmental Concerns: Breakthroughs in Research and Practice (pp. 525-540). Hershey, PA: IGI Global. doi:10.4018/978-1-5225-5487-5.ch027

Kanyamuka, J. S., Jumbe, C. B., & Ricker-Gilbert, J. (2018). Making Agricultural Input Subsidies More Effective and Profitable in Africa: The Role of Complementary Interventions. In A. Obayelu (Ed.), *Food Systems Sustainability and Environmental Policies in Modern Economies* (pp. 172–187). Hershey, PA: IGI Global. doi:10.4018/978-1-5225-3631-4.ch008

Karimi, H., & Gholamrezafahimi, F. (2017). Study of Integrated Coastal Zone Management and Its Environmental Effects: A Case of Iran. In R. Singh, A. Singh, & V. Srivastava (Eds.), *Environmental Issues Surrounding Human Overpopulation* (pp. 64–88). Hershey, PA: IGI Global. doi:10.4018/978-1-5225-1683-5.ch004

Kaya, I. R., Hutabarat, J., & Bambang, A. N. (2018). "Sasi": A New Path to Sustain Seaweed Farming From Up-Stream to Down-Stream in Kotania Bay, Molucass. *International Journal of Social Ecology and Sustainable Development*, *9*(2), 28–36. doi:10.4018/IJSESD.2018040103

Khader, V. (2018). Technologies for Food, Health, Livelihood, and Nutrition Security. In I. Management Association (Ed.), Food Science and Nutrition: Breakthroughs in Research and Practice (pp. 94-112). Hershey, PA: IGI Global. doi:10.4018/978-1-5225-5207-9.ch005

Kocadağlı, A. Y. (2017). The Temporal and Spatial Development of Organic Agriculture in Turkey. In W. Ganpat, R. Dyer, & W. Isaac (Eds.), *Agricultural Development and Food Security in Developing Nations* (pp. 130–156). Hershey, PA: IGI Global. doi:10.4018/978-1-5225-0942-4.ch006

Koundouri, P., Giannouli, A., & Souliotis, I. (2017). An Integrated Approach for Sustainable Environmental and Socio-Economic Development Using Offshore Infrastructure. In I. Management Association (Ed.), Renewable and Alternative Energy: Concepts, Methodologies, Tools, and Applications (pp. 1581-1601). Hershey, PA: IGI Global. doi:10.4018/978-1-5225-1671-2.ch056

Kumar, A., & Dash, M. K. (2017). Sustainability and Future Generation Infrastructure on Digital Platform: A Study of Generation Y. In N. Ray (Ed.), *Business Infrastructure for Sustainability in Developing Economies* (pp. 124–142). Hershey, PA: IGI Global. doi:10.4018/978-1-5225-2041-2.ch007

Kumar, A., Mukherjee, A. B., & Krishna, A. P. (2017). Application of Conventional Data Mining Techniques and Web Mining to Aid Disaster Management. In A. Kumar (Ed.), *Web Usage Mining Techniques and Applications Across Industries* (pp. 138–167). Hershey, PA: IGI Global. doi:10.4018/978-1-5225-0613-3.ch006

Kumar, C. P. (2017). Impact of Climate Change on Groundwater Resources. In I. Management Association (Ed.), Natural Resources Management: Concepts, Methodologies, Tools, and Applications (pp. 1094-1120). Hershey, PA: IGI Global. doi:10.4018/978-1-5225-0803-8.ch052

Kumari, S., & Patil, Y. (2017). Achieving Climate Smart Agriculture with a Sustainable Use of Water: A Conceptual Framework for Sustaining the Use of Water for Agriculture in the Era of Climate Change. In P. Rao & Y. Patil (Eds.), *Reconsidering the Impact of Climate Change on Global Water Supply, Use, and Management* (pp. 122–143). Hershey, PA: IGI Global. doi:10.4018/978-1-5225-1046-8.ch008

Kumari, S., & Patil, Y. (2018). Achieving Climate Smart Agriculture With a Sustainable Use of Water: A Conceptual Framework for Sustaining the Use of Water for Agriculture in the Era of Climate Change. In I. Management Association (Ed.), Climate Change and Environmental Concerns: Breakthroughs in Research and Practice (pp. 111-133). Hershey, PA: IGI Global. doi:10.4018/978-1-5225-5487-5.ch006

Kursah, M. B. (2017). Least-Cost Pipeline using Geographic Information System: The Limit to Technicalities. *International Journal of Applied Geospatial Research, 8*(3), 1–15. doi:10.4018/ijagr.2017070101

Lahiri, S., Ghosh, D., & Bhakta, J. N. (2017). Role of Microbes in Eco-Remediation of Perturbed Aquatic Ecosystem. In J. Bhakta (Ed.), *Handbook of Research on Inventive Bioremediation Techniques* (pp. 70–107). Hershey, PA: IGI Global. doi:10.4018/978-1-5225-2325-3.ch004

Lallo, C. H., Smalling, S., Facey, A., & Hughes, M. (2018). The Impact of Climate Change on Small Ruminant Performance in Caribbean Communities. In I. Management Association (Ed.), Climate Change and Environmental Concerns: Breakthroughs in Research and Practice (pp. 193-218). Hershey, PA: IGI Global. doi:10.4018/978-1-5225-5487-5.ch010

Laurini, R. (2017). Nature of Geographic Knowledge Bases. In S. Faiz & K. Mahmoudi (Eds.), *Handbook of Research on Geographic Information Systems Applications and Advancements* (pp. 29–60). Hershey, PA: IGI Global. doi:10.4018/978-1-5225-0937-0.ch002

Lawrence, J., Simpson, L., & Piggott, A. (2017). Protected Agriculture: A Climate Change Adaptation for Food and Nutrition Security. In I. Management Association (Ed.), Natural Resources Management: Concepts, Methodologies, Tools, and Applications (pp. 140-158). Hershey, PA: IGI Global. doi:10.4018/978-1-5225-0803-8.ch007

Leadbetter, A., Cheatham, M., Shepherd, A., & Thomas, R. (2017). Linked Ocean Data 2.0. In P. Diviacco, A. Leadbetter, & H. Glaves (Eds.), *Oceanographic and Marine Cross-Domain Data Management for Sustainable Development* (pp. 69–99). Hershey, PA: IGI Global. doi:10.4018/978-1-5225-0700-0.ch004

Leipnik, M. R., & Mehta, S. S. (2005). Geographic Information Systems (GIS) in E-Marketing. In I. Clarke III & T. Flaherty (Eds.), *Advances in Electronic Marketing* (pp. 193–210). Hershey, PA: IGI Global. doi:10.4018/978-1-59140-321-0.ch011

Lucas, M. R., Rego, C., Vieira, C., & Vieira, I. (2017). Proximity and Cooperation for Innovative Regional Development: The Case of the Science and Technology Park of Alentejo. In L. Carvalho (Ed.), *Handbook of Research on Entrepreneurial Development and Innovation Within Smart Cities* (pp. 199–228). Hershey, PA: IGI Global. doi:10.4018/978-1-5225-1978-2.ch010

Ma, X., Beaulieu, S. E., Fu, L., Fox, P., Di Stefano, M., & West, P. (2017). Documenting Provenance for Reproducible Marine Ecosystem Assessment in Open Science. In P. Diviacco, A. Leadbetter, & H. Glaves (Eds.), *Oceanographic and Marine Cross-Domain Data Management for Sustainable Development* (pp. 100–126). Hershey, PA: IGI Global. doi:10.4018/978-1-5225-0700-0.ch005

Mabe, L. K., & Oladele, O. I. (2017). Application of Information Communication Technologies for Agricultural Development through Extension Services: A Review. In T. Tossy (Ed.), *Information Technology Integration for Socio-Economic Development* (pp. 52–101). Hershey, PA: IGI Global. doi:10.4018/978-1-5225-0539-6.ch003

Malomo, B. I. (2018). A Review of Psychological Resilience as a Response to Natural Hazards in Nigeria. In A. Eneanya (Ed.), *Handbook of Research on Environmental Policies for Emergency Management and Public Safety* (pp. 147–165). Hershey, PA: IGI Global. doi:10.4018/978-1-5225-3194-4.ch008

Manchiraju, S. (2018). Predicting Behavioral Intentions Toward Sustainable Fashion Consumption: A Comparison of Attitude-Behavior and Value-Behavior Consistency Models. In I. Management Association (Ed.), Fashion and Textiles: Breakthroughs in Research and Practice (pp. 1-21). Hershey, PA: IGI Global. doi:10.4018/978-1-5225-3432-7.ch001

Manzella, G. M., Bartolini, R., Bustaffa, F., D'Angelo, P., De Mattei, M., Frontini, F., ... Spada, A. (2017). Semantic Search Engine for Data Management and Sustainable Development: Marine Planning Service Platform. In P. Diviacco, A. Leadbetter, & H. Glaves (Eds.), *Oceanographic and Marine Cross-Domain Data Management for Sustainable Development* (pp. 127–154). Hershey, PA: IGI Global. doi:10.4018/978-1-5225-0700-0.ch006

Mbonigaba, J. (2018). Comparing the Effects of Unsustainable Production and Consumption of Food on Health and Policy Across Developed and Less Developed Countries. In A. Obayelu (Ed.), *Food Systems Sustainability and Environmental Policies in Modern Economies* (pp. 124–158). Hershey, PA: IGI Global. doi:10.4018/978-1-5225-3631-4.ch006

McKeown, A. E. (2017). Nurses, Healthcare, and Environmental Pollution and Solutions: Breaking the Cycle of Harm. In I. Management Association (Ed.), Natural Resources Management: Concepts, Methodologies, Tools, and Applications (pp. 392-415). Hershey, PA: IGI Global. doi:10.4018/978-1-5225-0803-8.ch020

Mennecke, B. E., & West, L. A. Jr. (2001). Geographic Information Systems in Developing Countries: Issues in Data Collection, Implementation and Management. *Journal of Global Information Management*, *9*(4), 44–54. doi:10.4018/jgim.2001100103

Mili, B., Barua, A., & Katyaini, S. (2017). Climate Change and Adaptation through the Lens of Capability Approach: A Case Study from Darjeeling, Eastern Himalaya. In I. Management Association (Ed.), Natural Resources Management: Concepts, Methodologies, Tools, and Applications (pp. 1351-1365). Hershey, PA: IGI Global. doi:10.4018/978-1-5225-0803-8.ch064

Mir, S. A., Shah, M. A., Mir, M. M., & Iqbal, U. (2017). New Horizons of Nanotechnology in Agriculture and Food Processing Industry. In B. Nayak, A. Nanda, & M. Bhat (Eds.), *Integrating Biologically-Inspired Nanotechnology into Medical Practice* (pp. 230–258). Hershey, PA: IGI Global. doi:10.4018/978-1-5225-0610-2.ch009

Mujere, N., & Moyce, W. (2017). Climate Change Impacts on Surface Water Quality. In W. Ganpat & W. Isaac (Eds.), *Environmental Sustainability and Climate Change Adaptation Strategies* (pp. 322–340). Hershey, PA: IGI Global. doi:10.4018/978-1-5225-1607-1.ch012

Mukherjee, A. B., Krishna, A. P., & Patel, N. (2018). Geospatial Technology for Urban Sciences. In *Geospatial Technologies in Urban System Development: Emerging Research and Opportunities* (pp. 99–120). Hershey, PA: IGI Global. doi:10.4018/978-1-5225-3683-3.ch005

Nagarajan, S. K., & Sangaiah, A. K. (2017). Vegetation Index: Ideas, Methods, Influences, and Trends. In N. Kumar, A. Sangaiah, M. Arun, & S. Anand (Eds.), *Advanced Image Processing Techniques and Applications* (pp. 347–386). Hershey, PA: IGI Global. doi:10.4018/978-1-5225-2053-5.ch016

Naraine, L., & Meehan, K. (2017). Strengthening Food Security with Sustainable Practices by Smallholder Farmers in Lesser Developed Economies. In W. Ganpat, R. Dyer, & W. Isaac (Eds.), *Agricultural Development and Food Security in Developing Nations* (pp. 57–81). Hershey, PA: IGI Global. doi:10.4018/978-1-5225-0942-4.ch003

Naraine, L., & Meehan, K. (2017). Strengthening Food Security with Sustainable Practices by Smallholder Farmers in Lesser Developed Economies. In W. Ganpat, R. Dyer, & W. Isaac (Eds.), *Agricultural Development and Food Security in Developing Nations* (pp. 57–81). Hershey, PA: IGI Global. doi:10.4018/978-1-5225-0942-4.ch003

Naraine, L., & Meehan, K. (2017). Strengthening Food Security with Sustainable Practices by Smallholder Farmers in Lesser Developed Economies. In W. Ganpat, R. Dyer, & W. Isaac (Eds.), *Agricultural Development and Food Security in Developing Nations* (pp. 57–81). Hershey, PA: IGI Global. doi:10.4018/978-1-5225-0942-4.ch003

Neufeld, D. J., & Griffith, S. (2006). Isobord's Geographic Information System (GIS) Solution. In M. Khosrow-Pour, D.B.A. (Ed.), Cases on Database Technologies and Applications (pp. 64-80). Hershey, PA: IGI Global. doi:10.4018/978-1-59904-399-9.ch004

Nikolaou, K., Tsakiridou, E., Anastasiadis, F., & Mattas, K. (2017). Exploring Alternative Distribution Channels of Agricultural Products. *International Journal of Food and Beverage Manufacturing and Business Models*, *2*(2), 36–66. doi:10.4018/IJFBMBM.2017070103

Nishat, K. J., & Rahman, M. S. (2018). Disaster, Vulnerability, and Violence Against Women: Global Findings and a Research Agenda for Bangladesh. In N. Mahtab, T. Haque, I. Khan, M. Islam, & I. Wahid (Eds.), *Handbook of Research on Women's Issues and Rights in the Developing World* (pp. 235–250). Hershey, PA: IGI Global. doi:10.4018/978-1-5225-3018-3.ch014

O'Hara, S., Jones, D., & Trobman, H. B. (2018). Building an Urban Food System Through UDC Food Hubs. In A. Burtin, J. Fleming, & P. Hampton-Garland (Eds.), *Changing Urban Landscapes Through Public Higher Education* (pp. 116–143). Hershey, PA: IGI Global. doi:10.4018/978-1-5225-3454-9.ch006

Obayelu, A. E. (2018). Integrating Environment, Food Systems, and Sustainability in Feeding the Growing Population in Developing Countries. In A. Obayelu (Ed.), *Food Systems Sustainability and Environmental Policies in Modern Economies* (pp. 1–14). Hershey, PA: IGI Global. doi:10.4018/978-1-5225-3631-4.ch001

Othman, R., Nath, N., & Laswad, F. (2018). Environmental Reporting and Accounting: Sustainability Hybridisation. In G. Azevedo, J. da Silva Oliveira, R. Marques, & A. Ferreira (Eds.), *Handbook of Research on Modernization and Accountability in Public Sector Management* (pp. 130–158). Hershey, PA: IGI Global. doi:10.4018/978-1-5225-3731-1.ch007

Padigala, B. S. (2018). Traditional Water Management System for Climate Change Adaptation in Mountain Ecosystems. In I. Management Association (Ed.), Climate Change and Environmental Concerns: Breakthroughs in Research and Practice (pp. 630-655). Hershey, PA: IGI Global. doi:10.4018/978-1-5225-5487-5.ch033

Panda, C. K. (2018). Mobile Phone Usage in Agricultural Extension in India: The Current and Future Perspective. In F. Mtenzi, G. Oreku, D. Lupiana, & J. Yonazi (Eds.), *Mobile Technologies and Socio-Economic Development in Emerging Nations* (pp. 1–21). Hershey, PA: IGI Global. doi:10.4018/978-1-5225-4029-8.ch001

Pandian, S. L., Yarrakula, K., & Chaudhury, P. (2018). GIS-Based Decision Support System for Village Level: A Case Study in Andhra Pradesh. In S. Chaudhuri & N. Ray (Eds.), *GIS Applications in the Tourism and Hospitality Industry* (pp. 275–295). Hershey, PA: IGI Global. doi:10.4018/978-1-5225-5088-4.ch012

Quaranta, G., & Salvia, R. (2017). Social-Based Product Innovation and Governance in The Milk Sector: The Case of Carciocacio and Innonatura. In T. Tarnanidis, M. Vlachopoulou, & J. Papathanasiou (Eds.), *Driving Agribusiness With Technology Innovations* (pp. 293–310). Hershey, PA: IGI Global. doi:10.4018/978-1-5225-2107-5.ch015

Rahman, M. K., Schmidlin, T. W., Munro-Stasiuk, M. J., & Curtis, A. (2017). Geospatial Analysis of Land Loss, Land Cover Change, and Landuse Patterns of Kutubdia Island, Bangladesh. *International Journal of Applied Geospatial Research*, *8*(2), 45–60. doi:10.4018/IJAGR.2017040104

Rajack-Talley, T. A. (2017). Agriculture, Trade Liberalization and Poverty in the ACP Countries. In W. Ganpat, R. Dyer, & W. Isaac (Eds.), *Agricultural Development and Food Security in Developing Nations* (pp. 1–20). Hershey, PA: IGI Global. doi:10.4018/978-1-5225-0942-4.ch001

Rajamanickam, S. (2018). Exploring Landscapes in Regional Convergence: Environment and Sustainable Development in South Asia. In I. Management Association (Ed.), Sustainable Development: Concepts, Methodologies, Tools, and Applications (pp. 1051-1087). Hershey, PA: IGI Global. doi:10.4018/978-1-5225-3817-2.ch047

Rizvi, S. M., & Dearden, A. (2018). KHETI: ICT Solution for Agriculture Extension and Its Replication in Open and Distance Learning. In U. Pandey & V. Indrakanti (Eds.), *Open and Distance Learning Initiatives for Sustainable Development* (pp. 163–174). Hershey, PA: IGI Global. doi:10.4018/978-1-5225-2621-6.ch008

Roşu, L., & Macarov, L. I. (2017). Management of Drought and Floods in the Dobrogea Region. In I. Management Association (Ed.), Agri-Food Supply Chain Management: Breakthroughs in Research and Practice (pp. 372-403). Hershey, PA: IGI Global. doi:10.4018/978-1-5225-1629-3.ch016

Rouzbehani, K., & Rouzbehani, S. (2018). Mapping Women's World: GIS and the Case of Breast Cancer in the US. *International Journal of Public Health Management and Ethics*, *3*(1), 14–25. doi:10.4018/IJPHME.2018010102

Roy, D. (2018). Success Factors of Adoption of Mobile Applications in Rural India: Effect of Service Characteristics on Conceptual Model. In M. Khosrow-Pour, D.B.A. (Ed.), Green Computing Strategies for Competitive Advantage and Business Sustainability (pp. 211-238). Hershey, PA: IGI Global. doi:10.4018/978-1-5225-5017-4.ch010

Sajeva, M., Lemon, M., & Sahota, P. S. (2017). Governance for Food Security: A Framework for Social Learning and Scenario Building. *International Journal of Food and Beverage Manufacturing and Business Models*, *2*(2), 67–84. doi:10.4018/IJFBMBM.2017070104

Sallam, G. A., Youssef, T., Embaby, M. E., & Shaltot, F. (2011). Using Geographic Information System to Infollow the Fertilizers Pollution Migration. In I. Management Association (Ed.), Green Technologies: Concepts, Methodologies, Tools and Applications (pp. 564-586). Hershey, PA: IGI Global. doi:10.4018/978-1-60960-472-1.ch312

Sambhanthan, A., & Potdar, V. (2017). A Study of the Parameters Impacting Sustainability in Information Technology Organizations. *International Journal of Knowledge-Based Organizations*, *7*(3), 27–39. doi:10.4018/IJKBO.2017070103

Sanga, C., Kalungwizi, V. J., & Msuya, C. P. (2017). Bridging Gender Gaps in Provision of Agricultural Extension Service Using ICT: Experiences from Sokoine University of Agriculture (SUA) Farmer Voice Radio (FVR) Project in Tanzania. In I. Management Association (Ed.), Discrimination and Diversity: Concepts, Methodologies, Tools, and Applications (pp. 682-697). Hershey, PA: IGI Global. doi:10.4018/978-1-5225-1933-1.ch031

Santra, A., & Mitra, D. (2017). Role of Remote Sensing in Potential Fishing Zone Forecast. In A. Santra & S. Mitra (Eds.), *Remote Sensing Techniques and GIS Applications in Earth and Environmental Studies* (pp. 243–257). Hershey, PA: IGI Global. doi:10.4018/978-1-5225-1814-3.ch012

Schaap, D. (2017). SeaDataNet: Towards a Pan-European Infrastructure for Marine and Ocean Data Management. In P. Diviacco, A. Leadbetter, & H. Glaves (Eds.), *Oceanographic and Marine Cross-Domain Data Management for Sustainable Development* (pp. 155–177). Hershey, PA: IGI Global. doi:10.4018/978-1-5225-0700-0.ch007

Seckin-Celik, T. (2017). Sustainability Reporting and Sustainability in the Turkish Business Context. In U. Akkucuk (Ed.), *Ethics and Sustainability in Global Supply Chain Management* (pp. 115–132). Hershey, PA: IGI Global. doi:10.4018/978-1-5225-2036-8.ch006

Segbefia, A. Y., Barnes, V. R., Akpalu, L. A., & Mensah, M. (2018). Environmental Location Assessment for Seaweed Cultivation in Ghana: A Spatial Multi-Criteria Approach. *International Journal of Applied Geospatial Research*, *9*(1), 51–64. doi:10.4018/IJAGR.2018010104

Sen, Y. (2018). How to Manage Sustainability: A Framework for Corporate Sustainability Tools. In I. Management Association (Ed.), Sustainable Development: Concepts, Methodologies, Tools, and Applications (pp. 568-589). Hershey, PA: IGI Global. doi:10.4018/978-1-5225-3817-2.ch026

Shamshiry, E., Abdulai, A. M., Mokhtar, M. B., & Komoo, I. (2015). Regional Landfill Site Selection with GIS and Analytical Hierarchy Process Techniques: A Case Study of Langkawi Island, Malaysia. In P. Thomas, M. Srihari, & S. Kaur (Eds.), *Handbook of Research on Cultural and Economic Impacts of the Information Society* (pp. 248–282). Hershey, PA: IGI Global. doi:10.4018/978-1-4666-8598-7.ch011

Sharma, Y. K., Mangla, S. K., Patil, P. P., & Uniyal, S. (2018). Analyzing Sustainable Food Supply Chain Management Challenges in India. In M. Ram & J. Davim (Eds.), *Soft Computing Techniques and Applications in Mechanical Engineering* (pp. 162–180). Hershey, PA: IGI Global. doi:10.4018/978-1-5225-3035-0.ch008

Silvestrelli, P. (2018). The Impact of Events: To Which Extent Are They Sustainable for Tourist Destinations? Some Evidences From Expo Milano 2015. In M. Risso & S. Testarmata (Eds.), *Value Sharing for Sustainable and Inclusive Development* (pp. 185–204). Hershey, PA: IGI Global. doi:10.4018/978-1-5225-3147-0.ch008

Silvius, G. (2018). Sustainability Evaluation of IT/IS Projects. In I. Management Association (Ed.), Sustainable Development: Concepts, Methodologies, Tools, and Applications (pp. 26-40). Hershey, PA: IGI Global. doi:10.4018/978-1-5225-3817-2.ch002

Singh, R., Srivastava, P., Singh, P., Upadhyay, S., & Raghubanshi, A. S. (2017). Human Overpopulation and Food Security: Challenges for the Agriculture Sustainability. In R. Singh, A. Singh, & V. Srivastava (Eds.), *Environmental Issues Surrounding Human Overpopulation* (pp. 12–39). Hershey, PA: IGI Global. doi:10.4018/978-1-5225-1683-5.ch002

Singh, R., Srivastava, P., Singh, P., Upadhyay, S., & Raghubanshi, A. S. (2017). Human Overpopulation and Food Security: Challenges for the Agriculture Sustainability. In R. Singh, A. Singh, & V. Srivastava (Eds.), *Environmental Issues Surrounding Human Overpopulation* (pp. 12–39). Hershey, PA: IGI Global. doi:10.4018/978-1-5225-1683-5.ch002

Srivastava, N. (2017). Climate Change Mitigation: Collective Efforts and Responsibly. In I. Management Association (Ed.), Natural Resources Management: Concepts, Methodologies, Tools, and Applications (pp. 64-76). Hershey, PA: IGI Global. doi:10.4018/978-1-5225-0803-8.ch003

Stanganelli, M., & Gerundo, C. (2017). Understanding the Role of Urban Morphology and Green Areas Configuration During Heat Waves. *International Journal of Agricultural and Environmental Information Systems*, *8*(2), 50–64. doi:10.4018/IJAEIS.2017040104

Stewart, M. K., Hagood, D., & Ching, C. C. (2017). Virtual Games and Real-World Communities: Environments that Constrain and Enable Physical Activity in Games for Health. *International Journal of Game-Based Learning*, *7*(1), 1–19. doi:10.4018/IJGBL.2017010101

Stone, R. J. (2017). Modelling the Frequency of Tropical Cyclones in the Lower Caribbean Region. In W. Ganpat & W. Isaac (Eds.), *Environmental Sustainability and Climate Change Adaptation Strategies* (pp. 341–349). Hershey, PA: IGI Global. doi:10.4018/978-1-5225-1607-1.ch013

Syed, A., & Jabeen, U. A. (2018). Climate Change Impact on Agriculture and Food Security. In A. Eneanya (Ed.), *Handbook of Research on Environmental Policies for Emergency Management and Public Safety* (pp. 223–237). Hershey, PA: IGI Global. doi:10.4018/978-1-5225-3194-4.ch012

Tam, G. C. (2017). The Global View of Sustainability. In *Managerial Strategies and Green Solutions for Project Sustainability* (pp. 1–24). Hershey, PA: IGI Global. doi:10.4018/978-1-5225-2371-0.ch001

Tam, G. C. (2017). Understanding Project Sustainability. In *Managerial Strategies and Green Solutions for Project Sustainability* (pp. 110–139). Hershey, PA: IGI Global. doi:10.4018/978-1-5225-2371-0.ch005

Tam, G. C. (2017). Perspectives on Sustainability. In *Managerial Strategies and Green Solutions for Project Sustainability* (pp. 53–76). Hershey, PA: IGI Global. doi:10.4018/978-1-5225-2371-0.ch003

Tang, M., & Karunanithi, A. T. (2018). Visual Logic Maps (vLms). In *Advanced Concept Maps in STEM Education: Emerging Research and Opportunities* (pp. 108–149). Hershey, PA: IGI Global. doi:10.4018/978-1-5225-2184-6.ch005

Taşçıoğlu, M., & Yener, D. (2018). The Value and Scope of GIS in Marketing and Tourism Management. In S. Chaudhuri & N. Ray (Eds.), *GIS Applications in the Tourism and Hospitality Industry* (pp. 189–211). Hershey, PA: IGI Global. doi:10.4018/978-1-5225-5088-4.ch009

Tianming, G. (2018). Food Security and Rural Development on Emerging Markets of Northeast Asia: Cases of Chinese North and Russian Far East. In V. Erokhin (Ed.), *Establishing Food Security and Alternatives to International Trade in Emerging Economies* (pp. 155–176). Hershey, PA: IGI Global. doi:10.4018/978-1-5225-2733-6.ch008

Tiftikçigil, B. Y., Yaşgül, Y. S., & Güriş, B. (2017). Sustainability of Foreign Trade Deficit in Energy: The Case of Turkey. In N. Ray (Ed.), *Business Infrastructure for Sustainability in Developing Economies* (pp. 94–109). Hershey, PA: IGI Global. doi:10.4018/978-1-5225-2041-2.ch005

Tiwari, S., Vaish, B., & Singh, P. (2018). Population and Global Food Security: Issues Related to Climate Change. In I. Management Association (Ed.), Climate Change and Environmental Concerns: Breakthroughs in Research and Practice (pp. 41-64). Hershey, PA: IGI Global. doi:10.4018/978-1-5225-5487-5.ch003

Toujani, A., & Achour, H. (2018). A Data Mining Framework for Forest Fire Mapping. In I. Management Association (Ed.), Information Retrieval and Management: Concepts, Methodologies, Tools, and Applications (pp. 771-794). Hershey, PA: IGI Global. doi:10.4018/978-1-5225-5191-1.ch033

Tripathy, B., & K., S. B. (2017). Rough Fuzzy Set Theory and Neighbourhood Approximation Based Modelling for Spatial Epidemiology. In I. Management Association (Ed.), *Public Health and Welfare: Concepts, Methodologies, Tools, and Applications* (pp. 1257-1268). Hershey, PA: IGI Global. doi:10.4018/978-1-5225-1674-3.ch058

Trukhachev, A. (2017). New Approaches to Regional Branding through Green Production and Utilization of Existing Natural Advantages. In I. Management Association (Ed.), Advertising and Branding: Concepts, Methodologies, Tools, and Applications (pp. 1758-1778). Hershey, PA: IGI Global. doi:10.4018/978-1-5225-1793-1.ch081

Tsobanoglou, G. O., & Vlachopoulou, E. I. (2017). Social-Ecological Systems in Local Fisheries Communities. In G. Korres, E. Kourliouros, & M. Michailidis (Eds.), *Handbook of Research on Policies and Practices for Sustainable Economic Growth and Regional Development* (pp. 306–316). Hershey, PA: IGI Global. doi:10.4018/978-1-5225-2458-8.ch026

Tsou, M. (2010). Geographic Information Retrieval and Text Mining on Chinese Tourism Web Pages. *International Journal of Information Technology and Web Engineering*, *5*(1), 56–75. doi:10.4018/jitwe.2010010104

Tuydes-Yaman, H., & Karatas, P. (2018). Evaluation of Walkability and Pedestrian Level of Service. In I. Management Association (Ed.), Intelligent Transportation and Planning: Breakthroughs in Research and Practice (pp. 264-291). Hershey, PA: IGI Global. doi:10.4018/978-1-5225-5210-9.ch012

Uddin, S., Chakravorty, S., Ray, A., & Sherpa, K. S. (2018). Optimal Location of Sub-Station Using Q-GIS and Multi-Criteria Decision Making Approach. *International Journal of Decision Support System Technology*, *10*(2), 65–79. doi:10.4018/IJDSST.2018040104

Uddin, S., Chakravorty, S., Sherpa, K. S., & Ray, A. (2018). Power Distribution System Planning Using Q-GIS. *International Journal of Energy Optimization and Engineering*, *7*(2), 61–75. doi:10.4018/IJEOE.2018040103

Uzun, F. V. (2018). Natural Resources Management. In A. Eneanya (Ed.), *Handbook of Research on Environmental Policies for Emergency Management and Public Safety* (pp. 1–21). Hershey, PA: IGI Global. doi:10.4018/978-1-5225-3194-4.ch001

V, M., Agrawal, R., Sharma, V., & T.N., K. (2018). Supply Chain Social Sustainability and Manufacturing. In I. Management Association (Ed.), *Technology Adoption and Social Issues: Concepts, Methodologies, Tools, and Applications* (pp. 226-252). Hershey, PA: IGI Global. doi:10.4018/978-1-5225-5201-7.ch011 van der Vliet-Bakker, J. M. (2017). Environmentally Forced Migration and Human Rights. In C. Akrivopoulou (Ed.), *Defending Human Rights and Democracy in the Era of Globalization* (pp. 146–180). Hershey, PA: IGI Global. doi:10.4018/978-1-5225-0723-9.ch007

Vaskov, A. G., Lin, Z. Y., Tyagunov, M. G., Shestopalova, T. A., & Deryugina, G. V. (2018). Design of Renewable Sources GIS for ASEAN Countries. In V. Kharchenko & P. Vasant (Eds.), *Handbook of Research on Renewable Energy and Electric Resources for Sustainable Rural Development* (pp. 1–25). Hershey, PA: IGI Global. doi:10.4018/978-1-5225-3867-7.ch001

Vázquez, D. G., & Gil, M. T. (2017). Sustainability in Smart Cities: The Case of Vitoria-Gasteiz (Spain) – A Commitment to a New Urban Paradigm. In L. Carvalho (Ed.), *Handbook of Research on Entrepreneurial Development and Innovation Within Smart Cities* (pp. 248–268). Hershey, PA: IGI Global. doi:10.4018/978-1-5225-1978-2.ch012

Wahab, I. N., & Soonthodu, S. (2018). Geographical Information System in Eco-Tourism. In S. Chaudhuri & N. Ray (Eds.), *GIS Applications in the Tourism and Hospitality Industry* (pp. 61–75). Hershey, PA: IGI Global. doi:10.4018/978-1-5225-5088-4.ch003

Weiss-Randall, D. (2018). Cultivating Environmental Justice. In *Utilizing Innovative Technologies to Address the Public Health Impact of Climate Change: Emerging Research and Opportunities* (pp. 110–143). Hershey, PA: IGI Global. doi:10.4018/978-1-5225-3414-3.ch004

Weiss-Randall, D. (2018). Cultivating Resilience. In *Utilizing Innovative Technologies to Address the Public Health Impact of Climate Change: Emerging Research and Opportunities* (pp. 204–235). Hershey, PA: IGI Global. doi:10.4018/978-1-5225-3414-3.ch007

Weiss-Randall, D. (2018). Climate Change Solutions: Where Do We Go From Here? In *Utilizing Innovative Technologies to Address the Public Health Impact of Climate Change: Emerging Research and Opportunities* (pp. 236–268). Hershey, PA: IGI Global. doi:10.4018/978-1-5225-3414-3.ch008

Whyte, K. P., List, M., Stone, J. V., Grooms, D., Gasteyer, S., Thompson, P. B., . . . Bouri, H. (2018). Uberveillance, Standards, and Anticipation: A Case Study on Nanobiosensors in U.S. Cattle. In I. Management Association (Ed.), Biomedical Engineering: Concepts, Methodologies, Tools, and Applications (pp. 577-596). Hershey, PA: IGI Global. doi:10.4018/978-1-5225-3158-6.ch025

Wulff, E. (2017). Data and Operational Oceanography: A Review in Support of Responsible Fisheries and Aquaculture. In P. Diviacco, A. Leadbetter, & H. Glaves (Eds.), *Oceanographic and Marine Cross-Domain Data Management for Sustainable Development* (pp. 303–324). Hershey, PA: IGI Global. doi:10.4018/978-1-5225-0700-0.ch013

Yener, D. (2017). Geographic Information Systems and Its Applications in Marketing Literature. In S. Faiz & K. Mahmoudi (Eds.), *Handbook of Research on Geographic Information Systems Applications and Advancements* (pp. 158–172). Hershey, PA: IGI Global. doi:10.4018/978-1-5225-0937-0.ch006

About the Contributors

Ahmed Karmaoui started his academic career as Research Fellow at Cadi Ayyad University and awarded Ph.D. from Faculty of Semlalia; and as an associate editor at the International Journal of Climate Change: Impacts and Responses, editor at several International Journals such as Journal of Research in Chemistry & Environment (SRI: JRCE), the International Journal of Earth & Environmental Sciences, Nanomedicine and Nanoscience Research Journal, Environment and Pollution Journal, Asian Journal of Fisheries and Aquatic Research the Social Education Research (SER) journal, of the Global Economics Science (GES) journal; as International Advisory Board member of the International Journal of ICT Research and Development in Africa (IJICTRDA), and the International Journal of Social Ecology and Sustainable Development (IJSESD). He is a Member of the Asian Council of Science Editors. KARMAOUI has reviewed for 65 international journals, graduated from Publons Academy, and awarded Top 1% of international Reviewers (2018) by Publons (part of CLARIVATE, supported by WEB OF SCIENCE). He has published various peer-reviewed international journal articles and book chapters on a range of environmental topics. He is currently the president of the Southern Center for Culture and Science, Morocco. Broad areas of his research interests include ecosystem services assessment.

Kirby Barrick earned the Ph.D. in Agricultural Education from Ohio State University and joined the faculty where he taught courses in teaching and learning. He was awarded the OSU Distinguished Teaching Award, the Gamma Sigma Delta Teaching Award, and the NACTA Distinguished Educator Award. His research interests include teaching and learning theory and practice in domestic and international settings. He has authored/presented more than 275 articles, papers, posters, and workshops, and served as principal investigator for more than $8 million in grants. He is a Fellow in the AAAE and is a Fulbright Senior Specialist. Barrick was a department chair Ohio State, Associate Dean for Academic Programs at the University of Illinois, and Dean of the College of Agricultural and Life Sciences at the University of Florida. He has taught graduate-level courses in teacher education, instructional design, and university governance, and conducted professional development activities for practicing educators. He has worked extensively in Egypt, conducting teaching and learning and curriculum development workshops for university and technical school faculty. www.kirbybarrick.org.

* * *

Mohamed Ait El Mokhtar was born in 9 September 1985 in Charleville-Mézières, Ardennes state (France). He finished his high school studies in Marrakesh, Morocco, to obtain his Baccalaureate in "Experimental Sciences" in 2004. He continued his university studies at Cadi Ayyad University in Marrakesh to graduate with bachelor degree in Biology. In 2009, he was graduated from the same university with Master's degree in "Life and Health Sciences". Mr. Ait El Mokhtar started his professional carrier as professor of Biology in the high school since 2010. Four years later, he began his PhD in Plant biotechnology and physiology at Cadi Ayyad University. He's investigating the effect of organic and biological fertilizers on the tolerance of plants to biotic and abiotic stresses and on arbuscular mycorrhizal fungi infectivity of soils from Moroccan palm groves. He's also working on the impact of climate change on ecosystems and especially on the oasis ecosystem.

Mohamed Anli is a PhD student, Cadi Ayyad University, Faculty of Sciences Semlalia, Biology Department.

Marwane Bel-Lahcen is a PhD Student at Higher National School of Computer Science and Systems Analysis, Mohamed V University.

Banwari Dandotiya is a researcher at School of Studies in Environmental Science Jiwaji University, Gwalior, India. He works on Air Quality, Air Pollution Meteorology Environmental Pollution and Health Effects of air pollution and also working on Conservation Biology.

Fatima El Amerany is a PhD student in physiology and biochemistry of plant at Cadi Ayyad University Marrakech, Morocco; main research interests: Chitosan (functions in plant and fruits development) and arbuscular mycorrhiza.

Ismail Elhassnaoui is a Ph.D. student at the Civil Engineering Department at Ecole Mohammadia d'Ingénieurs (EMI). Mohamed V University. He is a rural engineer. His research interests: Hydrology, water resources management, disaster management.

Brema J. has an academic background with B.Tech in Civil Engineering, Postgraduate degree in Water Resources and Doctoral degree in Civil Engineering (Specialization in Water Resources). Apart from teaching, she has carried out research work and have published 90 research papers in national and international journals and conferences. Currently she is guiding 6 doctoral students. Her area of interest includes water resources management, climate modelling, flood modelling etc. She is a Fellow of Institution of Engineers and a member of Indian Society of Technical education and Indian Water Resources Society. She has published a book on Rainfall Variation with reference to Landuse change- Noyyal Basin, Lambert Academic Publishing in the year 2017.

Soufiane Lahbouki is a PhD student in the laboratory of Plant Biotechnology and Physiology, Faculty of Science Semlalia, University Cadi Ayad Marrakech, Morocco. Subject: Water management and agronomic evaluation of cactus plantations (Opuntia spp.).

Hieromin A. Lamtane is a senior lecturer working with the Department of Animal, Aquaculture and Range Sciences teaching aquatic science subjects including fisheries biology, limnology, aquaculture, fish systematic and taxonomy. He was awarded a BSc degree at the University of Dar es Salaam in 1997, MSc degree at the same university in 2001 and a PhD degree in Aquatic Ecology from the University of London, King's College, 2009. Before joined Sokoine University of Agriculture in 2006 as Assistant lecturer Dr. Lamtane was working with National Environmental Management Council since 1998 as an environmental officer. Apart from teaching Dr. Lamtane has been participated in several consultancy and outreach activities and; research projects in both aquaculture and fisheries. He has written and contributed to several publications and also has been supervising students at undergraduate and postgraduate levels.

Abdelilah Meddich is a professor at the Faculty of Science Semlalia, who works on the role of organic and biological fertilizers in improving tolerance of oasis crops including date palm to different biotic and abiotic stresses.

Mohammed Messouli has nearly 30 years of professional experience in the fields of environment, climate change and sustainable development. He is currently the Director of the Research Group on Impact, Vulnerability and Adaptation to Climate Change (GRIVAC) at the Faculty of Sciences of Marrakech. He has also acted as contributing author to the IPCC 5th Assessment Report, Working group III (2010-2014).

Renalda N. Munubi is a lecturer working with the Department of Animal, Aquaculture and Range Sciences at Sokoine University of Agriculture-Tanzania. She graduated a BSc. degree in Agriculture General at Sokoine University of Agriculturein (1998), MSc degree at Ghent University in Europe (2003) and a PhD degree in Environmental Sciences from Wright State University in the USA (2016). Before joined Sokoine University of Agriculture in 2007 as Assistant Lecturer, Dr. Munubi was working with Tanzanian Coastal Management Partnership-USAID in Dar es salaam as a training coordinator in aquaculture and Beekeeping. Apart from teaching Dr. Munubi has been participated in several outreach activities and research projects in both aquaculture. She has written and contributed to several publications, (Peer-review Journals, book chapters). In addition to research Dr. Munubi has been supervising research projects to undergraduate and postgraduate students.

K. S. Sastry Musti has obtained his BTech in Electrical Engineering from JNTU, Kakinada, MTech and PhD from NIT, Warangal., India. He is currently with Namibia University of Science and Technology as Associate Professor with the Department of Electrical and Computer Engineering. He has more than 30 years of professional experience in both academia and industry. His research interests are in the extended areas of Engineering Education, Power Systems, Software Engineering and Systems and Information Systems. Professor Sastry is a senior member IEEE, USA and a Life member IE, India.

Joan Nyika is an Environmental Science PhD student at the University of South Africa and a Tutorial Fellow at the Technical University of Kenya, Geoscience and the Environment Department. She obtained Msc and Bsc degrees in Land and Water Management and Biochemistry from the University of Nairobi and Jomo Kenyatta University of Agriculture and Technology, Kenya in 2017 and 2011, respectively. Her PhD work revolves around solid waste management and the effects of landfill leachate on soil and groundwater. She has interests in sustainable environmental management, water management and ecosystem management. Joan has published more than 15 articles and 5 book chapters in peer reviewed journals and books, respectively. She has and is mentoring both undergraduate and diploma students and has more than 5 years experience teaching and training in environmental science.

Suchismita Satapathy is currently working as an Associate Professor-1 in the School of Mechanical Sciences, KIIT University. She has published more than 25 articles, many national and international journals and conferences. She has more than 15 years of teaching and research experience. Her area of interest is production operation Management, Operation Research, accoustics, sustainability, supply chain mgmt., etc. She has guided many PhD, Mtech, and Btech students.

Said Wahbi is a professor at the Faculty of Science Semlalia. He is the head of the Plant Biotechnology and Physiology Unit and works on the role of organic and biological fertilizers in improving tolerance of oasis crops including date palm to different biotic and abiotic stresses.

Index

A

B

C

E

F

G

H

I

L

M

N

O

P

Q

R

S

T

U

V

W

www.ingramcontent.com/pod-product-compliance
Lightning Source LLC
LaVergne TN
LVHW081343300926
849161LV00008B/2049
* 9 7 8 1 7 9 9 8 3 3 4 3 7 *